Fundamentals of Condensed Matter and Crystalline Physics

This undergraduate textbook merges traditional solid state physics with contemporary condensed matter physics, providing an up-to-date introduction to the major concepts that form the foundations of condensed materials. The main foundational principles are emphasized, providing students with the knowledge beginners in the field should understand. The book is structured in four parts, and allows students to appreciate how the concepts in this broad area build upon each other to produce a cohesive whole as they work through the chapters. Illustrations work closely with the text to convey concepts and ideas visually, enhancing student understanding of difficult material, and end-of-chapter exercises, varying in difficulty, allow students to put into practice the theory they have covered in each chapter, and reinforce new concepts. Additional resources including solutions to exercises, lesson plans and pre-lecture reading quiz questions are available online at www.cambridge.org/sidebottom.

David L. Sidebottom is Associate Professor in the Physics Department at Creighton University. He is an experienced teacher and has taught a wide variety of courses at both undergraduate and graduate level in subject areas including introductory physics, thermodynamics, electrodynamics, laser physics and solid state physics. He has taught a course on solid state physics since 2003, adapting and revising its content to reflect the broader themes of condensed matter physics beyond those of the conventional solid state. This textbook stems from that course.

Fundamentals of Condensed Matter and Crystalline Physics

An Introduction for Students of Physics and Materials Science

DAVID L. SIDEBOTTOM

Creighton University, Omaha

CAMBRIDGE
UNIVERSITY PRESS

CAMBRIDGE UNIVERSITY PRESS
Cambridge, New York, Melbourne, Madrid, Cape Town,
Singapore, São Paulo, Delhi, Mexico City

Cambridge University Press
The Edinburgh Building, Cambridge CB2 8RU, UK

Published in the United States of America by Cambridge University Press, New York

www.cambridge.org
Information on this title: www.cambridge.org/9781107017108

First published 2012

Printed in the United Kingdom at the University Press, Cambridge

A catalog record for this publication is available from the British Library

Library of Congress Cataloging-in-Publication Data
Sidebottom, David L., 1961–
Fundamentals of condensed matter and crystalline physics: An Introduction for Students of
Physics and Materials Science / David L. Sidebottom, Creighton University, Omaha.
pages   cm
Includes bibliographical references and index.
ISBN 978-1-107-01710-8
1. Condensed matter–Textbooks.   2. Crystals–Textbooks.   I. Title.
QC173.454.S53 2012
530.4′1–dc23
2011046218

ISBN 978-1-107-01710-8 Hardback

Additional resources for this publication at www.cambridge.org/sidebottom

10 0669866 9

Contents

Preface   *page* xiii
Permission Disclosures   xvii

Part I: Structure

1 Crystal structure   3
  1.1 Crystal lattice   3
    1.1.1 Basis set   5
    1.1.2 Primitive cells   6
  1.2 Symmetry   8
    1.2.1 Conventional cells   10
  1.3 Bravais lattices   10
    1.3.1 Cubic lattices   12
    1.3.2 Hexagonal lattices   16
  Summary   17
  Exercises   18

2 Amorphous structure   20
  2.1 A statistical structure   20
    2.1.1 Ensemble averaging   21
    2.1.2 Symmetry   21
    2.1.3 The pair distribution function   22
  2.2 Two amorphous structures   26
    2.2.1 Random close packed structure   26
    2.2.2 Continuous random network   28
  Summary   31
  Exercises   32

3 Bonds and cohesion   35
  3.1 Survey of bond types   35
    3.1.1 The van der Waals bond   35
    3.1.2 Ionic, covalent and metallic bonds   38
    3.1.3 The hydrogen bond   41
  3.2 Cohesive energy   41
    3.2.1 Crystals   41
    3.2.2 Amorphous materials   46

Summary   47
Exercises   47

4 Magnetic structure   50
4.1 The ordering process   50
    4.1.1 Correlations and pattern formation   51
4.2 Magnetic materials   53
    4.2.1 Magnetic moments   53
    4.2.2 Diamagnetism   55
    4.2.3 Paramagnetism   57
    4.2.4 Ferromagnetism   60
Summary   64
Exercises   65

Part II: Scattering

5 Scattering theory   69
5.1 The dipole field   69
    5.1.1 The scattering cross section   71
5.2 Interference   72
    5.2.1 Scattering from a single atom   73
5.3 Static structure factor   76
    5.3.1 A relevant scattering length scale   77
    5.3.2 A Fourier relationship: the density–density correlation
        function   79
Summary   80
Exercises   80

6 Scattering by crystals   82
6.1 Scattering by a lattice   82
    6.1.1 A set of allowed scattering wave vectors   84
6.2 Reciprocal lattice   85
6.3 Crystal planes   86
    6.3.1 Miller indices   86
    6.3.2 Bragg diffraction   88
    6.3.3 Missing reflections   89
Summary   92
Exercises   92

7 Scattering by amorphous matter   95
7.1 The amorphous structure factor   95
    7.1.1 Equivalence for liquids and glasses   97
    7.1.2 Investigating short-range order   97
    7.1.3 Rayleigh scattering   100

7.2 Light scattering by density fluctuations 101
   7.2.1 The van Hove space correlation function 103
   7.2.2 Intermediate-range order: SAXS
       and SANS 105
Summary 107
Exercises 107

8 Self-similar structures and liquid crystals 109
8.1 Polymers 109
   8.1.1 The random walk 110
   8.1.2 Swollen polymers: self-avoiding walks 114
8.2 Aggregates 115
   8.2.1 Fractals 117
   8.2.2 Example: soot formation 118
8.3 Liquid crystals 124
   8.3.1 Thermotropic liquid crystals 125
   8.3.2 Lyotropic liquid crystals: micelles
       and microemulsions 129
Summary 132
Exercises 132

Part III: Dynamics

9 Liquid dynamics 139
9.1 Dynamic structure factor 139
   9.1.1 The van Hove correlation function 141
   9.1.2 Brownian motion: the random walk
       revisited 142
   9.1.3 Hydrodynamic modes in liquids 145
9.2 Glass transition 150
   9.2.1 Kauzmann paradox 151
   9.2.2 Structural relaxation 153
9.3 Polymer liquids 154
   9.3.1 Rouse model 155
   9.3.2 Reptation 157
Summary 160
Exercises 160

10 Crystal vibrations 163
10.1 Monatomic basis 163
   10.1.1 Dispersion relation 166
   10.1.2 Brillouin zone 167
   10.1.3 Boundary conditions and allowed modes 169
   10.1.4 Phonons 170

10.2 Diatomic basis   173
    10.2.1 Long wavelength limit   174
    10.2.2 Waves near the Brillouin zone   176
    10.2.3 Acoustical waves, optical waves and energy gaps   176
10.3 Scattering from phonons   177
    10.3.1 Elastic (Bragg) scattering: The Debye–Waller factor   178
    10.3.2 Inelastic scattering by single phonons   179
Summary   180
Exercises   181

11 Thermal properties   182
11.1 Specific heat of solids   182
    11.1.1 Einstein model   184
    11.1.2 Debye model   186
11.2 Thermal conductivity   189
    11.2.1 Phonon collisions   191
11.3 Amorphous materials   193
    11.3.1 Two-level systems   194
    11.3.2 Phonon localization   197
Summary   199
Exercises   200

12 Electrons: the free electron model   201
12.1 Mobile electrons   201
    12.1.1 The classical (Drude) model   202
12.2 Free electron model   204
    12.2.1 Fermi level   206
    12.2.2 Specific heat   207
    12.2.3 Emission effects   210
    12.2.4 Free electron model in three dimensions   210
    12.2.5 Conduction in the free electron model   212
    12.2.6 Hall effect   214
Summary   216
Exercises   216

13 Electrons: band theory   218
13.1 Nearly free electron model   218
    13.1.1 Bloch functions   218
    13.1.2 Bragg scattering and energy gaps   220
13.2 Kronig–Penney model   221
    13.2.1 Energy bands and gaps   223
    13.2.2 Mott transition   226

13.3 Band structure   226
13.4 Conductors, insulators, and
      semiconductors   230
      13.4.1 Holes   232
      13.4.2 Intrinsic semiconductors   233
      13.4.3 Extrinsic semiconductors   235
13.5 Amorphous metals: the Anderson transition   241
Summary   243
Exercises   244

14 Bulk dynamics and response   246
14.1 Fields and deformations   246
      14.1.1 Mechanical deformations   247
      14.1.2 Electric and magnetic deformations   249
      14.1.3 A generalized response   251
14.2 Time-dependent fields   252
      14.2.1 Alternating fields and response functions   253
      14.2.2 Energy dissipation   256
14.3 The fluctuation–dissipation theorem   257
Summary   261
Exercises   261

Part IV: Transitions

15 Introduction to phase transitions   267
15.1 Free energy considerations   267
15.2 Phase diagrams for fluids   269
      15.2.1 PT diagram   269
      15.2.2 PV diagram   270
      15.2.3 TV diagram   272
      15.2.4 Order parameter   272
15.3 Supercooling/heating and nucleation   274
15.4 Critical phenomena   276
      15.4.1 A closer look: density fluctuations   278
15.5 Magnetic phase transitions   282
      15.5.1 Exchange interaction   283
      15.5.2 Magnetic phase diagrams   284
15.6 Universality: the law of corresponding
      states   285
Summary   287
Exercises   287

16 Percolation theory   289
    16.1 The percolation scenario   289
        16.1.1 Percolation threshold: the spanning cluster   292
        16.1.2 A closer look: cluster statistics   293
    16.2 Scaling relations   297
        16.2.1 Finite-sized scaling   298
        16.2.2 Renormalization   300
        16.2.3 Universality and the mean field limit   303
    16.3 Applications of percolation theory   305
        16.3.1 Orchard blight and forest fires   305
        16.3.2 Gelation   305
        16.3.3 Fractal dynamics: anomalous diffusion   307
    Summary   310
    Exercises   310

17 Mean field theory and renormalization   313
    17.1 Mean field theory   313
        17.1.1 The mean field approximation   313
    17.2 The mean field equation of state   316
        17.2.1 Fluids: the van der Waals model   316
        17.2.2 Magnets: the Ising model   318
    17.3 Law of corresponding states   319
    17.4 Critical exponents   321
        17.4.1 Compressibility and susceptibility   323
        17.4.2 Order parameter   323
    17.5 Landau theory   324
    17.6 Renormalization theory   326
        17.6.1 A matter of perspective   327
        17.6.2 Kadanoff spin renormalization   327
        17.6.3 Scaling relations   330
    Summary   332
    Exercises   333

18 Superconductivity   334
    18.1 Superconducting phenomena   334
        18.1.1 Discovery   334
        18.1.2 Meissner effect   335
        18.1.3 Critical field   340
        18.1.4 Specific heat   340
        18.1.5 Energy gap   343
        18.1.6 Isotope effect   344
    18.2 Cooper pairs and the BCS theory   345
        18.2.1 Cooper pairs   346
        18.2.2 Flux quantization   348

18.3 Thermodynamics: Ginzburg–Landau theory   349
  18.3.1 Mean field theory   349
  18.3.2 Type II superconductors   351
  18.3.3 The Ginzburg–Landau equations   352
  18.3.4 Type II critical fields   356
  18.3.5 High-T_c superconductors   360
Summary   360
Exercises   361

Appendix: Toolbox   364
A.1 Complex notation   364
  A.1.1 Trigonometric identities   365
  A.1.2 Other items   365
A.2 Wave notation   366
A.3 Fourier analysis   367
  A.3.1 Fourier series   368
  A.3.2 Fourier transforms   368
  A.3.3 Fourier transforms expressed in complex notation 370
  A.3.4 Extension of Fourier transforms to higher dimensions 373
A.4 The Dirac delta function   373
  A.4.1 Dirac delta functions and Fourier transforms 375
A.5 Elements of thermodynamics   376
  A.5.1 First and second laws   376
  A.5.2 The free energies   378
  A.5.3 Free energy and the second law   378
A.6 Statistical mechanics   380
  A.6.1 Microstates and macrostates   380
  A.6.2 The Boltzmann factor   381
A.7 Common integrals   382

Glossary   383
References   390
Index   394

Preface

Purpose and motivation

This textbook was designed to accompany a one-semester, undergraduate course that itself is a hybridization of conventional solid state physics and "softer" condensed matter physics.

Why the hybridization? Conventional (crystalline) solid state physics has been pretty much understood since the 1960s at a time when non-crystalline physics was still a fledgling endeavour. Some 50 years later, many of the foundational themes in condensed matter (scaling, random walks, percolation) have now matured and I believe the time is ripe for both subjects to be taught as one. Moreover, for those of us teaching at smaller liberal arts institutions like my own, the merging of these two subjects into one, better accommodates a tight curriculum that is already heavily laden with required coursework outside the physics discipline.

Why the textbook? For some years now I have taught a one-semester course, originally listed as "solid state physics", which evolved through each biannual reincarnation into a course that now incorporates many significant condensed matter themes, as well as the conventional solid state content. In past offerings of the course, a conventional solid state textbook was adopted (Kittel's *Introduction to Solid State Physics*) and students were provided with handouts for the remaining material. This worked poorly. Invariably, the notation and style of the handouts clashed with that of the textbook and the disjointed presentation of the subject matter was not only annoying to students, but a source of unnecessary confusion. Students were left with the impression that solid state and condensed matter were two largely unrelated topics being crammed into a single course. Frustrated, I opted to spend a portion of a recent sabbatical assembling all of the material into a single document that would better convey the continuity of these two fields by threading both together into a seamless narrative.

So if you are looking for a reference-style textbook that provides a comprehensive coverage of the entire field of condensed matter, read no further because this is not it. This textbook was not written for practitioners, but rather for novices. It was designed to help students comprehend, not so much the details, but the major concepts that form the foundations of condensed matter and crystalline physics. At the very least I want students to leave the course able to comprehend the meaning behind terminology used by solid state physicists (e.g., "symmetry operations", "Brillouin zones", "Fermi sufaces") and condensed matter physicists (e.g, "mean field theory", "percolation", "scaling laws", "structure factors") so that they might rapidly acclimate to current research in either field.

Layout and use

I confess that my inspiration for the textbook style was Kittel's *Introduction to Solid State Physics*, which has been a valuable guide for maintaining the development at a level appropriate for an undergraduate audience. Although criticized by some, his text is now in its eighth edition and has remained a popular choice for many undergraduate courses on solid state physics (including my own). Those familiar with Kittel, will find that this hybrid textbook incorporates most of the same subject matter (albeit abbreviated in places and arranged in a different order due to the way it is now interwoven with other non-crystalline topics) as is found in the first twelve chapters of Kittel.

Students will need a limited exposure to both quantum mechanics and statistical mechanics. The level of quantum mechanics that is provided in an introductory sophomore-level course on modern physics (1D wave mechanics, particle in a box, harmonic oscillators) should be sufficient. Beyond that, statistical mechanics and thermodynamics (specifically, Boltzmann statistics and free energies) are introduced periodically throughout the text and this is more likely to be the deficiency for some students. In an effort to help alleviate this and other potential deficiencies, an appendix is included which provides an introduction to such things as statistical mechanics, Fourier transforms and the use of Dirac delta functions.

The text is divided into four major parts: Structure, Scattering, Dynamics, and Transitions. Within each part are anywhere from four to six chapters designed more to delineate topics than to represent equal amounts of material. Although a common rule of thumb would be to allot three, 50-minute lecture periods per chapter, several chapters (e.g., 2, 3, 7, 10, 14, 15) could be adequately discussed in just two periods and Chapter 5 could likely be addressed in a single period. The lesson plan that I have adopted looks something like this:

	Lecture #1 (50 min)	Lecture #2 (50 min)	Lecture #3 (50 min)
Week	Chapter	Chapter	Chapter
1	1	1	1
2	2	2,3	3
3	4	4	5
4	6	6	6,7
5	7	7,8	8
6	8	9	9
7	10	10	11

	Lecture #1 (50 min)	Lecture #2 (50 min)	Lecture #3 (50 min)
Week	Chapter	Chapter	Chapter
8	11	11	12
9	12	12,13	13
10	13	13	13
11	14	14	15
12	15	16	16
13	16,17	17	17
14	18	18	18

Can all these topics be covered in a semester? Maybe. In my experience, I have so far only managed to cover about 85%. Topics to skip are really a matter of preference. I had no reservations about skipping the subject of bonds and cohesion (Chapter 3) and only modest discomfort at skipping the subject of bulk dynamics (Chapter 14). Others that are less interested in amorphous solids could skip glass structure (Chapter 2), but I would advise not to skip the material on scattering from self-similar objects (Chapter 8), as this contributes an important conceptual foundation for much of the materials in the last four chapters (Chapters 15–18) of the text.

Some might be tempted to skip the development of scattering theory presented in Chapter 5, so let me petition against this. In my experience, students struggle with the concept of reciprocal space primarily because of how most conventional solid state textbooks mysteriously introduce it directly after discussing Bragg's law. Students rarely grasp the significance of this abstract space and probably question why it is introduced at all, given how Bragg's law seems sufficient. By first introducing the fundamentals of scattering in Chapter 5, the reciprocal space appears more naturally as the discrete set of scattering wave vectors for which non-zero scattering occurs. Bragg's law is only presented as a consequence.

Anywhere from five to ten exercises can be found at the end of each chapter. These come in a variety of difficulty levels and are designed mostly to help students digest the material and develop skills. Many of the easier problems are derived from the text itself and ask students to complete the missing steps in a derivation. Although some may see this as aimless "busy work", for many undergraduate students (in my experience) these exercises represent a challenging skill yet to be mastered.

For students

Good luck and I hope this textbook helps you. Please let me know what you do and don't like about the textbook so that I can improve it in the future.

Acknowledgements

Let me start by thanking the many students that have taken my course in the past several years, and in particular the 2011 class (David, Jamison, Clifford, Nathan, Stan, Tri and Yuli), who braved an early prototype of the textbook and provided many valuable suggestions for revision and improvement. I thank also several close colleagues, Chris Sorensen, Jeppe Dyre and Per Jacobsson whose positive feedback on an early draft inspired me considerably and eventually prompted me to seriously consider publication. I am especially indebted to Chris who has been a mentor to me throughout the years and who helped immensely by giving an early draft of the textbook a thorough read.

Naturally, the support of Creighton University in the form of employment and a sabbatical leave is gratefully acknowledged. Grateful too am I for support from the faculty and staff in my department who have tolerated my erratic behavior during the past two years. I am especially grateful for advice given to me by Robert Kennedy about the publishing process.

And last, but certainly not least, I want to acknowledge the support and encouragement from my best friend and wife, Lane. Thank you, my love.

Permission Disclosures

The following figures are reprinted with permission:

Fig. 2.7b adapted from J. L. Finney (1970) "Random Packings and the Structure of Simple Liquids. I. The Geometry of Random Close Packing," *Proc. Roy. Soc. (London)* **A319**, 479–493 (Copyright (1970) with permission from The Royal Society).

Fig. 2.8 adapted from E. H. Henniger, R. C. Buschert and L. Heaton (1967) "Atomic structure and correlation in vitreous silica by X-ray and neutron diffraction," *J. Phys. Chem. Solids* **28**(3), 423–432 (Copyright (1967) with permission from Elsevier).

Fig. 4.5 adapted from W. E. Henry (1952) "Spin paramagnetism of Cr^{3+}, Fe^{3+} and Gd^{3+} at liquid helium temperatures and in strong magnetic fields," *Phys. Rev.* **88**(3), 559–562 (Copyright (1952) with permission from The American Physical Society).

Fig. 4.6 adapted from L. C. Jackson (1936) "The paramagnetism of the rare earth sulphates at low temperatures," *Proc. Royal Soc. (London)* **48**, 741–746 (Copyright (1936) with permission from The Royal Society).

Fig. 4.7 adapted from H. E. Nigh, S. Legvold and F. H. Spedding (1963) "Magnetization and electrical resistivity of gadolinium single crystals," *Phys. Rev.* **132**(3), 1092–1097 (Copyright (1963) with permission from The American Physical Society).

Figs. 7.1b, 7.2 adapted from R. J. Temkin, W. Paul, and G. A. N. Connell (1973) "Amorphous germanium II. Structural properties," *Adv. Phys.* **22**(5), 581–641 (Copyright (1973) with permission from Taylor and Francis Group, http://www.informaworld.com).

Fig. 7.7 adapted from S. Susman, K. J. Volin, D. L. Price, M. Grimsditch, J. P. Rino, R. K. Kalia, P. Vashishta, G. Gwanmesia, Y. Wang, and R. C. Liebermann (1991) "Intermediate-range order in permanently densified vitreous SiO_2: A neutron-diffraction and molecular-dynamics study," *Phys. Rev. B* **43**, 1194–1197 (Copyright (1991) with permission from The American Physical Society).

Fig. 12.2 (inset) adapted from D. K. C. MacDonald and K. Mendelssohn (1950) "Resistivity of Pure Metals at Low Temperatures I. The Alkali Metals," *Proc. Royal Soc. (London)* **A202**, 103–126 (Copyright (1950) with permission from The Royal Society).

Fig. 13.13 adapted from F. J. Moran and J. P. Maita (1954) "Electrical properties of silicon containing arsenic and boron," *Phys. Rev.* **96**(1), 28–35 (Copyright (1954) with permission from The American Physical Society).

Fig. 15.10 adapted from J. E. Thomas and P. W. Schmidt (1963) "X-ray Study of Critical Opalescence in Argon," *J. Chem. Phys.* **39**, 2506–2516 (Copyright (1963) with permission from The American Institute of Physics).

Fig. 15.13 adapted from E. A. Guggenheim (1945) "The Principle of Corresponding States," *J. Chem. Phys.* **13**, 253–261 (Copyright (1945) with permission from The American Institute of Physics).

Fig. 18.1 adapted from H. K. Onnes (1911) *Comm. Leiden* **124c** (1911) (Courtesy of the Kamerlingh Onnes Laboratory, Leiden Institute of Physics).

Figs. 18.5, 18.7 adapted from M. A. Bondi, A. T. Forrester, M. P. Garfunkel and C. B. Satterthwaite (1958) "Experimental Evidence for and Energy Gap in Superconductors," *Rev. Mod. Phys.* **30**(4), 1109–1136 (Copyright (1958) with permission from The American Physical Society).

Fig. 18.8 adapted from P. Townsend and J. Sutton (1962) "Investigation by Electron Tunneling of the Superconducting Energy Gaps in Nb, Ta, Sn, and Pb," *Phys. Rev.* **128**(2), 591–595 (Copyright (1962) with permission from The American Physical Society).

Fig. 18.9 adapted from E. Maxwell (1952) "Superconductivity of the Isotopes of Tin," *Phys. Rev.* **86**(2), 235–242 (Copyright (1952) with permission from The American Physical Society).

PART I

STRUCTURE

Picture with me an old cottage nestled in the woods. There is a small house built of clay bricks that were thoughtfully stacked and interlaced by a master bricklayer so as to produce a repeated interlocking pattern. The house has a thatched roof consisting of bundles of straw. The straws in each bundle are oriented in a common direction to direct rainwater off the roof, and are lashed together with twine. Around the house is a garden enclosed by a stone wall. Like the brick walls of the house, the stones in the wall are bonded together with mortar. But unlike the bricks, the stones lack any sense of a repeating pattern.

In this part of the textbook, we examine the basic structures that are found in condensed matter as well as the forces (the mortar and twine) that maintain these structures over long time periods. For our purposes, structures are divided into two main categories: ordered (like the bricks and the straw of the house) and disordered (like the stones in the garden wall).

We begin in Chapter 1 with an examination of the structure of crystals whose periodic arrangement of atoms is a prime example of ordered matter. Particle positions in the crystal are well-defined and the periodic structure is seen to extend for very long distances. As a result of this ordering, crystal structures are rather easy to describe mathematically and provide an excellent introduction to the concept of symmetry. All of this simplicity and symmetry is lost for amorphous materials and in Chapter 2 we examine alternative means for quantifying structures in which particle positions are aperiodic. In the third chapter, we pause to examine the inter-particle forces that provide the mortar necessary for condensed matter to form. There we survey the fundamental types of bonds and discuss how each can influence the resulting structure. In our final chapter on the topic of structures, we look at magnetic materials. Although the atoms that compose these materials may be arranged in an ordered manner, their magnetic moments can either be oriented randomly or, like the aligned straws of a thatched roof, assume an ordered configuration.

1   Crystal structure

Introduction

We often think of crystals as the gemstones we give to a loved one, but most metals (e.g. copper, aluminum, iron) that we encounter daily are common crystals too. In this chapter, we will examine the structure of crystalline matter in which particles are arranged in a repeating pattern that extends over very long distances. This long-range order is formally described by identifying small local groupings of particles, known as a *basis* set, that are identically affixed to the sites of a regularly repeating space *lattice*. As it happens, most crystals found in nature assume one of a limited set of special space lattices known as Bravais lattices. These lattices are special by virtue of their unique symmetry properties wherein only discrete translations and rotations allow the lattice to appear unchanged. Chief among these Bravais lattices are the cubic and hexagonal lattice structures that appear most frequently in nature. We focus extra attention on both to provide a useful introduction to coordination properties and packing fractions.

1.1 Crystal lattice

Crystals have a decided advantage because of the inherent repeating pattern present in their structure. In an ideal (perfect) crystal, this repeating pattern extends indefinitely. However, for real crystals found in nature, the pattern is often interrupted by imperfections known as *defects* that can include vacancies, in which a single particle is missing, and dislocations in which the repeating pattern is offset. These defects are important for some crystal properties, but for now we restrict ourselves to only ideal structures. Besides, even in real crystals large regions containing substantial numbers of particles exist in which a perfectly repeating pattern is maintained.

Let's start with an imaginary, two-dimensional example of a crystal that contains two types of particles (say, large A atoms and small B atoms) as illustrated in Fig. 1.1. It is clear from inspection that this collection of particles displays a well-ordered repeating pattern of A and B atoms that can be

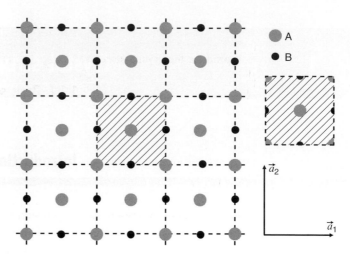

Figure 1.1   The repeating pattern of atoms A (gray circles) and B (black circles) is mapped onto a lattice (dashed lines) that is defined by two lattice vectors ($\vec{a}_1$ and $\vec{a}_2$). The pattern of atoms can be viewed as the result of attaching tiles (hashed area that contains a total of two A atoms and two B atoms) onto the lattice.

arranged neatly on the square grid that is superimposed. How can we best describe this repeating pattern? We could simply establish an arbitrary origin and then list the position vectors for every particle of each type. But that would be unnecessarily cumbersome given that there is an obvious repeating pattern. Instead, consider the square grid. The points formed by the intersections of these grid lines can be referenced from any other point by any combination of translations of the form:

$$\vec{T} = h\vec{a}_1 + k\vec{a}_2, \tag{1.1}$$

where h and k are the complete set of integer numbers. The complete set of these translations define what is known as a *space lattice* – an abstract set of points in space that convey the inherent repeating pattern behind the crystal's structure.

In Fig. 1.1, we see that some of the larger A atoms are located directly on the points of the space lattice (grid) and their positions can be referenced by the set of translations in Eq. (1.1) alone. But other A atoms, as well as the smaller B atoms, reside off the lattice. To completely describe the particle positions of all the atoms of the crystal, we must combine with the space lattice a small subset of atoms (known as a *basis*) that are repeatedly attached to each lattice site so as to produce the entire structure. This is much like flooring your kitchen with linoleum tiles. Imagine that each linoleum tile has a pattern stamped onto it corresponding to one of the squares in Fig. 1.1. This particular tile would have two of each type of atom: a complete A atom at the center, one-quarter of an A atom at each corner, and one-half of a B atom at the middle of each side. When each such tile is positioned with its lower left-hand corner coincident with a space lattice point, the completed assembly of tiles would reproduce the crystal structure of Fig. 1.1 as a whole.

1.1.1 Basis set

Thus, to describe the entire structure of a crystal we combine a space lattice, described by the translations of Eq. (1.1), with a set of basis vectors (referenced to, say, the lower left-hand corner of the tile) to describe the contents of each tile:

$$\vec{R}_i = x_i\vec{a}_1 + y_i\vec{a}_2, \qquad (1.2)$$

where x_i and y_i are fractions. For the particular tile illustrated in Fig. 1.1, the basis vectors would include:

single central A atom: $\vec{R}_1 = \dfrac{1}{2}\vec{a}_1 + \dfrac{1}{2}\vec{a}_2$

four corner A atoms: $\left.\begin{cases} \vec{R}_2 = 0\vec{a}_1 + 0\vec{a}_2 \\ \vec{R}_3 = 1\vec{a}_1 + 0\vec{a}_2 \\ \vec{R}_4 = 0\vec{a}_1 + 1\vec{a}_2 \\ \vec{R}_5 = 1\vec{a}_1 + 1\vec{a}_2 \end{cases}\right\} \dfrac{1}{4}$ of an A atom each

four side B atoms: $\left.\begin{cases} \vec{R}_6 = \frac{1}{2}\vec{a}_1 + 0\vec{a}_2 \\ \vec{R}_7 = 0\vec{a}_1 + \frac{1}{2}\vec{a}_2 \\ \vec{R}_8 = 1\vec{a}_1 + \frac{1}{2}\vec{a}_2 \\ \vec{R}_9 = \frac{1}{2}\vec{a}_1 + 1\vec{a}_2 \end{cases}\right\} \dfrac{1}{2}$ of a B atom each

This is still more cumbersome than necessary. Consider, as shown in Fig. 1.2, an alternative space lattice composed of diagonal grid lines. Notice that we

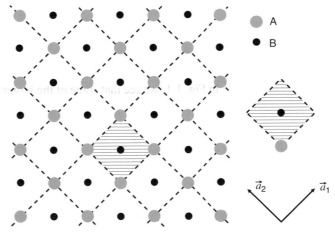

Figure 1.2   The same pattern of two atoms found in Fig. 1.1 are referenced to an alternative, diagonal lattice with a corresponding redefinition of the tile (hashed area) to contain only one each of each atom.

have not repositioned any of the particles, only redefined the space lattice we choose to associate with them. Our diamond-shaped tiles now contain only one atom of each type. This sort of tile is known as a *primitive cell*. It is the smallest-sized tile that can be used together with the space lattice to fill the space with our desired repeating pattern. Our basis set now requires only two vectors:

$$\text{A atom: } \vec{R}_1 = 0\vec{a}_1 + 0\vec{a}_2$$
$$\text{B atom: } \vec{R}_2 = \frac{1}{2}\vec{a}_1 + \frac{1}{2}\vec{a}_2. \tag{1.3}$$

Note here that the entire A atom is now being associated with the tile (even though three quarters of it sticks outside). Tiles affixed to neighboring lattice sites will then provide the other three A atoms.

1.1.2 Primitive cells

Primitive cells can be identified by several properties. A primitive cell:

(1)   contains only one lattice point,
(2)   has the smallest size (area, $A = |\vec{a}_1 \times \vec{a}_2|$) that can just fill the space by repetition, and
(3)   has a basis set containing only one molecular unit (in our case: AB).

Primitive cells are *not* unique. As shown in Fig. 1.3, yet another alternative space lattice has been chosen to describe our AB system. The shaded cell shown has the same smallest size area as our diamonds in Fig. 1.2 and contains

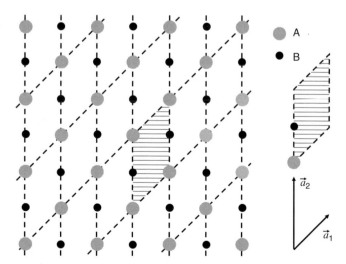

Figure 1.3   The same pattern of two atoms found in Fig. 1.1 and Fig. 1.2 are referenced to yet another alternative lattice with an alternative primitive cell (hashed area).

(a)

(b)

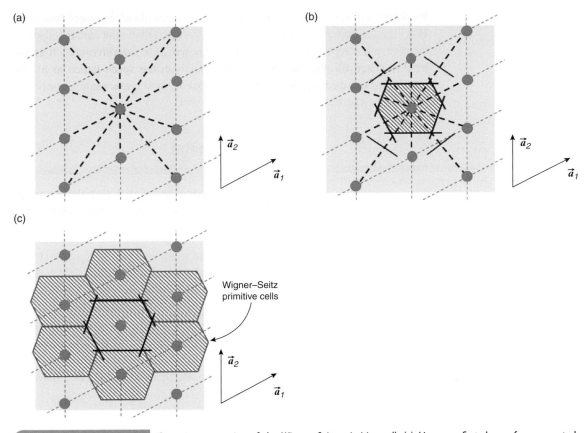

(c)

Wigner–Seitz
primitive cells

Figure 1.4   Steps in construction of the Wigner–Seitz primitive cell. (a) Lines are first drawn from a central lattice site to all neighboring sites (heavy dashed lines). (b) Each of these lines is then bisected by a perpendicular plane (heavy solid lines) and the volume enclosed becomes the Wigner–Seitz cell. (c) The cell is capable of tiling the entire space and is a primitive cell because it contains one lattice site (at its center).

one of each atom type. However, the basis vectors for this situation would need to be revised as:

$$\begin{aligned} \text{A atom: } \vec{R}_1 &= 0\vec{a}_1 + 0\vec{a}_2 \\ \text{B atom: } \vec{R}_2 &= 0\vec{a}_1 + \frac{1}{2}\vec{a}_2. \end{aligned} \tag{1.4}$$

Wigner–Seitz primitive cell

Although there are many choices for the primitive cell as illustrated above, there is one alternative known as the *Wigner–Seitz cell*, which will have special relevance later on in our discussions of solid state physics. Construction of the Wigner–Seitz cell is illustrated in a series of panels in Fig. 1.4 and begins by drawing lines from any arbitrary lattice site to neighboring lattice sites (see Fig. 1.4a). Next, each line is bisected by a perpendicular line (or plane in the case of a 3D lattice), as illustrated in Fig. 1.4b. The interior region bounded by these perpendicular lines is then the Wigner–Seitz cell. The cell is seen to be primitive because it contains just one lattice point (namely, the one at its center) and can successfully tile the entire space.

1.2 Symmetry

Aside from its repeating pattern, the space lattice possesses another important characteristic known as *symmetry*. Consider yourself as a (very small) observer located on one of the A atoms in Fig. 1.5. When you look around, you observe nearby B atoms (to the north, south, east and west) and nearby A atoms (to the NE, NW, SE and SW). If you now move to another point of the space lattice (atop another A atom), by a translation, $\vec{T} = h\vec{a}_1 + k\vec{a}_2$, you will experience no sense that your surroundings have changed in any way. In this way the space lattice is said to possess *translational symmetry* – if the entire space lattice is shifted by any of the translation vectors that describe it, the resulting pattern is unchanged in any observable manner.

In addition to this translational symmetry, which all space lattices possess by virtue of their repeating nature, there are other important symmetry operations that define different space lattices. For example, consider yourself again atop an A atom in Fig. 1.5. If you rotate by $90°$ you again see the same surroundings as before you rotated. The space lattice is said to possess a certain *rotational symmetry*. Note that this symmetry appears only for specific angles of rotation in

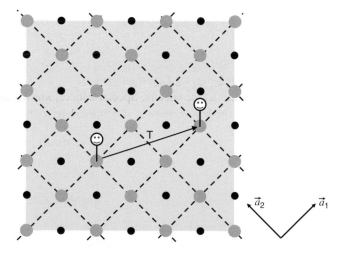

Figure 1.5   An observer situated on a lattice undergoes a translation to any other lattice site and finds his/her surroundings unaltered. The system of particles is then said to possess translational symmetry.

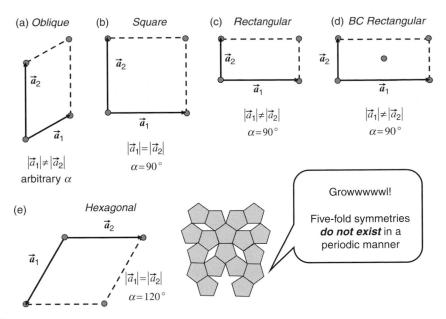

Figure 1.6

The complete set of Bravais lattices (nets) in two dimensions. In addition to the oblique, there are four other lattices possessing distinct symmetry properties. Of these, only the BC rectangular is a conventional lattice. Rotational symmetries include 2, 3 and 4-fold but do not allow for 5-fold symmetries.

the crystal. For example, a rotation by 45° on the lattice of Fig. 1.5 will not return your surroundings to their original state. Only rotations by a multiple of 90° will do this. Because there are four 90° increments in a full circle, this particular case of rotational symmetry is referred to as '4-fold' rotational symmetry.

For the two-dimensional situations we are currently discussing, there are an unlimited number of possible space lattices owing to the fact that any lengths of the two lattice vectors ($\vec{a}_1$ and $\vec{a}_2$) can be chosen as well as any angle between them. However, these generic, *oblique* lattices like that shown in Fig. 1.6a will only have 2-fold rotational symmetry unless special restrictions are applied to the lattice vectors. Special lattices, known as *Bravais lattices*, can be obtained with higher degrees of rotational symmetry by placing restrictions on the lengths and angles between the two lattice vectors $\vec{a}_1$ and $\vec{a}_2$. For 2D, there are just four other lattices that can be constructed with other than 2-fold symmetry. These are shown in Fig. 1.6. Note that 5-fold symmetry is not possible. As one can see in Fig. 1.6, primitive cells based on pentagons do not correctly fill space.

Additional symmetry operations under which certain space lattices will return to their original situation include:

(1) Mirror symmetry: reflection about a plane.
(2) Inversion symmetry: rotation by 180° about an axis followed by reflection through a plane normal to the rotation axis.

(3) Glide symmetry: a combination of reflection and translation.

(4) Screw symmetry: a combination of rotation and translation.

1.2.1 Conventional cells

One of the lattices presented in Fig. 1.6 is not a primitive lattice. The lattice shown in Fig. 1.6d has lattice vectors identical with those in Fig. 1.6c, but has an additional lattice point at the center of the cell. In this instance, the two lattice vectors mark off a *conventional unit cell* (non-primitive) referred to as a 'body-centered' (BC) rectangular lattice. Conventional cells are often introduced as an alternative to their primitive lattices as they afford a better visualization of the geometrical structure.

1.3 Bravais lattices

Our discussion of 2D lattices has laid much of the groundwork for discussing lattices in three dimensions. The structures of 3D crystals are again defined by the combination of a space lattice, described by a set of translation vectors:

$$\vec{T} = h\vec{a}_1 + k\vec{a}_2 + l\vec{a}_3, \tag{1.5}$$

where h, k and l are the complete set of integers, and an appropriate set of basis vectors:

$$\vec{R}_i = x_i\vec{a}_1 + y_i\vec{a}_2 + z_i\vec{a}_3, \tag{1.6}$$

that locate the contents of each unit cell in relation to any given lattice point. The volume of a 3D cell is now given by

$$V = |\vec{a}_1 \cdot \vec{a}_2 \times \vec{a}_3| \tag{1.7}$$

and is smallest for any of the possible primitive cells that can be constructed.

While any sort of generic lattice could be created with appropriate choice of the lengths of the three lattice vectors ($\vec{a}_1$, $\vec{a}_2$ and $\vec{a}_3$) as well as the angle between them, symmetry considerations lead to only 13 other, special or *Bravais lattices*. All 14 lattice types are illustrated in Fig. 1.7. The generic lattice (with arbitrary lengths and angles between $\vec{a}_1$, $\vec{a}_2$ and $\vec{a}_3$) is known as the triclinic, and the other 13 are grouped into six sub-categories based on how the lattice vectors are restricted to produce a unique symmetry: monoclinic, orthorhombic, tetragonal, cubic, trigonal and hexagonal. In addition to the primitive cell forms, some of these categories also contain conventional cell forms. These are non-primitive cells in which more than one lattice point is included in the cell. As the majority of crystals found in nature assume either a cubic or a hexagonal lattice structure, we focus next on the detailed properties of these two lattice types.

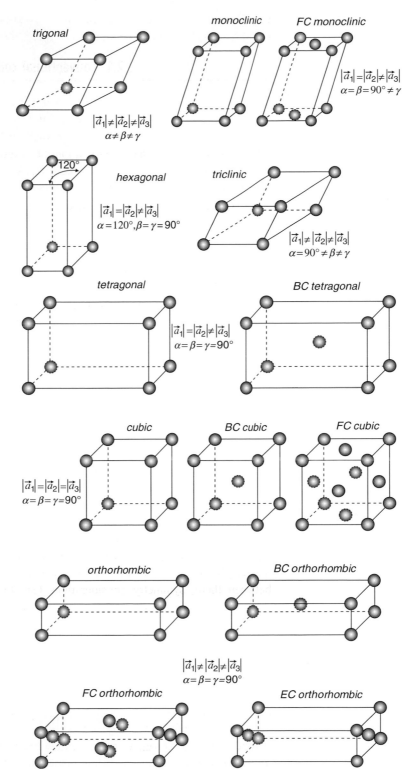

Figure 1.7

The complete set of 14 Bravais lattices in three dimensions grouped into seven sets: trigonal, monoclinic, hexagonal, triclinic, tetragonal, cubic and orthorhombic.

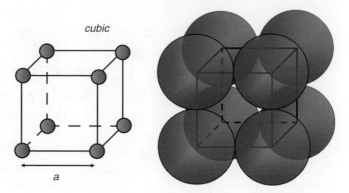

cubic

a

Figure 1.8   The simple cubic lattice. The ideal packing fraction is determined by imagining that balloons at each lattice site are inflated to just touch one another, as shown in the right-hand figure.

1.3.1 Cubic lattices

The cell of a simple cubic (SC) lattice, shown in Fig. 1.8, is a primitive cell because it contains just one lattice site. To see this, you may need to pretend that the lattice points in each corner of the cube are actually finite-sized balls. One-eighth of a ball at each corner resides inside the cube while the remaining seven-eighths resides in other adjacent cells. Thus, in spite of it having eight corners, any given cell contains just one total lattice site. Because the lengths of the three lattice vectors are equivalent and orthogonal, the cell volume is simply $V = |\vec{a}_1 \cdot \vec{a}_2 \times \vec{a}_3| = a^3$.

This simple cubic structure is to be contrasted with the two conventional cell structures of the body-centered cubic (BCC) and face-centered cubic (FCC) types. The body-centered cubic (BCC) has *two* lattice points per cell (one in the center and one-eighth in each of eight corners) and the face-centered cubic (FCC) has *four* lattice points per cell (one-eighth in each of the corners and one-half in each of six faces). While the FCC and BCC cells are conventional, each can alternatively be described by corresponding primitive cells affixed to a non-cubic lattice. Since a primitive cell must contain only one lattice point, a direct way of constructing these primitive cells would be to assign an origin to any one of the lattice sites, and choose lattice vectors ($\vec{a}_1'$, $\vec{a}_2'$ and $\vec{a}_3'$) that correspond to the shortest distance to three neighboring lattice sites which are oriented so as to trace out a rhombohedral with the minimum volume. The result is shown in Fig. 1.9 for both situations. With a little effort, one can show (Ex. 1 and Ex. 2) that the volumes of these two primitive cells are $a^3/2$ for the BCC and $a^3/4$ for the FCC, as expected.

Packing fractions

Another important property of any structural arrangement of particles is the *packing fraction*. The ideal packing fraction is a measure of how much space is

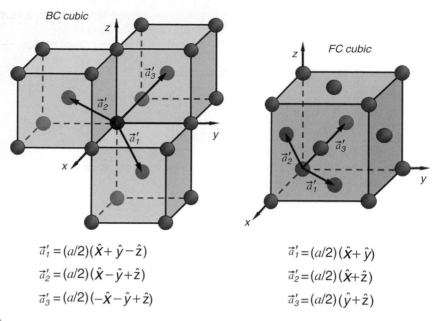

$$\vec{a}_1' = (a/2)(\hat{x} + \hat{y} - \hat{z})$$
$$\vec{a}_2' = (a/2)(\hat{x} - \hat{y} + \hat{z})$$
$$\vec{a}_3' = (a/2)(-\hat{x} - \hat{y} + \hat{z})$$

$$\vec{a}_1' = (a/2)(\hat{x} + \hat{y})$$
$$\vec{a}_2' = (a/2)(\hat{x} + \hat{z})$$
$$\vec{a}_3' = (a/2)(\hat{y} + \hat{z})$$

Figure 1.9   Lattice vectors for the corresponding primitive lattice of both the BC cubic and FC cubic conventional lattices.

occupied by identical spherical atoms when they are placed on the lattice sites such that they just touch one another. Imagine that inflatable balloons are located at the corners of the cube in Fig. 1.8 and are inflated at equal rates until they just begin to touch. Each balloon will have a radius equal to half the lattice spacing ($a/2$) and one-eighth of its volume will reside inside the cube. Hence the total space inside the cube that is physically occupied by the inflated balloons is $4\pi(a/2)^3/3 = 0.524a^3$, and the fraction of *occupied* space (the packing fraction) would be 0.524. A similar analysis (Ex. 3 and Ex. 4) of the BCC and FCC lattices, results in packing fractions of 0.680 and 0.740, respectively. These higher packing fractions are to be anticipated since, in each case, the BCC and FCC structures represent an attempt to compensate for the empty space of the SC lattice, which is seen in Fig. 1.8 to be concentrated in the cube center and at the center of each face.

Coordination spheres

Yet another characteristic of lattice structure is its coordination properties. This concerns the number of nearest (or next-nearest) neighboring lattice points and their distance. Consider again the SC lattice in Fig. 1.8. For a given lattice point, the shortest distance to another lattice site is the lattice spacing a. If we search around any given lattice point at this distance we will encounter six other lattice sites. Hence, the nearest neighbor *coordination number* for the SC

Table 1.1 Properties of cubic lattices.			
	SC	BCC	FCC
Conventional cell volume	a^3	a^3	a^3
Lattice points per cell	1	2	4
Primitive cell volume	a^3	$a^3/2$	$a^3/4$
Nearest neighbors	6	8	12
Nearest neighbor separation	a	$\sqrt{3}a/2$	$a/\sqrt{2}$
Next-nearest neighbors	12	6	6
Next-nearest neighbor separation	$\sqrt{2}a$	a	a
Packing fraction	$\pi/6 = 0.524$	$\sqrt{3}\pi/8 = 0.680$	$\sqrt{2}\pi/6 = 0.740$

(a)   (b)

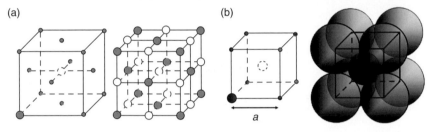

Figure 1.10   (a) The FC cubic structure of NaCl. Left-hand figure highlights the diatomic basis set consisting of one Cl anion (large solid circle) and one Na cation (large open circle in cube center). Right-hand figure shows the result when this basis set is attached to the sites of a FC cubic lattice.
(b) The simple cubic structure of CsCl. Left-hand figure highlights the diatomic basis set consisting of one Cl anion (large solid circle) and one Cs cation (large open circle in cube center). Right-hand figure illustrates how the smaller size of the Cs cation is comfortably fitted to the void space present in the SC center.

lattice is six. Likewise, the next largest distance to another lattice site is $\sqrt{2}a$ and we will find a *next-nearest* neighbor coordination number of 12. A summary of the coordination properties and ideal packing fractions of cubic lattices is provided in Table 1.1.

Rocksalt and diamond

Now let's consider some common examples of cubic crystals found in nature to see how their lattice structures arise. First we consider NaCl (rocksalt) whose Na^+ and Cl^- atoms are arranged as shown in Fig. 1.10a. The structure is built from a FCC space lattice containing a basis of two atoms:

$$\text{Cl atom: } \vec{R}_{Cl} = 0\vec{a}_1 + 0\vec{a}_2 + 0\vec{a}_3$$
$$\text{Na atom: } \vec{R}_{Na} = \frac{1}{2}\vec{a}_1 + \frac{1}{2}\vec{a}_2 + \frac{1}{2}\vec{a}_3. \tag{1.8}$$

Because the FCC conventional cell contains four lattice points, we should find four units of the chemical formula (NaCl) present in each cell. Let's check. Each one of eight corners provides one-eighth of a Cl anion and each one of six faces provides one-half of a Cl anion inside the cell, resulting in a total of four Cl anions per cell. Likewise, there is one entire Na cation in the cell center and each of 12 edges provide one-quarter of a Na cation, resulting in a total of four Na cations per cell.

Compare this NaCl structure with that of a chemically equivalent salt, CsCl, shown in Fig. 1.10b. Interestingly enough, the structure here is *not* built on the FCC lattice, but on the SC lattice with a basis of two atoms:

$$\text{Cl atom: } \vec{R}_{Cl} = 0\vec{a}_1 + 0\vec{a}_2 + 0\vec{a}_3$$
$$\text{Cs atom: } \vec{R}_{Cs} = \frac{1}{2}\vec{a}_1 + \frac{1}{2}\vec{a}_2 + \frac{1}{2}\vec{a}_3. \tag{1.9}$$

Because the SC is a primitive cell, it should contain just one unit of the chemical formula CsCl, and indeed it does. Why though does CsCl not assume the FCC structure like NaCl? This difference stems from the differing sizes of the ions and an inherent tendency for nature to favor efficient packing of space. In NaCl, the Na^+ and Cl^- are nearly equal in diameter and, in this instance, space is best filled by using the FCC structure. However, Cs^+ is much smaller in diameter and space is more efficiently occupied using the SC lattice. Thus the crystalline structure realized in nature is a consequence of many factors including the size of the particles, their bonding requirements, and a desire to minimize empty space.

As another example, we consider the structure of diamond. Diamond is composed entirely of C atoms that are bonded covalently. Because of the discrete nature of the covalent bond, each C atom must form a single covalent bond with four other C atoms in order to satisfy the requirement of a closed electronic shell configuration. This bonding requirement promotes a *tetragonal* aspect of the diamond structure, which can be seen in Fig. 1.11a. Here the diamond structure is composed of an FCC lattice with a basis of two identical carbon atoms located at

(a)   (b)

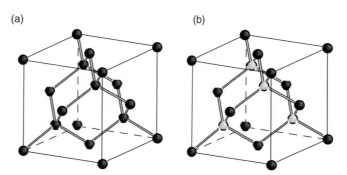

Crystal structure (FC cubic) of (a) diamond and (b) zincblende (ZnS).

$$\vec{R}_1 = 0\vec{a}_1 + 0\vec{a}_2 + 0\vec{a}_3$$
$$\vec{R}_2 = \frac{1}{4}\vec{a}_1 + \frac{1}{4}\vec{a}_2 + \frac{1}{4}\vec{a}_3 \qquad (1.10)$$

Zincblende (ZnS), shown in Fig. 1.11b, is identical in structure to that of diamond, except that the basis contains two dissimilar atoms:

$$\vec{R}_S = 0\vec{a}_1 + 0\vec{a}_2 + 0\vec{a}_3$$
$$\vec{R}_{Zn} = \frac{1}{4}\vec{a}_1 + \frac{1}{4}\vec{a}_2 + \frac{1}{4}\vec{a}_3 \qquad (1.11)$$

1.3.2 Hexagonal lattices

The primitive cell of the hexagonal space lattice, shown in Fig. 1.12a, looks nothing at all like a hexagon. Instead it resembles a tall rectangular box that has been squished from four 90° angles to a pair of 120° and 60° angles, respectively. The hexagonal appearance only emerges when three or more of these boxes are combined.

Hexagonal close packed (HCP)

By far the most prominent occurrence of hexagonal structure in nature appears in the form of the *hexagonal close packed* (HCP) structure in which the

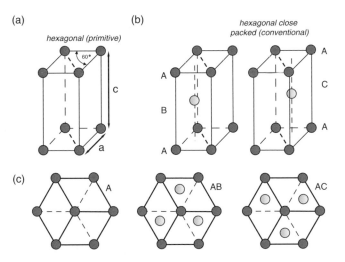

Figure 1.12   (a) Primitive cell of the hexagonal lattice and (b) corresponding conventional cells for the hexagonal close packed structure. The two conventional cells differ only in the location of the central lattice site (B versus C). (c) Three conventional cells are combined to form a hexagonal base (A) with two possibilities for the central layer (AB versus AC).

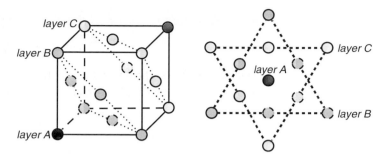

Figure 1.13   The lattice sites of a conventional FC cubic lattice (left-hand figure) are shown to be equivalent to the lattice sites of the hexagonal layering ABCABC . . ..

primitive cell contains a two particle basis. As shown in Fig. 1.12b, the second particle is located on one side or the other of the squished box midway up. The simplest way to view the HCP structure is to consider that it is constructed by alternating layers of the 2D hexagonal lattice. When we combine three primitive cells each with two basis particles, our bottom layer (layer A) appears as a hexagonal net (see Fig. 12c). The next layer (layer B) is formed by the second particle of the basis and again forms another hexagonal net which is *offset* from layer A such that the particles in layer B minimize waste space by fitting into some of the shallows of layer A. If the second particle in the basis set happened to be located on the other side of the primitive cell, an alternative layer (layer C) could likewise be positioned atop layer A. In either case, the second layer is then covered by another layer A, directly over the first layer, which corresponds to the top of the three primitive cells.

Layering of the form ABABAB . . . or ACACAC . . . makes up the HCP structure. But layering of the form ABCABC . . . does not! As shown in Fig. 1.13, this third layering pattern just reproduces the FCC structure. The packing fractions of the HCP and FCC are identical and both correspond to the best packing efficiency possible for an ordered arrangement of spheres.

Summary

- The arrangement of atoms in any ideal crystal can be described by a combination of a space lattice (defined by the set of translations, $\vec{T} = h\vec{a}_1 + k\vec{a}_2 + l\vec{a}_3$) and a basis set (defined in reference to a lattice site by $\vec{R}_i = x_i\vec{a}_1 + y_i\vec{a}_2 + z_i\vec{a}_3$) affixed to each lattice site.

- A primitive cell contains only one lattice site and has the smallest volume needed to fill space by the translations, $\vec{T}$. Conventional cells contain more than one lattice site.

- Space lattices are categorized according to their special symmetry properties into 14 distinct types known as Bravais lattices.

- The ideal packing fraction is the ratio of cell space occupied by a monatomic basis with atoms of maximal diameter to the volume of the cell itself.

- The coordination number refers to the number of neighboring lattice sites located at a common distance from a central site.

Exercises

1.1. Show that the volume of the primitive cell of a BCC crystal lattice is $a^3/2$, where a is the lattice constant of the conventional cell.

1.2. Show that the volume of the primitive cell of a FCC crystal lattice is $a^3/4$, where a is the lattice constant of the conventional cell.

1.3. Show that the packing fraction of a BCC crystal lattice is $\sqrt{3}\pi/8 = 0.680$.

1.4. Show that the packing fraction of a FCC crystal lattice is $\sqrt{2}\pi/6 = 0.740$.

1.5. The 2D crystal shown in Fig. 1.14 contains three atoms with a chemical formula ABC_2. Illustrated in the figure are several possible tiles. (a) Identify which of the tiles are primitive cells. (b) Identify which of the tiles are conventional cells. (c) Identify any tiles that are unable to correctly fill the space. (d) For each primitive cell, provide expressions for the appropriate basis vectors describing the basis set of atoms.

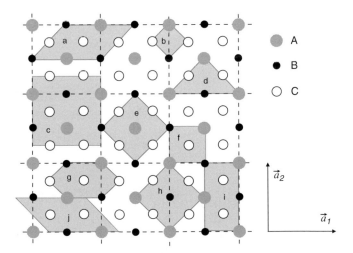

Figure 1.14

1.6. Consider again the 2D crystal shown in Fig. 1.14. Describe all the basic symmetry operations (translation, rotation and mirror only) satisfied by this lattice.

1.7. For the HCP crystal structure, show that the ideal c/a ratio is 1.633.

1.8. Bromine has an orthorhombic lattice structure with $|\vec{a}_1| = 4.65$ Å, $|\vec{a}_1| = 6.73$Å, $|\vec{a}_1| = 8.70$Å. (a) The atomic weight of bromine is 79.9 g/mol. If it has a density of 3.12 g/cc, how many bromine atoms reside in a single unit cell? (b) Which type of orthorhombic lattice (i.e, BC, FC, etc.) is suggested by your finding in part (a)? Explain. (c) If the atomic radius of bromine is 1.51Å, determine the packing fraction.

1.9. Shown in Fig. 1.15 is the unit cell of a monatomic crystal. (a) How would you describe this particular crystal structure? (b) What is the maximum packing fraction you should expect for this specific structure?

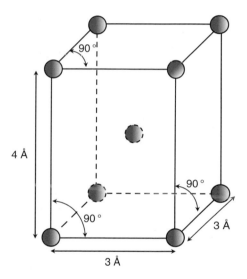

Figure 1.15

Suggested reading

There are many good introductory textbooks that develop crystal structure. These are just a few favorites:

C. Kittel, *Introduction to Solid State Physics*, 8th Ed. (John Wiley and Sons, 2005).

J. S. Blakemore, *Solid State Physics*, 2nd Ed. (W. B. Saunders Co., Philadelphia, 1974).

N. W. Ashcroft and N. D. Mermin, *Solid State Physics* (Holt, Rinehart and Winston, New York, 1976).

M. A. Omar, *Elementary Solid State Physics: Principles and Applications* (Addison-Wesley, Reading, MA, 1975).

Amorphous structure

Introduction

In the previous chapter, we saw how the long-range order of particle positions in a crystal could be described in a rather elegant manner using a space lattice that extends indefinitely. In this chapter, we examine instead disordered matter such as liquids or glasses in which a long-range repeated pattern is absent. These amorphous materials might not seem as glamorous as their crystalline counterparts, but they are increasingly prevalent in our world as they comprise the windows, computer screens and vast array of plastic components that surround us on a daily basis. In comparison with crystalline structures, these amorphous materials pose a challenge to describe, and their structure can only be defined in a statistical sense by introducing an ensemble-averaged, *pair distribution function*. In spite of their disordered nature, a robust pattern of particle positions emerges over short distances. This *short-range order* reflects the local coordination of particles and we briefly review the *random close pack* and the *continuous random network* systems as common examples of amorphous structure.

2.1 A statistical structure

Disordered or amorphous condensed matter has a clear disadvantage in that particle positions lack any long-range repeating pattern akin to that found in crystals. This is evident in Fig. 2.1, which illustrates the typical particle positions of either a glass or a liquid captured at a particular instant in time.

We face a dilemma. The absence of a repeating pattern means that the crystal description based upon a lattice described by a set of translation vectors $(\vec{T} = h\vec{a}_1 + k\vec{a}_2 + l\vec{a}_3)$ will not work in an amorphous system. Looking at Fig. 2.1, one might be inclined to believe that there is no useful pattern of particle positions at all. This, however, is not the case. Patterns do emerge, in a statistical sense, that provide for a robust description of the amorphous structure.

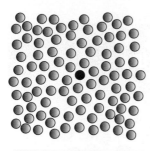

Figure 2.1

Typical, non-repeating structure for a disordered monatomic liquid or glass. The central atom (black) represents a possible point of observation.

2.1.1 Ensemble averaging

Let us return to a thought experiment we performed in the last chapter. When we examined crystalline structures we imagined sitting on any given lattice site (or particle associated with it) and observing our surroundings. For the crystal, we observed the surroundings to be unaffected for certain symmetry operations (e.g. translation, rotation, reflection). But what happens if we attempt this same experiment in an amorphous structure like that depicted in Fig. 2.1?

Imagine we choose a particle at random to sit on and then survey our surroundings. Our record of observations would include a handful of particles found nearby and a larger number of others in the distance. If we sat on some other particle, the details might be a little different, but some aspects would be retained. We still see a handful of nearby particles and larger numbers off in the distance. Now, if we desire to characterize the amorphous structure as a whole, we cannot just use an observation taken when sitting on a single given particle. Rather, we need to *average* the observations taken when we have sat on each and every particle in turn. Or, alternatively, if our amorphous system is a liquid in which the particles are moving about, we could sit on a given particle and average the observations of our surroundings as they evolve in time from one configuration to the next.

These sorts of averages are known in statistical mechanics as *ensemble averages*. In each situation, the properties of a large number of thermodynamically equivalent configurations (ensemble members) of a given system are averaged to produce a *robust* result that is characteristic of the system as a whole. As an example, imagine that a large bag of marbles is vigorously shaken and the density is then measured. The particular density found will depend upon the actual configuration of the marbles, but if the shaking and measuring is repeated many times, a robust and reproducible value of density will be found that is characteristic of a bag of marbles. For amorphous systems, this sort of ensemble averaging must be performed as part of any robust description of the structure as a whole.

2.1.2 Symmetry

Now that we have carried out our observations, along with some appropriate averaging, what do we find? Firstly, if we sit at *any point* in space (either on a particle or elsewhere) and rotate about, we will find (on average) the same surroundings, regardless of the angle through which we turn or the axis about which we turn. Unlike our cubic crystal that had a 4-fold rotational symmetry in which only multiples of 90° turns could be executed to return the surroundings to their original configuration, our disordered system is *isotropic* and has an *infinite* rotational symmetry.

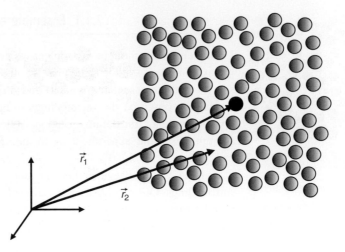

Figure 2.2 Amorphous structure is described in reference to correlations between particle positions. Given that a particle is present at $\vec{r}_1$, we inquire as to the probability that another is present at some arbitrary $\vec{r}_2$.

Secondly, if we consider making a translation from any arbitrary point in space to some other location, we will find that (1) the direction we choose to move does not matter (i.e. the structure is isotropic), and (2) that the distance we move is (on average) irrelevant. Thus, with appropriate ensemble averaging, amorphous materials are both rotationally and translationally invariant and possess a high degree of symmetry.

Given this invariance, one might speculate that there is nothing useful to conclude about the structure. However, while translations from any arbitrary point to another are invariant, there do emerge robust patterns in the surroundings when viewed from the perspective of the center of any particle. These patterns arise from the pairwise forces acting between the particles and produce an inherent structure in the form of *correlations* between particle positions.

2.1.3 The pair distribution function

In the crystal structure, the presence of a repeating pattern produced a strong correlation between particle positions. That is, given the position of any one particle we could predict with 100% accuracy where other particles would be located. This correlation is formally expressed in terms of a *pair distribution function*, $g(\vec{r}_1, \vec{r}_2)$, which is defined (see Fig. 2.2) to represent the conditional probability that, given that there is a first particle center located at $\vec{r}_1$, a second particle center will be located at $\vec{r}_2$. For a crystal lattice, $g(\vec{r}_1, \vec{r}_2)$ would be zero, except for the condition when $\vec{r}_2 - \vec{r}_1$ equals any one of the lattice translations $\vec{T}$. In an amorphous structure, such a 'hit or miss' probability is

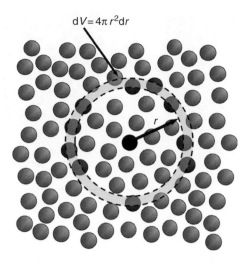

Figure 2.3 The local particle density, $\langle n(r) \rangle_{excl}$, is defined by the number of particle centers (black circles) resident within the spherical volume element $dV = 4\pi r^2 dr$ located a distance r from the central particle divided by the volume element itself.

not present. Nevertheless, we can explore similar statements about the *probability* that two particles will (on average) be located relative to one another.

Because an amorphous material is isotropic, the only feature of $\vec{r}_2 - \vec{r}_1$ that matters is the magnitude, $r = |\vec{r}_2 - \vec{r}_1|$. Consequently, for amorphous materials:

$$g(\vec{r}_1, \vec{r}_2) \rightarrow g(r). \qquad (2.1)$$

Again, *g(r)* is the conditional probability that another particle center will be found a distance r from the center of the first particle. As depicted in Fig. 2.3, this probability is proportional to the average number of particle centers found in a small shell of radius r and thickness dr:

$$g(r) \propto \left\langle \frac{\text{\# particle centers in dV}}{4\pi r^2 dr} \right\rangle, \qquad (2.2)$$

where the brackets ($\langle \cdots \rangle$) indicate that this quantity is appropriately ensemble averaged. This quantity is just a *local* density of particle centers (excluding the first particle), $\langle n(r) \rangle_{excl}$, with units of inverse volume. To make this look more like a dimensionless probability, we will divide it by the overall (*global*) average density, $\langle n \rangle$, so as to define the pair distribution function as:

$$g(r) = \frac{\langle n(r) \rangle_{excl}}{\langle n \rangle} = \frac{1}{\langle n \rangle} \left\langle \frac{\text{\# particle centers in dV}}{4\pi r^2 dr} \right\rangle. \qquad (2.3)$$

Since the volume element ($dV = 4\pi r^2 dr$) grows in size as r increases, the local density at large values of r will eventually mimic that of the global density. Thus at large r, *g(r)* tends to a value of unity, implying that the

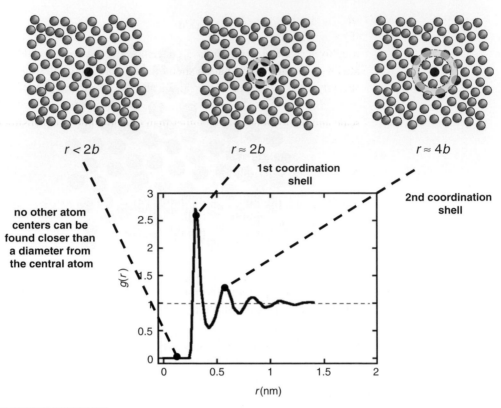

$r < 2b$

no other atom centers can be found closer than a diameter from the central atom

$r \approx 2b$

1st coordination shell

$r \approx 4b$

2nd coordination shell

Figure 2.4 The pair distribution function for a monatomic liquid of particles of diameter $2b$. For particles with a hard interaction, no two particle centers can be located at less than $2b$ apart. The distribution function exhibits a sharp peak at the first coordination shell (near $r = 2b$), where a relatively large density of neighboring particles are located. Higher-order coordination shells are marked by weaker peaks in $g(r)$.

probability of finding a second particle center is no better or worse than the random odds associated with an average density of particles.

At short distances, however, one finds correlations (undulations) in $g(r)$ caused by pairwise interactions between the central particle and its nearest neighbors. Because of these interactions, the local density near the first particle varies rapidly in response to the coordination shells of nearest and next-nearest neighboring particles. A typical example of the pair distribution function for a glass or liquid is shown in Fig. 2.4, and the following features should be noted.

(1) **$g(r) = 0$ very near $r = 0$.**

Because $g(r)$ is a *conditional* probability, it only refers to the location of a second particle. If the particles are taken to be hard spheres of radius b, then a second particle center cannot be located any nearer the first than $r = 2b$, without the two particles penetrating into one another.

(2) **g(r) is maximum near r = 2b.**

If we travel outwards from the center of the first particle by a distance of roughly $r = 2b$, we encounter a concentration of particle centers associated with the first coordination shell of neighboring particles that are most strongly attracted to, and thus in a sense "pressed up against", the first particle. There is a strong correlation between the particles at this length scale and the local density is higher than the global density causing $g(r)$ to attain a value far in excess of unity.

(3) **g(r) is minimum near r = 3b.**

As we continue past the first coordination shell but not yet to the second (next-nearest neighbors) coordination shell, we find a region of space that is generally void of particle centers, causing the local density to fall below that of the global density.

(4) **g(r) displays weak maxima and minima at r ≈ 4b, 5b, 6b, etc.**

Beyond the first coordination shell, we observe evidence for weakened correlations associated with higher-level coordination shells in a series of damped oscillations in $g(r)$.

(5) **g(r) → 1 as r → ∞.**

These correlations eventually vanish and at large length scales the likelihood of finding a second particle center becomes indistinguishable from random chance.

The undulations in $g(r)$ for an amorphous system are mainly found only for distances corresponding to a few layers of coordination and, for this reason, are described as *short-range order*, unlike the long-range correlations found in crystals.

Radial distribution function and coordination numbers

In the case of crystals, we also characterize the lattice structure in terms of coordination numbers and distances. Each peak in $g(r)$ represents a coordination shell whose coordination distance corresponds to the location of the peak. We can likewise determine a corresponding coordination number by integration. If we rearrange Eq. (2.3), we can express the number of particle centers in a small volume element as:

$$\langle \text{\# particle centers in dV} \rangle = \langle n \rangle g(r) 4\pi r^2 dr. \qquad (2.4)$$

The quantity, $\langle n \rangle g(r) 4\pi r^2$, is often referred to as the *radial distribution function* and, when it is integrated over a range of r corresponding to a specific peak in $g(r)$, as illustrated in Fig. 2.5, we obtain the coordination number for the corresponding shell:

$$\langle \text{\# particle centers in shell} \rangle = \int_{\text{peak in } g(r)} \langle n \rangle g(r) 4\pi r^2 dr. \qquad (2.5)$$

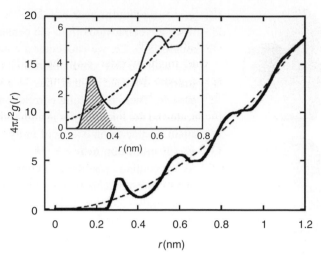

Figure 2.5   Radial distribution function based on $g(r)$ presented in Fig. 2.4. Note how the undulations in $g(r)$ are now superimposed on $4\pi r^2$ (dashed curve). Inset illustrates how integration of the area under the first peak of the radial distribution function (hashed area) produces the coordination number of the first shell.

2.2 Two amorphous structures

Amorphous materials possess short-range order as evidenced by correlations in $g(r)$ at short distances. However, there are two distinct variants in the short-range order that depend upon the nature of the bonds that are present between the particles. In some materials, the bonding is isotropic: a given particle can form multiple bonds with other particles at arbitrary directions and distances. This is typical of van der Waals bonds and ionic bonds. In other, covalently bonded, materials the bonding is discrete: a given particle forms only a limited number of bonds with neighboring particles. These two extremes of bonding produce differences in the nature of the short-range order of the amorphous structure. Isotropic bonds result in what is known as a *random close packed structure*, while discrete bonds result in a *continuous random network*.

2.2.1 Random close packed structure

The random close packed (RCP) structure corresponds to the structure illustrated in Fig. 2.1 and is that which forms when hard spheres (marbles, ball bearings) are packed together. Indeed much of the investigation of the RCP structure is derived from studies (using either real balls or simulations on a computer) of the packing of hard spheres. These studies reveal that random packing leads to a poorer filling of space than that of hexagonal close packing

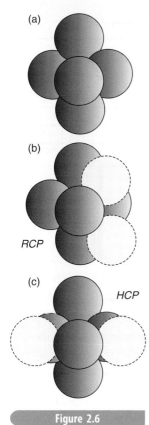

(a)

(b)

RCP

(c)

HCP

(a) Five spherical particles are assembled while maximizing the packing fraction. (b) When the sixth particle is added it can either lead to the RCP structure, if the minimization of wasted space is enforced on a local scale, or to the HCP structure, if the minimization of wasted space is only enforced on the global scale.

(HCP), discussed in the previous chapter. The RCP structure averages a packing fraction of about 64% in comparison with the 74% found in the crystalline HCP.

One might wonder why the RCP structure is less efficient at filling space, given that we are no longer constrained by a repeating pattern but are free to jam balls together in any fashion we wish. There is an insightful moral underlying the differing packing efficiencies of the (ordered) HCP and (disordered) RCP. In the HCP, it appears that Nature "plans ahead" in adopting this structure of long-range order, knowing that this entire structure will more efficiently fill space in the end. Contrast this with what happens when particles are assembled without the benefit of a blueprint in a haphazard "one-at-a-time" process, guided only by a desire to minimize wasted space at each step. If we took marbles and glued them together one at a time, all the while striving for the most compact assembly, by the time we had glued five together we would have obtained a bipyramid arrangement, like that shown in Fig. 2.6a.

So far, this arrangement of particles is identical to ones found in the HCP structure. However, when we add the sixth marble (Fig. 2.6b), our narrow-minded recipe dictates that we place it on the recess of any one of the six faces of the bipyramid, and this is where we begin to deviate from the HCP blueprint. Although we have worked to maintain a compact, space-saving arrangement, as we continue to add marbles to our assembly, we will quickly find that we are developing large voids of empty space! In the case of the HCP, Nature, by planning ahead, accepts a small amount of empty space on a local scale to achieve an overall conservation of empty space on a larger scale. In the case of the RCP, there is no blueprint to follow and the tendency to fill space on a local scale eventually results in less efficient packing.

In addition to differences in packing, the HCP and RCP structures have other characteristic differences. Consider again our bag of marbles and suppose we construct about the center of each marble the Wigner–Seitz cell in which a perpendicular face is placed at the midpoint of the line connecting the marble center with every other nearby marble center, as shown in Fig. 2.7a. The resulting Wigner–Seitz cell will assume the form of a polyhedron known as a *Voronoi polyhedron*, and will have variations in the number of faces per cell and the number of edges per face, depending on the marble in question. If we had constructed Wigner–Seitz cells for the crystalline HCP structure, all the polyhedra would be identical with 12 faces per cell and four edges per face. However, for the RCP structure, a *distribution* of polyhedra arises, as shown in Fig. 2.8, with a high occurrence of pentagonal faces and a larger number of faces per cell than found for the HCP.

The random packing of M&M ellipsoids

For many years, it was largely assumed that any and all forms of random packing were limited to a packing fraction no greater than the 64% seen for

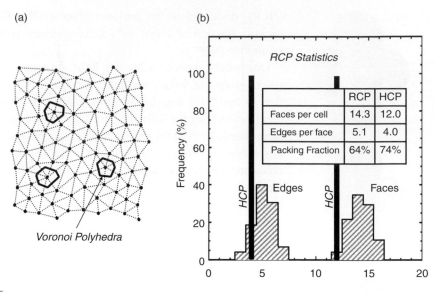

(a)

(b)

Voronoi Polyhedra

Figure 2.7 (a) The RCP lattice is decomposed into Wigner–Seitz cells (also known as Voronoi polyhedra). Like snowflakes, no two polyhedra are exactly alike. (b) The RCP structure can, however, be characterized by the distribution of Voronoi polyhedra types according to the number of faces and number of edges they possess. Solid vertical lines locate the corresponding properties for the Wigner–Seitz cell of the HCP structure for comparison. (Adapted from Finney (1970).)

packing of hard spheres. However, in 2004, students working in the lab of David Weitz surprised the condensed matter community when they observed a packing fraction of 72% for the random packing of ellipsoids (actually M&M candies).

Why would ellipsoids pack better than spheres? The answer seems to be that, unlike spheres, ellipsoids experience both a force and a torque due to contact with neighbors. When a sphere is pressed by other spheres, the forces act only along the radius without producing any torque about the sphere's center. Hence, the sphere is stabilized against translation and rotation by balancing the forces alone. In order to stabilize an ellipsoid, both the forces and the torques must be balanced. It seems that the added requirement of balancing the torque necessitates an increased number of neighbors (on average), and that this is responsible for the larger packing fraction.

2.2.2 Continuous random network

The RCP structure is most often encountered in amorphous materials that interact by weak, isotropic forces such as ionic bonds and van der Waals bonds. In the case where covalent bonds produce the condensed phase, a *continuous random network* (CRN) of covalent bonds is formed. This arises because the

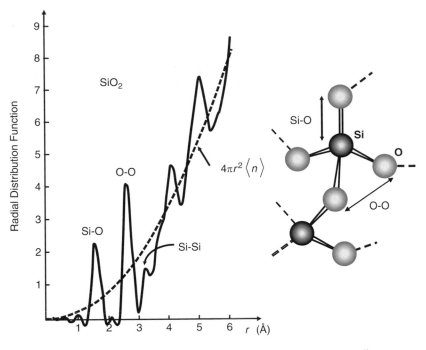

Figure 2.8   The radial distribution function of amorphous silica obtained by neutron scattering. Short-range tetrahedral ordering is indicated by the series of peaks corresponding to increasing particle separations: the Si-to-O bond length, the O-to-O separation and the Si-to-Si separation. (Adapted from Henniger, Buschert and Heaton (1967).)

covalent bonds form discrete linkages between particles and thus lead to a restriction on the number of particles present in the first coordination shell. As an example, consider the structure of amorphous SiO_2. Silicon is a group 4 element of the periodic table and so requires four shared electrons to mimic the closed electron configuration of an inert gas element. It obtains these four additional electrons (one each) by forming a single covalent bond with each of four oxygen atoms. Oxygen resides in group 6 of the periodic table, and so needs two additional electrons to mimic the electron configuration of an inert gas. To do this, each oxygen forms a single covalent bond with each of two Si atoms. Thus, a continuous network of Si and O atoms is produced in which each Si bonds to four O and each O bonds to two Si, as shown in Fig. 2.8.

Short-range order

In its crystalline form, SiO_2 (quartz) is constructed of SiO_4 tetrahedra, each with identical bond angles of $109°$ for the O–Si–O linkage and identical bond lengths. In the amorphous form, these angles and bond lengths vary somewhat

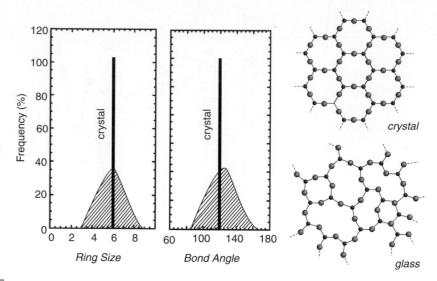

Figure 2.9 A schematic comparison of the covalent network structure of a two-dimensional crystal and glass. The disorder results in a distribution of bond angles and ring sizes that characterize the continuous random network structure.

from one tetrahedral unit to the next, forming a *distribution* of bond angles and lengths, as illustrated for a 2D analog in Fig. 2.9. What emerges is a network that possesses short-range order (SRO) but no long-range order (LRO). That is, while orderly on short distances in the sense that there is a discrete set of atoms in the first coordination shell owing to the tetrahedral units, variations in bond angles and lengths generate a network that becomes progressively more random at larger length scales. This SRO is evident in the radial distribution function for SiO_2, shown in Fig. 2.8, where the first three peaks correspond to the coordination of Si–O, O–O and Si–Si atoms, in increasing order of their separation.

Another insightful example of CRNs comes from so-called *chalcogenide* glasses (e.g. Se, S) that are based on elements from column 6 of the periodic table. Because Se and S reside in column 6 of the periodic table, each desires two electrons in covalent bonding to complete a closed electron configuration. This means that amorphous Se (or S) would be composed of chains of atoms in which each Se (or S) is connected to two others (Se or S). In the case of S, these chains tend to close up to form eight-membered rings. But for Se, they tend to form long extended chains or linear *polymers*. Disorder arises because the bond angles at each linkage can be altered slightly causing the chain to bend and fold to resemble something like a strand of spaghetti in a bowl of pasta (see Fig. 2.10a). The bending and folding, compounded with the interweaving of multiple chains, renders an absence of long-range order, despite a well-defined short-range order, with a coordination number of 2.

Figure 2.10   (a) The polymeric structure of amorphous Se is formed by long chains of covalently bonded atoms with two bonds per atom. (b) Additions of either As or Ge allow for crosslinks to form with sites that have either three or four covalent bonds, respectively.

The addition of small amounts of elements from nearby columns of the periodic table, such as As and/or Ge, produces crosslinks between the Se chains. For example, when As (column 5) is added, it will form three covalent bonds with Se to create an $AsSe_3$ unit. Likewise, if Ge (column 4) is added, it will form four covalent bonds with Se to produce a $GeSe_4$ unit. The effect of these additions, as shown in Fig. 2.10b, is to create randomly placed attachments between the polymer Se chains in a process known as *gelation*.

Summary

- The structural features of amorphous matter emerge only in a statistical sense via appropriate ensemble averaging, and only in reference to particles correlations about any arbitrarily chosen central particle.

- The pair distribution function, $g(r)$, is a ratio of the average local density found at some distance, r, from a central particle to the global density overall. Strong correlations appear as undulations in $g(r)$ associated with the first few coordination spheres.

- Short-range order (SRO) is characterized by $g(r)$ for $r \approx$ particle spacing and is sensitive to the specific nature of the pairwise forces between particles. The random closed packed (RCP) structure arises in situations of weak or non-existent bonding (e.g. in van der Waals liquids and systems of hard spheres), while stronger and discrete covalent bonds produce a continuous random network (CRN) structure.

- In comparison with their crystalline counterparts, amorphous materials are characterized by a distribution of bond angles and bond lengths.

Exercises

2.1. An ideal gas consists of non-interacting, point particles that move about in rapid and incessant fashion. Sketch the form of $g(r)$ for this ideal gas and discuss its features.

2.2. Find a copy machine and make a reproduction of Fig. 2.11 below that represents the atoms in an amorphous solid and, using a compass, manually calculate $g(r)$ for a single ensemble using the dark particle as the central particle. Do this with a dr no larger than the particle radius, b. Plot your result and identify the first and second coordination spheres.

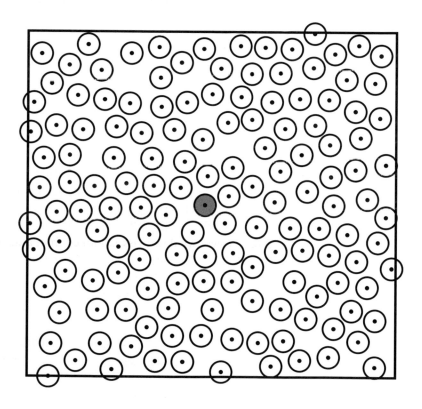

Figure 2.11

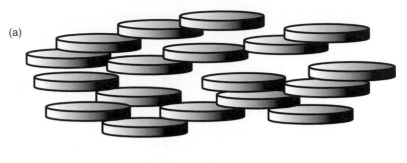

(a)

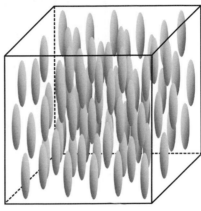

(b)

Figure 2.12

2.3. Figure 2.12 shows the nematic phases of two liquid crystals: (a) a discotic liquid crystal and (b) a lipid liquid crystal. For each of these partially amorphous systems, discuss the symmetry properties including both translational and rotational symmetries.

2.4. The pair distribution function for a bag of marbles is shown in Fig. 2.13. From the figure, (a) determine the nearest and next-nearest separation distances corresponding to the first and second coordination shells, and (b) estimate the coordination number for the first and second coordination shells.

2.5. A common chalcogenide glass is As_2Ge_3. Determine the average coordination number for this system.

2.6. Typical window glass is formed from a mixture of approximately 70% SiO_2, 20% Na_2O and 10% CaO, known as soda-lime-silicate. How does the addition of Na_2O and CaO affect the CRN of SiO_2 if the O donated by either is to end up bonded with a Si atom?

Suggested reading

For a good introduction to amorphous materials, I recommend the first three texts. Those interested in more details about the packing of ellipsoids are directed to the article listed last.

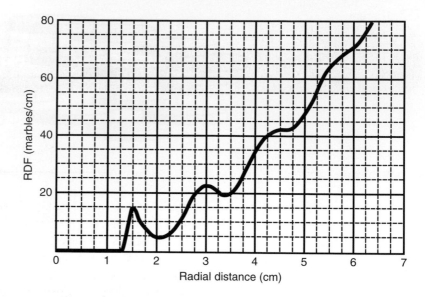

R. Zallen, *The Physics of Amorphous Materials* (John Wiley and Sons, New York, 1983).

J. Zarzycki, *Glasses and the Vitreous State* (Cambridge University Press, New York, 1991).

S. R. Elliott, *Physics of Amorphous Materials*, 2nd Ed. (John Wiley and Sons, New York, 1990).

D. A. Weitz, Packing in the Spheres, *Science* **303**, 968–969 (2004).

3 Bonds and cohesion

Introduction

What makes condensed matter "condensed"? The answer is stickiness (i.e. an attractive interaction) that exists between the particles that make up a liquid or solid. In the last two chapters we have looked at two extremes of particle arrangements in matter: ordered (crystalline) and disordered (amorphous). In this chapter, we now examine the nature of the forces that form between particles and which promote the formation of a condensed phase of matter. We begin by reviewing the five major bonds (van der Waals, covalent, ionic, metallic and hydrogen bonds) and conclude by considering the overall cohesive energy in a crystal. As thermodynamics generally favors the system with lowest energy, this cohesive energy is part of what determines why a particular crystal structure is adopted in Nature, rather than another structure.

3.1 Survey of bond types

In order for matter to condense, there must be an attractive force between the particles to promote their mutual gathering together. Of the four fundamental forces in Nature, the two nuclear forces (strong and weak) play no role in the condensation process and the gravitational force is far too weak to drive the process at ordinary terrestrial temperatures and pressures. Instead, the fundamental force that binds particles together in condensed matter arises from electrostatic interactions.

In spite of the common origin for the interaction, the bonds that form between particles are divided into five main types depending upon how the attractive interaction forms as a result of the distribution of charge on the particles themselves. These five bond types are: the van der Waals bond, the ionic bond, the covalent bond, the metallic bond and the hydrogen bond. Their relative strengths are indicated by a handful of examples presented in Fig. 3.1.

3.1.1 The van der Waals bond

Weak in comparison to other bond types, van der Waals bonds are always present in the condensed phase and are the only means by which the noble

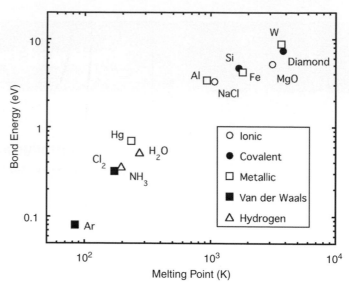

Figure 3.1   Plot of bond energy (per molecule) versus melting point temperature for a selection of materials that encompass all the five bond types (data from Callister (2000)).

elements in column 8 of the periodic table can be condensed from a gas. All of the atoms in this column have exactly the required electrons to completely fill an electron energy level. All the quantum numbers are balanced, resulting in zero net spin and a spherically symmetric electron charge distribution. Consequently, the atom as a whole produces no net external electric field and appears neutral.

Given this neutral appearance, one might then wonder how these elements are able to condense from a gas at all. They nevertheless do condense, albeit only at very low temperatures as indicated in Fig. 3.1. To understand where the attractive force between these elements comes from, we must recognize that the electron cloud surrounding the nucleus is not rigidly attached, but can be distorted slightly (say by small thermal fluctuations). When the electron cloud is displaced slightly off-center from the nucleus, a non-zero electric dipole moment results whose external electric field is no longer zero. Other atoms nearby can couple with the field produced by the first atom in such a way as to produce a weak, mutually attractive force whose pairwise potential energy varies inversely with the separation as:

$$u_{vdW}(r_{ij}) = -4\varepsilon \left(\frac{\sigma}{r_{ij}}\right)^{6}, \qquad (3.1)$$

where r_{ij} is the separation between the atom centers, and ε and σ are appropriately chosen energy and length scales, respectively.

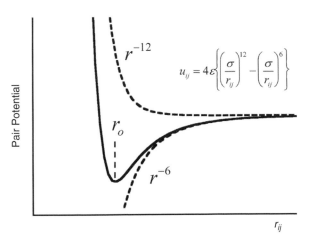

Figure 3.2 The van der Waals pair interaction (solid curve) showing the separate attractive and repulsive contributions (dashed curves).

Pauli repulsion

Equation (3.1) indicates that the potential energy of a pair of atoms interacting via the van der Waals force will be lowered as the two atoms approach one another, and would diverge should their positions coincide. This, however, does not occur because as the two atoms approach, there develops a stronger repulsive force due to the Pauli *exclusion principle*. As the atoms approach, first contact is made by the outer electrons of the electron clouds, and the electrons in the outer shell of one atom attempt to occupy the same space as electrons from the other atom. Because of Pauli exclusion, this situation is not tolerated and one of the electrons would need to be excited to a higher energy level. Since such an excitation is not thermodynamically favored (it requires input of external energy), the Pauli exclusion principle in effect produces a repulsive force whose pair potential varies approximately as:

$$u_{\text{repul}}(r_{ij}) = 4\varepsilon \left(\frac{\sigma}{r_{ij}}\right)^{12}. \qquad (3.2)$$

Together with the attractive van der Waals interaction, the total pair interaction potential, also known as the Lennard-Jones potential, is:

$$u(r_{ij}) = 4\varepsilon \left\{ \left(\frac{\sigma}{r_{ij}}\right)^{12} - \left(\frac{\sigma}{r_{ij}}\right)^{6} \right\}. \qquad (3.3)$$

A sketch of this potential is presented in Fig. 3.2, which shows how the potential develops a minimum at some finite separation distance known as the bond length.

3.1.2 Ionic, covalent and metallic bonds

The next three bond types occur for elements whose outer electron shell is not completely filled, but wishes to be. In these bonds, the excess outer electrons are akin to a commodity: something each element has which it either wishes to trade away or collect more of. This trading occurs in an effort to attain (or at least attain the appearance of) the happy existence of a closed electron shell configuration, such as is enjoyed by the noble gas elements. The differences between these three bond types lies in both the willingness of the element to trade electrons and the manner in which the electrons are shared by others.

Ionic bond

Ionic bonds form between elements that are aggressively interested in trading electrons. These are dominated by the combination of elements in group 1 of the periodic table (the alkalides), whose excess electron is rather easily ionized, with elements in group 7 (halides) which have a large electron affinity. A good example is rocksalt (NaCl), in which the Na atom gives away its electron to the Cl atom. As a result, Na^+ attains the closed shell configuration it desires as does Cl^-. Two happy consumers emerge after the trade is made and the result is the formation of two dissimilar charges that are attracted by the traditional Coulomb interaction:

$$u_{Coul}(r_{ij}) = -\frac{kZ_iZ_je^2}{r_{ij}}, \tag{3.4}$$

where $k = (4\pi\varepsilon_o)^{-1} = 8.99 \times 10^9$ Nm^2/C^2 and Z_i and Z_j are the magnitudes of the ion charges.

Again, this attractive interaction is overshadowed at short separation distances by the repulsive Pauli exclusion interaction to produce a net pair interaction potential of the form:

$$u(r_{ij}) = 4\varepsilon\left(\frac{\sigma}{r_{ij}}\right)^{12} - \frac{kZ_iZ_je^2}{r_{ij}}. \tag{3.5}$$

Note that, since the attractive potential for the ionic force varies as just $1/r_{ij}$, it is far more effective at large separations than was the van der Waals force and results in stronger bonds, as evidenced in Fig. 3.1.

Covalent bond

In comparison with the ionic bond, covalent bonds generally form between elements that are less inclined to trade away their electrons completely, but which are willing to share them with others. Details of the interaction potential for the covalent bond require a quantum mechanical treatment that is beyond

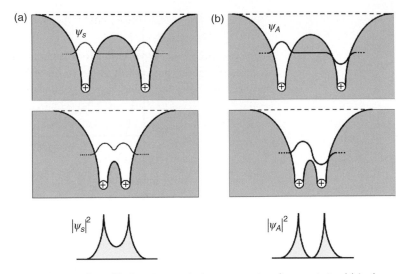

Figure 3.3 Diatomic hydrogen is formed by bringing two hydrogen atoms into close proximity. (a) In the case of a symmetric space wave function, a covalent bond is formed with finite sharing of the electron density. (b) A bond is not formed in the case of an anti-symmetric space wave function.

the scope of this textbook. However, some insight into how covalent bonds form can be gleaned from a simple quantum mechanical consideration of the bonding that occurs in H_2.

The quantum mechanical wave function that describes the two electrons in H_2 is formed by a combination of both a space wave function (describing the electron density) and a spin wave function:

$$\Psi = \psi_{\text{space}} \times \psi_{\text{spin}} \tag{3.6}$$

Since the two H nuclei represent a double-well potential, the space wave function has two possible states: one symmetric and the other anti-symmetric, as illustrated in Fig. 3.3. Similarly, the spin wave function can be either symmetric (with both spins aligned) or anti-symmetric (with both spins opposed).

Now, one feature of the Pauli exclusion principle is the requirement that the *total* wave function for the two electrons must be anti-symmetric. Thus, of the four possible combinations in Eq. (3.6), only two:

$$\Psi_{\text{A}} = \begin{cases} \psi_{\text{space, A}} \times \psi_{\text{spin, S}} \\ \psi_{\text{space, S}} \times \psi_{\text{spin, A}} \end{cases} \tag{3.7}$$

are permitted.

Consider what happens to the space wave function as the two nuclei approach each other. For the case of an anti-symmetric space wave function (Fig. 3.3b), the electron density remains localized around each nucleus separately with zero density at the midpoint in between. In this instance, the electrons are not being shared and a covalent bond is not formed. However,

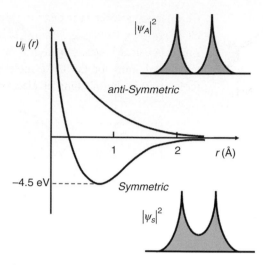

Figure 3.4 The pair interaction for diatomic hydrogen is purely repulsive in the case of an anti-symmetric space wave function, but exhibits a potential well (i.e. a covalent bond) in the case of a symmetric space wave function.

in the other situation (Fig. 3.3a) in which the space wave function is symmetric and the spin wave function is anti-symmetric, the electron density at the midpoint between the two nuclei increases on approach, giving rise to sharing of the two electrons and formation of a covalent bond. This interpretation is confirmed by exact calculation of the interaction potential, as shown in Fig. 3.4, which indeed reveals a potential minimum forming in the case of a symmetric space wave function with spins anti-aligned.

For our purposes, the origin of the covalent bond is far less important than its consequences. Because the covalent bond involves the sharing of electrons, there arises a limit to the number of such bonds that can be formed with any given element. As a result, covalent bonds form only *discrete*, directional bonds to a limited number of neighboring atoms. This is unlike the van der Waals and ionic bonds, which in principle form *isotropic* bonds with all other atoms in the vicinity.

Consider the ammonia molecule (NH_3) as a simple example of covalent bonds. Nitrogen, from group 5 of the periodic table has $v = 5$ excess (valence) electrons in its outer electron shell, while each H atom is just shy by one electron of completing a noble gas configuration. Because N has $v = 5$ valence electrons, it can form a maximum of only $8 - v = 3$ single covalent bonds with other elements. In the case of NH_3, three of the electrons from N are shared with each of three H atoms and each electron of the H atoms is shared by N. By sharing these electrons, each atom in the molecule attains the appearance of a closed electron configuration. The directional nature of the covalent bonds in NH_3 is evidenced by the highly symmetric molecular shape it assumes, as shown in Fig. 3.5.

Other examples of covalent bonding include CCl_4, CO_2 and N_2. For C (group 4) our $8 - v$ rule limits covalent bonding to a maximum of four bonds.

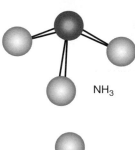

NH₃

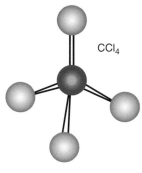

CCl₄

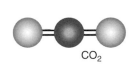

CO₂

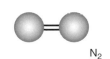

N₂

Figure 3.5

Illustrations of the 8 − v rule for covalent bonding in NH₃ and CCl₄. Exceptions to this rule occur in the case of CO₂ and N₂ as a result of double and triple covalent bonds, respectively.

Chlorine, which like H is just shy of a full electron shell, contributes one of its electrons to the bond and, together with four electrons shared by C, produces four single covalent bonds. Again, the discrete bonding gives rise to a symmetric tetrahedral form for the CCl_4 molecule (see Fig. 3.5). In CO_2, the O atom (column 6) is shy of filling its electron shell by two electrons. In bonding with C, two electrons of C are shared by each O (a double covalent bond) to attain the noble gas configuration for all the atoms. In N_2, our $8 - v$ rule suggests a maximum of three covalent bonds. However, N_2 forms a solitary triple covalent bond in which three electrons are mutually shared between each N atom.

Metallic bond

The metallic bond is commonly found in monatomic materials, including the many metals found in the central portion of the periodic table. Like the ionic bond, the atom is eager to trade away its excess valence electron in order to achieve a closed shell configuration. The only problem is that there is no other atom present that desires to collect these unwanted electrons, and so they are cast off to produce a continuous gas of electrons surrounding the, now positively charged, ion cores, as shown in Fig. 3.6. The ion cores, which remain after ejecting their excess electron, would normally be repelled by electrostatic forces, but because they are surrounded by a gas of electrons, they actually experience an overall attraction.

3.1.3 The hydrogen bond

With a single electron, the hydrogen atom should form a covalent bond with only one other atom. However, when it forms covalent bonds with highly electronegative atoms like oxygen, nitrogen or fluorine, the electron cloud distribution of the hydrogen atom is severely distorted and the positive charge of the hydrogen nucleus is left exposed on the side opposite the covalent bond, as illustrated in Fig. 3.7. This exposed nuclear charge forms a hydrogen bond *donor* site that can "bond" to another nearby electronegative atom (a hydrogen bond *acceptor*) so as to form a *hydrogen bond*. The hydrogen bond is intermediate in strength between that of a covalent bond and a van der Waals interaction, and is responsible for the unique properties of many associated liquids including water.

3.2 Cohesive energy

3.2.1 Crystals

Having examined the five bond types, we now consider how the multiplicity of bonds present in a large collection of particles might influence the resulting

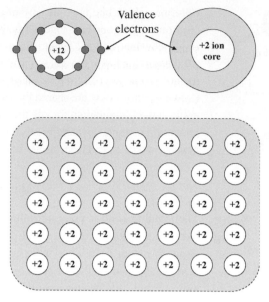

The metallic bond in Mg represented as a gas of dissociated electrons and residual ion cores. From the second column of the periodic table, Mg has two valence electrons that become fully dissociated in the crystal. The nucleus and remaining inner electrons form ion cores that are arranged in a periodic fashion and held together by a surrounding gas of free electrons.

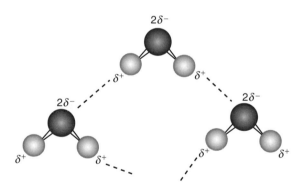

Hydrogen bonding in H_2O. In forming a covalent bond with O, the single electron of H is severely dislocated, leaving the positive charge (δ^+) of the H nucleus exposed to provide a donor site for hydrogen bonding. Charge neutrality requires the O atom to be dominated by a net negative charge ($2\delta^+$) to form an acceptor site.

structure in the crystalline phase. In certain cases (van der Waals and ionic bonds), we can analytically determine the cohesive energy of a proposed crystalline form starting from the pair potential (Eq. (3.3) and Eq. (3.5) above). The *cohesive energy* is defined as the energy required to disassemble the particles of the crystal and relocate all so that they are at rest and infinitely separated (so the attractive

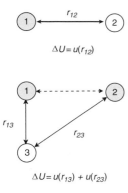

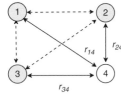

$\Delta U = u(r_{14}) + u(r_{24}) + u(r_{34})$

Figure 3.8

Contributions to the electrostatic energy during piecewise assembly of a van der Waals crystal.

interaction between any two particles vanishes), and to return them to neutral atoms, if necessary. As a result, the cohesive energy is a measure of ultimate stability of the particles when they are assembled in a given crystalline form. It provides a benchmark for the mechanical strength of the material and a measure of the melting point at which thermal energy would be likely to disrupt the structure. Furthermore, Nature is predisposed to seek out the crystalline form that will achieve the lowest overall potential energy or greatest cohesive energy. Thus, a comparison (Ex. 6) of the cohesive energy in possible crystalline forms can help us to understand why one form might occur instead of another.

We begin with the crystalline forms taken by the noble gas elements of column 8 in the periodic table, whose pair potential is dominated by the van der Waals interaction of Eq. (3.3). Inherent in the development of Eq. (3.3) was the notion that these inert, monatomic atoms prefer to minimize their mutual separation, provided that their electron clouds do not severely overlap. Thus we can think of the atoms roughly as spherical particles pressed up in contact with one another and we can anticipate that the minimum potential energy will coincide with crystal structures that afford the greatest packing efficiency. Indeed, the crystalline forms seen for the noble gas elements are either the HCP or FCC structures for which the packing fraction (74%) is greatest amongst ordered structures.

To determine the cohesive energy of a van der Waals crystal, we consider assembling the structure atom by atom. We begin with neutral atoms infinitely separated and consider the changes in the total potential energy as atoms are brought into proximity with one another. In Fig. 3.8, we show how the potential energy changes as the first four atoms are assembled. When a second atom is placed next to the first, the change in potential energy is $u(r_{12})$. The third atom changes that potential energy by an amount $u(r_{13}) + u(r_{23})$ and the fourth by an amount $u(r_{14}) + u(r_{24}) + u(r_{34})$, and so on. The total potential energy for N assembled atoms is thus:

$$U_{\text{tot}} = u(r_{12}) + u(r_{13}) + u(r_{23}) + u(r_{14}) + u(r_{24}) + u(r_{34}) + \cdots$$

$$= \frac{1}{2} \sum_{i=1,N} \sum_{\substack{j=1,N \\ j \neq i}} u(r_{ij}). \tag{3.8}$$

Because we are dealing with a monatomic basis set associated with an ordered lattice structure, our surroundings as viewed from any atom should appear the same. Thus, the second sum in Eq. (3.8) should not depend upon which ith atom is chosen in the first sum, and so the double sum should be the same as N repetitions of the second sum alone as referenced to any arbitrary $i = 1$ atom:

$$U_{\text{tot}} = \frac{N}{2} \sum_{\substack{j=1,N \\ j \neq i}} u(r_{ij}) = \frac{N}{2} \sum_{j=2,N} u(r_{1j}), \tag{3.9}$$

where N is the number of atoms in the crystal.

van der Waals crystals

We can now insert the van der Waals pair potential (Eq. (3.3)) to obtain:

$$U_{tot} = \frac{N}{2} \sum_{j=2,N} 4\varepsilon \left\{ \left(\frac{\sigma}{r_{1j}}\right)^{12} - \left(\frac{\sigma}{r_{1j}}\right)^{6} \right\}. \tag{3.10}$$

The distances r_{1j} represent the distances between the central ($i = 1$) atom and all other atoms in the crystal. Since the crystal has some unique form, it is convenient to express these distances as fractions of the nearest neighbor distance, r_o, such that:

$$U_{tot} = 2N\varepsilon \left\{ \left(\frac{\sigma}{r_o}\right)^{12} \left[\sum_{j=2,N} \left(\frac{1}{p_{1j}}\right)^{12} \right] - \left(\frac{\sigma}{r_o}\right)^{6} \left[\sum_{j=2,N} \left(\frac{1}{p_{1j}}\right)^{6} \right] \right\}, \tag{3.11}$$

where $r_{1j} = p_{1j}r_o$. Eq. (3.11) can be cleaned up to read:

$$U_{tot} = 2N\varepsilon \left\{ A\left(\frac{\sigma}{r_o}\right)^{12} - B\left(\frac{\sigma}{r_o}\right)^{6} \right\}, \tag{3.12}$$

where $A = \sum_{j=2.N} p_{1j}^{-12}$ and $B = \sum_{j=2.N} p_{1j}^{-6}$ are constants that are unique to a specific lattice type.

Equation (3.12) only expresses the dependence of the total potential energy on the nearest neighbor distance. To obtain the potential energy of the crystal *at equilibrium*, we thus seek the value of r_o that minimizes U_{tot}. Hence, we must first differentiate Eq. (3.12) with respect to r_o to find the equilibrium nearest neighbor distance, and then substitute that result back into Eq. (3.12). We obtain the equilibrium nearest neighbor distance as

$$r_{eq} = \left(\frac{2A}{B}\right)^{1/6} \sigma, \tag{3.13}$$

and equilibrium energy as,

$$U_{tot,eq} = -N\varepsilon \left(\frac{B^2}{2A}\right). \tag{3.14}$$

For the FCC lattice, $A = 12.13188$ and $B = 14.45392$, for which $r_{eq}/\sigma = 1.09$ and the total cohesive energy/atom is given by:

$$\left| U_{tot,eq}/N \right| = 8.6102\varepsilon. \tag{3.15}$$

A similar calculation (Ex. 4) carried out for the BCC structure results in a lower cohesive energy, and thus supports the adoption of FCC by the noble gas elements.

Ionic crystals

We can work out the cohesive energy of an ionic crystal in a similar fashion. However, a small complication arises because the pair interaction potential (Eq. (3.5)) contains both attractive and repulsive electrostatic contributions. To see this, consider constructing NaCl into its FCC structure. We might begin with a Cl^- anion at one corner of the lattice and proceed to place a Na^+ next to it. The two dissimilar charges are attracted and the potential energy is lowered. However, when we approach with the next atom (either Na^+ or Cl^-) it will be both attracted and repelled by our NaCl molecule. Indeed, it is unclear whether a fully assembled NaCl crystal will be stable at all!

Another issue for our ionic crystal situation is that the basis is diatomic. This means that, unlike the case of the noble gas elements, the two ions generally will not have similar sizes. Therefore, it is possible that structures other than FCC and HCP may be favored when energy considerations are taken into account. Recall, for example, that while NaCl favors the FCC structure, CsCl favors the BCC structure.

To compute the cohesive energy, we start as before, but with *neutral* Na and Cl spread out infinitely apart. Some amount of energy must be input to ionize the Na atoms and some energy released when these electrons are received to form Cl^-. We will need to keep account of this net energy, as it contributes to our definition of the cohesive energy. Having formed the ions, we again assemble the lattice and find:

$$U_{\text{tot}} = \frac{N}{2} \sum_{j=2,N} u(r_{1j}) = \frac{N_{Na}}{2} \sum_{j=2,N} u(r_{1j}) + \frac{N_{Cl}}{2} \sum_{j=2,N} u(r_{1j}) \qquad (3.16)$$

where N is the total number of ions, and N_{Na} and N_{Cl} are the numbers of Na and Cl ions, respectively. The two sums in Eq. (3.16) are equivalent for a diatomic basis set. This is because we can think of NaCl as either an FCC lattice with Cl^- at the corner, or an equivalent lattice with Na^+ at the corner. Consequently, the total potential energy is:

$$U_{\text{tot}} = N_{\text{pairs}} \sum_{j=2,N} u(r_{1j}) \qquad (3.17)$$

where $N_{\text{pairs}} = N/2$ is the number of Na^+ and Cl^- ion pairs.

Inserting the ionic pair potential (Eq. (3.5)) and again scaling the distances to that of the nearest neighbor distance, r_o, we obtain:

$$U_{\text{tot}} = N_{\text{pairs}} \left\{ A \left(\frac{\sigma}{r_o} \right)^{12} - \alpha \left(\frac{kZ^2 e^2}{r_o} \right) \right\}, \qquad (3.18)$$

where

$$\alpha = \sum_{j=2,N} \mp \frac{1}{p_{1j}}, \qquad (3.19)$$

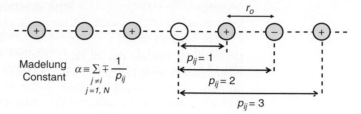

Figure 3.9 A one-dimensional ionic crystal illustrating the various terms that contribute to the Madelung constant.

is known as the *Madelung constant* and is unique to a given lattice type. Note that the minus sign is chosen for pairs of similar charge and the plus sign for pairs of dissimilar charge.

Again, we can solve for the cohesive energy (aside from the ionization energy component discussed earlier) by first solving Eq. (3.18) for the equilibrium nearest neighbor distance at which U_{tot} is minimized (Ex. 2) and then back substituting to obtain:

$$U_{tot,\,eq} = -N_{pairs} \frac{\alpha k Z^2 e^2}{r_{eq}} \left(1 - \frac{1}{12} \right). \qquad (3.20)$$

We find that the cohesive energy of an ionic crystal is determined chiefly by the *Madelung energy*, $\alpha k Z^2 e^2 / r_{eq}$, of which α depends on the specific lattice type. The Madelung constant involves an infinite sum of alternating sign and, unlike the parameters A and B discussed for the van der Waals case, its convergence is more gradual. Nevertheless, methods have been developed to insure the convergence (Ex. 7) and obtain an accurate result.

As an example of calculating the Madelung constant, we consider a simple one-dimensional ionic crystal, shown in Fig. 3.9. From the figure, we find the first several terms in the summation to be of the form:

$$\alpha = \sum_{j=2,N} \mp \frac{1}{p_{1j}} = 2 \left(1 - \frac{1}{2} + \frac{1}{3} - \frac{1}{4} + \cdots \right). \qquad (3.21)$$

This can be compared with the series identity

$$\ln(1 + x) = \left(x - \frac{x^2}{2} + \frac{x^3}{3} - \frac{x^4}{4} + \cdots \right),$$

to obtain $\alpha = 2 \ln 2 \approx 1.386$.

3.2.2 Amorphous materials

We saw that for the van der Waals crystals the double summation in Eq. (3.8) could be reduced to a single sum (Eq. (3.9)) because the result of the second summation did not depend upon the particle chosen in the first sum. This arose because of the crystal symmetry in which the surroundings appear identical from the perspective of any particle. A similar situation arises for a monatomic liquid or glass, wherein the rotational invariance of the amorphous structure produces surroundings that are (on average) identical from the perspective of any randomly chosen central particle. Hence we again should find that the average total potential energy of a van der Waals liquid (or glass) would be given by:

$$\langle U_{tot} \rangle = \frac{N}{2} \sum_{j=2,N} \langle u(r_{1j}) \rangle. \tag{3.22}$$

We can now employ the pair distribution function, $g(r)$, to replace the discrete summation by an integration. Recall (Eq. (2.4)) that the number of particle centers in a spherical shell a distance r from a central particle is given by the radial distribution function:

$$\langle \# \text{ particle centers at } r \rangle = \langle n \rangle 4\pi r^2 g(r) \mathrm{d}r.$$

Thus the average potential energy is

$$\langle U_{tot} \rangle = \frac{N}{2} \int \langle n \rangle 4\pi r^2 u(r) g(r) \mathrm{d}r. \tag{3.23}$$

Summary

- Attraction between inert gas elements arises from a dipolar interaction known as the van der Waals force.

- Repulsion between particles is a consequence of Pauli exclusion which occurs when neighboring electron clouds begin to overlap.

- Ionic, covalent and metallic bonds all involve a redistribution of valence electrons, motivated by the stability of a closed electronic configuration.

- The cohesive energy is the energy required to disassemble the constituent atoms of matter, displacing each to infinity and returning each to a neutral charge state if necessary. It is a measure of the ultimate stability of a given condensed structure.

Exercises

3.1. Show that the equilibrium separation of two particles interacting through the Lennard-Jones potential is $r_{eq} = 2^{1/6}\sigma$.

3.2. Show that the equilibrium nearest neighbor separation for an ionic crystal is $r_{eq} = \left(12A\sigma^{12}/\alpha kZ^2 e^2\right)^{1/11}$.

3.3. Ice floats in water because when water is frozen it expands into a lower density state of matter. Explain this anomalous behavior on the basis of what you know about the shape of a H_2O molecule and the nature of the hydrogen bond.

3.4. Demonstrate that, energetically, a noble gas element is slightly more stable in the FCC crystal structure than it would be in a BCC structure, by calculating the ratio of these two cohesive energies. The lattice sums for the BCC structure are:

$$A = \sum_{j=2.N} p_{1j}^{-12} = 9.11418$$

$$B = \sum_{j=2.N} p_{1j}^{-6} = 12.2533$$

3.5. Imagine that a 1D chain of ions (with $q = \pm Ze$) like that shown in Fig. 3.9 is compressed, so that the separation distance decreases from its equilibrium value by some fractional amount δ (i.e., $r_o \to r_{eq}(1 - \delta)$). For the ionic potential of Eq. (3.18), show that the work done (per unit length of the chain) in compressing the crystal is quadratic (in leading order) and given as: $\dfrac{11kq^2 \ln 2}{2r_{eq}} \delta^2$.

3.6. Calcium oxide has the FCC structure for which the Madelung constant is $\alpha = 1.7476$. Determine the cohesive energies per molecule of the hypothetical crystals Ca^+O^- and $Ca^{2+}O^{2-}$. Be sure to account also for energy required to return the ions to neutral atoms. Assume $r_o = 2.40$ Å is the same for both forms, and neglect the repulsive energy. The energy needed to ionize the first and second electrons of Ca is 6.11 and 11.87 eV, respectively. The energy released when a first and second electron is added to neutral oxygen is 1.46 and -9.0 eV, respectively. (Note that energy is actually required for adding the second electron.) Which of the two hypothetical crystals is most likely to occur in Nature and why?

3.7. Figure 3.10 shows a hypothetical 2D square lattice of ions. Estimate the Madelung constant in the following way: First sum only the contributions to the Madelung constant arising from the ions (or partial ions) contained within the innermost square. Note that an ion on one edge of a square is counted as 50% inside and 50% outside and one on a corner is counted as 25% inside and 75% outside. Show that this first contribution amounts to $+1.2939$. Continue on to the next layer (indicated by the hashed region). Show that this region alone contributes to the Madelung constant by an amount $+0.3140$. Lastly, compute the contribution from the outermost layer and show that out to this level the Madelung constant is $+1.6116$.

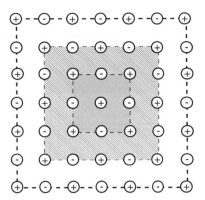

Figure 3.10

Suggested reading

The book by Callister, Jr. is an introductory-level materials engineering textbook I happened across which does a nice job of describing bonds and crystal structures.

C. Kittel, *Introduction to Solid State Physics*, 8th Ed. (John Wiley and Sons, 2005).

W. D. Callister, Jr., *Science and Engineering: An Introduction*, 5th Ed. (John Wiley & Sons, New York, 2000).

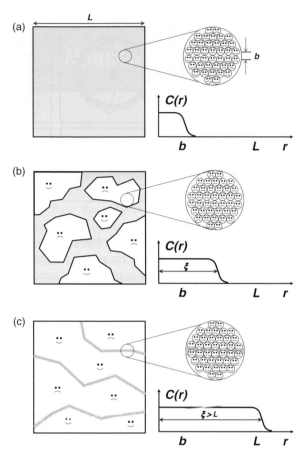

Figure 4.2 The possible development of correlated patterns of happy and sad communities in a population of people. In each successive panel, the size of correlated regions is increasing until it rivals the size of the system itself.

communities grow over time due to an inherent "feedback" built into the biasing. The larger a community becomes, the more strongly it biases those potential converts on the periphery. This spontaneous growth would result in regions of correlated mood that coarsen and eventually merge into one another, as illustrated in Fig. 4.2c.

But how do we characterize this growing pattern of correlated regions? A simple average of the property over the entire system would not necessarily reveal the developing correlations taking place. As we saw in our example above, correlated regions may be substantial in size, but could still be divided into nearly equal numbers of happy and sad communities. To illuminate the presence of these correlated patterns, we need to consider an appropriate *correlation function* of the form

$$C_X(\vec{r}_1, \vec{r}_2) = \langle X(\vec{r}_1) X(\vec{r}_2) \rangle, \tag{4.1}$$

where $X(\vec{r})$ represents the value of the property for a particle located at $\vec{r}$, and the brackets again signify the required ensemble averaging. Conceptually, this correlation function is much like the pair distribution function discussed in Chapter 2. It expresses a conditional probability that given that some central particle at $\vec{r}_1$ has a certain value of the property X, another particle at $\vec{r}_2$ will have the same value. In practice, the correlation function would be computed by selecting each particle in the system, in turn, as the central particle and sampling the similarity of those neighbors a distance $\Delta\vec{r} = \vec{r}_2 - \vec{r}_1$ away. Over distances shorter than a characteristic length, ξ, known as the *correlation length*, the property is similar and said to be correlated. At longer separation distances, the similarity vanishes and the property becomes uncorrelated.

Although our example involves people with mood swings, we could readily replace it with particles, each of which has some property whose value can match or differ from that of its neighbor. Magnetic particles with their intrinsic magnetic moment are one example.

4.2 Magnetic materials

When we speak of magnetic materials, most often we think of materials like iron that can be either magnetized or de-magnetized by an external field and retains the final state of magnetization when the field is removed. Iron is a *ferromagnetic* material and, on a microscopic level, each atom of iron contains a miniature magnetic moment (a particle property) whose individual orientation is influenced, not only by the external field that is applied, but is also biased by the orientation of its neighbors. Materials that possess a miniature moment, but which lack the neighboring interaction, are known as paramagnets. Although the moments of a paramagnet can be aligned with an applied field, paramagnets are unable to retain their magnetization when the field is removed.

In turn, we will look at both of these magnetic materials, as well as considering the *diamagnetic* response, seen mainly in non-magnetic materials that do not possess a permanent moment. We begin by inquiring into the origin of the miniature magnetic moment that endows magnetic materials with magnetization.

4.2.1 Magnetic moments

To understand the origin of the magnetic moment of an atom, consider for a moment the simplistic Bohr model illustrated in Fig. 4.3, in which an electron orbits the nucleus with speed v on a circular path of radius ρ.

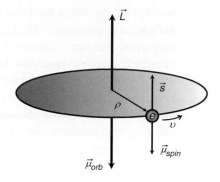

Bohr model of an orbiting electron indicating the orbital and spin contributions to the magnetic moment.

The steady motion of the orbiting electron constitutes a continuously flowing loop of current,

$$I = \frac{\text{charge}}{\text{period}} = -e/(2\pi\rho/v) = -ev/2\pi\rho, \tag{4.2}$$

which together with the area of the loop produces a magnetic moment,

$$\mu_{\text{orb}} = IA = (-ev/2\pi\rho)\pi\rho^2 = -ev\rho/2. \tag{4.3}$$

Meanwhile, the orbiting mass of the electron produces an angular momentum, $L = mv\rho$, directed opposite the magnetic moment. Combining this with Eq. (4.3), we find that

$$\vec{\mu}_{\text{orb}} = -(e/2m)\vec{L}, \tag{4.4}$$

and conclude that the magnetic moment ultimately arises from the angular momentum of the electron. This conclusion is unchanged by a quantum mechanical approach, and since quantum mechanics requires that the angular momentum should be quantized in discrete units of $\hbar$, the orbital magnetic moment must appear in discrete units of a *Bohr magneton*,

$$\mu_B = e\hbar/2m = 9.27 \times 10^{-24} \text{Am}^2.$$

In addition to the orbital angular momentum, there is also electron *spin* that contributes to the overall angular momentum of an atom. For the electron, spin is quantized and the quantum number can assume either $\vec{s} = \pm\frac{1}{2}\hbar$. Without going into the detailed quantum mechanical treatment, the electron spin of a free electron is found to contribute to the magnetic moment by an amount,

$$\vec{\mu}_{\text{spin}} = -g\mu_B\vec{s}, \tag{4.5}$$

where the so-called splitting factor, $g = 2.0023$.

For completeness, we should also acknowledge that there is yet another contribution to the magnetic moment that arises from the intrinsic spin of the nucleus. Again, the nuclear magnetic moment appears in quantized units, but of a smaller size, $\mu_n = e\hbar/2m_p = 5.05 \times 10^{-27} \text{Am}^2$, known as a *nuclear magneton*. Although this nuclear magnetic moment plays an important role in nuclear magnetic resonance (NMR), in magnetic materials it is completely masked by the much larger magnetic moment arising from the electron.

We now see that the magnetic moment has its origin in the combined orbital and spin angular momentum of the electron, and we begin to get a glimpse of why not all materials are magnetic. Consider any of the noble gas elements that form closed electron shells. Since each shell is filled, the total orbital angular momentum vanishes. Likewise, since each electron is paired to another with opposite spin (due to Pauli exclusion), the net spin angular momentum is also zero. Consequently, the noble gas elements have no net angular momentum and no permanent magnetic moment. They are non-magnetic. A similar fate is found for many elements of the periodic table in which electrons appear in pairs so that the vector sum of the angular momentum vanishes. Magnetic materials then develop only in those fortuitous instances when an outer electron shell is only partly filled and contains an odd number of electrons.

Determining the magnetic moment of an atom thus boils down to determining the total angular momentum of its constituent electrons, and presents a complicated vector addition problem, which is further compounded by rules of quantization. Luckily, this vector addition problem has already been tackled and is commonly treated using the Russell–Saunders scheme of spin–orbit coupling, which can be found in almost any introductory quantum mechanics textbook. In this scheme, the atom's total orbital angular momentum, $\hbar\vec{L}_{\text{atom}} = \hbar \sum_i \vec{L}_i$, and total spin angular momentum, $\hbar\vec{S}_{\text{atom}} = \hbar \sum_i \vec{s}_i$, are combined to obtain a total angular momentum,

$$\hbar\vec{J} = \hbar\vec{L}_{\text{atom}} + \hbar\vec{S}_{atom}. \tag{4.6}$$

The magnetic moment of the atom remains proportional to the total angular momentum and is given by

$$\vec{\mu}_{\text{atom}} = -g\mu_B\vec{J}, \tag{4.7}$$

where the splitting factor is now replaced by

$$g = 1 + \frac{J(J+1) + S(S+1) - L(L+1)}{2J(J+1)}. \tag{4.8}$$

4.2.2 Diamagnetism

We are keenly interested in those magnetic materials that fortuitously have a non-zero angular momentum. But, let us pause momentarily to consider how the remaining non-magnetic materials respond to an applied field. Although they lack a permanent moment, these non-magnetic materials do develop a weak magnetization that is *opposite* to the direction of the field. That is, they exhibit *dia*magnetic behavior as opposed to a *para*magnetic response.

To understand the origin of the diamagnetic response, consider again the example of the single electron atom illustrated in Fig. 4.3. Suppose we apply a magnetic field, $\vec{B} = \mu_o \vec{H}$, directed upwards, perpendicular to the orbital plane. This new field introduces an upward magnetic flux through the loop which, by virtue of Lenz's law (the one expressing Nature's displeasure with changing flux), causes the electron to react in such a way as to produce a counter flux. Assuming that the orbital radius is fixed, a quick check of our right hand rule reveals that the needed counter flux is achieved by having the electron speed up (or slow down, if the electron were orbiting in the opposite direction). One can show (Ex. 2) that the electron suffers a change in its angular velocity by an amount,

$$\Delta\omega = eB/2m, \tag{4.9}$$

known as the *Larmor* frequency. This speed change is common to all electrons in the orbital and, for an atom with Z electrons, the applied magnetic field thus induces a net change in the current by an amount,

$$I_{\text{induced}} = -Ze(\Delta\omega/2\pi) = -Ze^2 B/4\pi m. \tag{4.10}$$

This current change in turn induces an average magnetic moment,

$$\langle \mu_{\text{induced}} \rangle = -\left(Ze^2 R/4m\right)\langle \rho^2 \rangle \tag{4.11}$$

directed opposite to the applied magnetic field. Collectively, this produces a net magnetization per unit volume of

$$\vec{M} = n\langle \vec{\mu} \rangle, \tag{4.12}$$

where n is the particle number density. Because diamagnetic materials respond to an applied field by producing an induced magnetic moment proportional to the field but opposite in direction, the magnetic susceptibility of a diamagnet,

$$\chi_m = \frac{\mathrm{d}M}{\mathrm{d}H} = n\mu_o \frac{\mathrm{d}\langle \mu_{\text{induced}} \rangle}{\mathrm{d}B} = -\left(Ze^2 n\mu_o/4m\right)\langle \rho^2 \rangle, \tag{4.13}$$

is always less than zero. While the diamagnetic response is only measureable in non-magnetic materials, it is nevertheless found universally in all materials, including those magnetic materials that have a permanent magnetic moment due to a net angular momentum. The size of the induced moment is however much smaller than that of the permanent moments and in magnetic materials the diamagnetic contribution is masked by the paramagnetic response. Values for the susceptibility for a handful of common elements are listed in Table 4.1.

Table 4.1 Susceptibility of several elements at standard temperature and pressure (values obtained from *Handbook of Chemistry and Physics* (1983)).

Diamagnetic		Paramagnetic	
Element	Susceptibility	Element	Susceptibility
Bismuth	-1.6×10^{-4}	Sodium	8.5×10^{-6}
Gold	-3.4×10^{-5}	Aluminum	2.1×10^{-5}
Silver	-2.4×10^{-5}	Tungsten	7.8×10^{-5}
Copper	-9.7×10^{-6}	Platinum	2.8×10^{-4}
Hydrogen	-2.2×10^{-9}	Gadolinium	4.8×10^{-1}

4.2.3 Paramagnetism

As we learned earlier, magnetic materials are those whose constituent atoms possess a net, non-zero, magnetic moment in the absence of any applied magnetic field. Such materials are generally referred to as paramagnetic because the effect of an applied field produces a torque that tends to align moments in the direction of the field, with a corresponding lowering of internal energy given by,

$$U = -\vec{\mu} \cdot \vec{B}. \qquad (4.14)$$

Unlike ferromagnetic materials, which we will discuss in the next section, this alignment of moments with the field disintegrates at any finite temperature when the field is removed due to incessant thermal agitation. Consequently, the magnetization vanishes as the moments return to a disordered pattern of random orientations.

Let us consider then a paramagnetic material with total angular momentum quantum number J. How does the magnetization of this material depend on both the applied field (acting to align the moments) and the temperature (acting to randomize the orientations)? The answer to this question requires a thermodynamic approach. Imagine, as illustrated in Fig. 4.4, that the field is applied along the z-axis. Because angular momentum is quantized, its component along the z-direction is also restricted to values $J_z = m_J \hbar$, where the magnetic quantum number, m_J, ranges from $m_J = J$, $(J-1)$, ..., 0, ... $-(J-1)$, $-J$. From Eq. (4.7), this then implies that the component of $\vec{\mu}$ along the field is similarly quantized as $\mu_z = -g\mu_B m_J$.

Since we also anticipate no net magnetization to appear in either the x- or y-directions, the magnetization we seek is given by Eq. (4.12) using only the average value of μ_z consistent with conditions of thermodynamic equilibrium. Using Boltzmann statistics, this average can be expressed as

$$\langle \mu_z \rangle = \frac{\sum\limits_{-J}^{J} (-g\mu_B m_J) e^{-U_j/k_B T}}{\sum\limits_{-J}^{J} e^{-U_j/k_B T}}, \qquad (4.15)$$

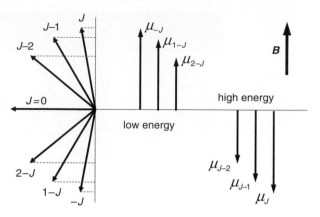

Figure 4.4 The total angular momentum, J, is restricted to a discrete set of allowed projections along a given direction (here taken to be parallel to an applied magnetic field). For each allowed projection there is a corresponding magnetic moment. States with the moment aligned with the magnetic field are energetically favored (see Eq. (4.14)).

where

$$U_J = -\mu_z B = -(-g\mu_B m_J)B = m_J(g\mu_B B).\tag{4.16}$$

One can show (Ex. 5) that this relation reduces to

$$\langle \mu_z \rangle = (g\mu_B J)B_J(y),\tag{4.17}$$

where $y = g\mu_B JB/k_B T$, and

$$B_J(y) = \left\{ \left(\frac{2J+1}{2J}\right) \coth\left[\frac{(2J+1)y}{2J}\right] - \left(\frac{1}{2J}\right)\coth\left(\frac{y}{2J}\right) \right\}\tag{4.18}$$

is known as the *Brillouin* function.

Good agreement of Eq. (4.17) with experiment is demonstrated in Fig. 4.5 for several materials of differing J. In all instances, the magnetization increases with increasing field, reaching a point of saturation in which all the moments are aligned with the field. In the weak field regime, the magnetization is roughly proportional to the applied field. Here, we can approximate the Brillouin function for small argument as

$$\lim_{y \ll 1} B_J(y) = \frac{y(J+1)}{3J},\tag{4.19}$$

and obtain the susceptibility as

$$\chi_m = \frac{dM}{dH} = n\mu_o \frac{g^2\mu_B^2 J(J+1)}{3k_B T} = \frac{C}{T},\tag{4.20}$$

a result known as the *Curie law*.

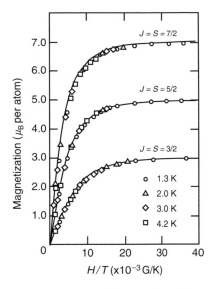

Figure 4.5 Magnetization curves for three paramagnets of differing total angular momentum. The solid lines are a fit to the Brillouin function of Eq. (4.17). (Adapted from Henry (1952).)

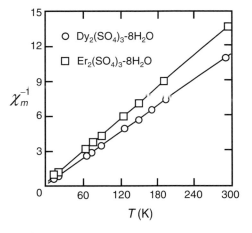

Figure 4.6 The inverse magnetic susceptibility of two rare earth sulphates, demonstrating a linear temperature dependence consistent with the Curie law of Eq. (4.20). (Adapted from Jackson (1936).)

An example of paramagnets exhibiting Curie behavior is shown in Fig. 4.6. In paramagnets, there is no interaction between neighboring moments, and the ability to align moments using an external applied field is countered only by the thermal agitation present at finite temperatures. When this thermal agitation vanishes, at absolute zero, the alignment of magnetic moments then occurs without competition and, as the diverging susceptibility suggests, any tiny field will be capable of aligning all the moments.

4.2.4 Ferromagnetism

Unlike paramagnets, ferromagnets are able to retain their magnetization when the external field is removed. In actuality, this is only true if they are held at a temperature below the so-called Curie temperature, T_c. Above the Curie temperature, the material responds like a usual paramagnet and the magnetization vanishes when the field is removed.

To interpret the ferromagnetic behavior it is somewhat natural to suppose that there exists in these materials some internal interaction between magnetic moments that biases neighboring moments to adopt a common orientation. Indeed, early theories of the ferromagnetic state advanced by Pierre Weiss fashioned this interaction in the form of an internal magnetic field, self-generated by the net alignment of moments, and thus proportional to the net magnetization,

$$\vec{H}_{\text{int}} = \lambda \vec{M} \quad \text{or} \quad \vec{B}_{\text{int}} = \mu_o \lambda \vec{M}. \tag{4.21}$$

This so-called *molecular field* was seen as being both a result of the aligned moments and yet also a stimulus for additional alignment. Inherent in this dual nature is a sort of feedback mechanism that can promote the rapid growth of regions of correlated magnetic moment below the Curie temperature.

Since the molecular field is proportional to the magnetization, the magnetization can then be expressed as a modified form of the Curie law,

$$M = (C/T)(H + H_{\text{int}}) = (C/T)(H + \lambda M), \tag{4.22}$$

where the molecular field is merely adding to that which is applied externally. Evident in this modified expression is the feedback alluded to earlier: the applied field promotes an incipient magnetization which in turn increases the effective field to promote even more magnetization. We can rearrange Eq. (4.22) to obtain

$$M = \frac{C}{(T - C\lambda)} H, \tag{4.23}$$

and obtain the susceptibility as

$$\chi_m = \frac{C}{(T - T_c)}, \tag{4.24}$$

where $T_c = C\lambda$ is identified with the *Curie temperature*. This temperature variation for ferromagnetic materials is known as the Curie–Weiss law and is illustrated in Fig. 4.7. Unlike a paramagnetic material whose susceptibility diverges only at absolute zero, the susceptibility of a system of interacting magnetic moments diverges at the finite Curie temperature. It is at this temperature that the molecular field, acting to align moments, achieves dominion over the thermal agitation working to disorder the moments.

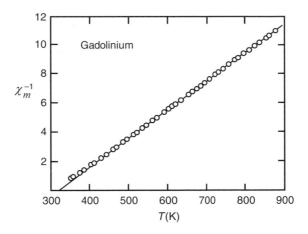

Figure 4.7 The inverse magnetic susceptibility of gadolinium demonstrating a linear temperature dependence consistent with the Curie–Weiss law of Eq. (4.24). (Adapted from Nigh, Legvold and Spedding (1963).)

Exchange interaction

Prior to the development of quantum mechanics, the only field known to exist within a magnetized specimen was a dipolar field collectively produced by the aligned magnetic moments themselves. At the location of each moment, this dipolar field has a magnitude given roughly as

$$B_{\text{dip}} \approx \frac{\mu_o}{4\pi}\left(\frac{2\langle\mu\rangle}{a^3}\right), \tag{4.25}$$

where a is the spacing between particles. Unfortunately, this dipolar field is too weak to account for most ferromagnetic materials, including iron whose Curie temperature ranges near 1000 K. Combining Eq. (4.20) and Eq. (4.25), we can express the Curie temperature for the case of a dipolar field as

$$T_c = C\lambda \approx \left[\mu_o \frac{g^2\mu_B^2 J(J+1)}{3k_B}\right]\left(\frac{1}{2\pi a^3}\right). \tag{4.26}$$

For iron (see Ex. 3) this yields only about 3 K, which is some two and a half orders of magnitude smaller than the experimental result.

The resolution to this problem came eventually from quantum mechanics. In a fashion analogous to our discussion of the covalent bond in Chapter 3, a quantum mechanical treatment of the overlapping wave function of two neighboring atoms produces an additional contribution to the potential energy of the form

$$u(r_{ij}) = -2J_{ex}(r_{ij})\,\vec{S}_i \cdot \vec{S}_j, \tag{4.27}$$

where $\vec{S}_i$ and $\vec{S}_j$ are the respective spins of the two atoms. This interaction is known as the *exchange interaction*, and the energy parameter J_{ex}, whose dependence on

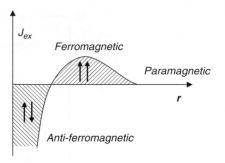

Figure 4.8 The exchange integral, J_{ex}, varies with separation between neighboring spins. For small separations, the interaction favors anti-alignment resulting in an anti-ferromagnetic state. Only for a limited range of larger separations is the ferromagnetic state favored.

the atom separation is illustrated in Fig. 4.8, is often referred to as the exchange integral. As seen in the figure, for a range of intermediate atomic separations, the exchange integral is positive and results in a lowering of internal energy if the two neighboring spins are aligned. This is the source of the mysterious biasing of moments that stabilizes the magnetization in a ferromagnetic material. Although the source is not really a molecular field but rather a pairwise interaction between adjacent spins, it is nevertheless convenient to treat the interaction as though it were the result of a mean field as described by Eq. (4.21).

Correlated domains

Now that we understand the origins of ferromagnetic behavior, we can finally return to our original theme concerning orientational pattern formation in magnetic materials, by considering what happens when we cool a ferromagnet toward its T_c in the absence of any field. Far above T_c, the moments are orientated randomly, because the thermal energy far exceeds that of the exchange interaction and serves to thwart its biasing effect. As we cool, we would expect to find small communities of similar spin orientation developing, as illustrated in Fig. 4.9.

The specific orientation of a region is random and reflects only that orientation which happened to be dominant when the region first began to form. To characterize this developing pattern, we would naturally introduce a correlation function, the *moment–moment correlation function*

$$\Gamma(\vec{r}) = \langle \vec{\mu}(0) \cdot \vec{\mu}(\vec{r}) \rangle, \tag{4.28}$$

which would be zero at very high temperatures (reflecting the random orientations), but would develop a finite, non-zero value over an extent characterized by a correlation length $\xi \gg a$ at lower temperatures. Arriving at T_c, we would observe the correlation span over great distances, comparable to the size of the system itself. The correlated regions would develop into so-called *domains* with a common orientation of the magnetic moment.

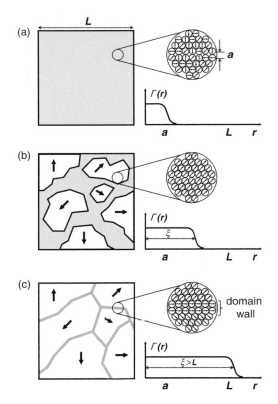

Figure 4.9 A series of panels illustrating how regions of ordered magnetic moments develop in a ferromagnet as the system is cooled to the Curie temperature in the absence of any external magnetic field. Just at the Curie point, the correlation length approaches the system size. Here the system exhibits no net magnetization, but is composed of many domains each containing commonly oriented spins.

Hysteresis loop

What happens if we now apply an external field to this zero-field cooled specimen? Our magnetization would follow a path as illustrated in Fig. 4.10, known as a *hysteresis* loop. Our specimen starts off in an unmagnetized state. Although there exist large domains of uniformly magnetized regions, these domains are individually oriented in random directions resulting in little or no initial magnetization. As we apply a field, those domains with orientation in the field direction begin to grow at the expense of other domains. These favored domains expand their boundaries by converting spins near the periphery of the domain. Consequently, the ill-favored domains shrink away. At some very large applied field, the favored domains have grown to macroscopic proportions and the magnetization saturates. As the field is then removed, the new domain structure adjusts slightly, but the favored domains remain dominant due to the internal biasing provided by the exchange interaction. When the field is completely removed, the finite magnetization that remains is known as the *remanence*, M_R.

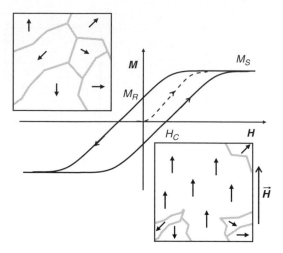

Figure 4.10   The hysteresis loop of a zero-field cooled ferromagnet (solid curve). Dashed line indicates the initial magnetization caused by an applied field. Applying the field grows favorably aligned domains at the expense of unfavored domains.

In order to de-magnetize the specimen, a reverse field must be applied. In applying this reverse field, the dominant domains are now shrunken while domains of opposite orientation (those aligned with the reversed field) begin to grow. Eventually, a situation is reached at a field of magnitude H_c, known as the coercive field (or *coercivity*), where the net magnetization vanishes. Further increase of the reverse field beyond H_c drives the system again to a state of saturation.

Summary

- Magnetic behavior appears in atoms with a net total (orbital plus spin) angular momentum, $\hbar \vec{J}$.

- The permanent magnetic moment of an atom is proportional to J and appears in units of a Bohr *magneton*, $\mu_B = e/2m$.

- All materials exhibit a diamagnetic response. For non-magnetic materials, the response is observed as a weak magnetization opposite to the applied field.

- Paramagnets lack any interaction between neighboring moments and lose their magnetization whenever a field is removed. Ferromagnets experience an interaction between neighboring moments and can retain their magnetization if the field is removed at temperatures below the *Curie temperature*, T_c.

- The interaction between moments in a ferromagnet arises from a quantum mechanical interaction between neighboring atomic spins, $u(r_{ij}) = -2J_{ex}(r_{ij})\, \vec{S}_i \cdot \vec{S}_j$, known as the *exchange interaction*.

- Near T_c, ferromagnets develop regions of correlated moments which increase in size as T_c is approached.

Exercises

4.1. A power transformer consists of two windings around a common ferromagnetic core and is used in ac circuits to step up (or down) an ac voltage. During each cycle, the ferromagnetic core repeats the process of growing and shrinking domains. (a) Show that the work done in one cycle of a hysteresis loop equals the area enclosed by the loop. (b) Which would be better for a power transformer: a core with a large coercivity or a small coercivity? Why? (Hint: for part (a), you might make use of the following thermodynamic relation: $dW = H dM$.)

4.2. Consider the centripetal force on the orbiting electron of the classical Bohr model illustrated in Fig. 4.3. Consider both a situation with and without a magnetic field present and show that when the field is present and directed upwards (perpendicular to the orbital plane), the electron speeds up by $\Delta v \approx \rho e B / 2m$, and that the change in its angular velocity then equals the Larmor frequency given in Eq. (4.9). As a follow-up to this, determine how large a magnetic field is needed to cause a 1% change in the orbital speed.

4.3. In the text, the T_c for iron was estimated to be only about 3 K, based on a classical molecular field due to the dipolar interaction. Verify this by direct substitution into Eq. (4.26) using for iron the angular momentum $J = S = 1$ and a particle spacing $a = 1$ Å.

4.4. Compute the diamagnetic susceptibility for a gas of hydrogen atoms (at standard temperature and pressure). Recall that the ground state wave function for a hydrogen atom is $\psi(r) = (\pi a^3)^{-1/2} e^{-r/a}$, where $a = \hbar^2/me^2 = 0.529$ Å. Note also that $\langle \rho^2 \rangle = \langle x^2 \rangle + \langle y^2 \rangle = \frac{2}{3}\langle r^2 \rangle$ for a spherically symmetric orbital. Compare your result with the experimental value listed in Table 4.1.

4.5. Obtain the form of the Brillouin function given in Eq. (4.18) starting from Eq. (4.15). By way of a hint, notice that: $\dfrac{\sum m e^{mx}}{\sum e^{mx}} = \dfrac{d}{dx}\left\{ \ln\left(\sum e^{mx} \right) \right\}$.

4.6. Consider the simple two-state paramagnet with $J = S = 1/2$. Show that for such a paramagnet, the magnetization is given simply as: $M = n\mu_B \tanh(\mu_B B / k_B T)$. Using this result, determine the magnitude of magnetic field required to produce a magnetization of half its saturation value at (a) 300 K and (b) 1 K.

Suggested reading

Chaikin and Lubensky provide additional depth on the subject of correlations, but at an advanced level that might not appeal to all. For a good introduction to magnetic materials I recommend reading the sixth chapter of Griffiths.

D. J. Griffiths, *Introduction to Electrodynamics*, 3rd Ed. (Prentice Hall, New Jersey, 1999).

P. M. Chaikin and T. C. Lubensky, *Principles of Condensed Matter Physics* (Cambridge University Press, New York, 2003).

C. Kittel, *Introduction to Solid State Physics*, 8th Ed. (John Wiley and Sons, 2005).

PART II

SCATTERING

Most of the light that enters our eyes has been scattered and when we see objects we see them because of the diffuse scattering of light they produce. Even the sky is blue because of how it scatters sunlight. But scattering is also an important mechanism for observing very small objects. As a classic example, recall how Lord Rutherford unveiled the internal structure of the atom by studying the scattering pattern of alpha particles directed at gold atoms. The abnormally large number of particles backscattered by these gold atoms pointed to the existence of a small, but very dense, center which we now refer to as the nucleus.

In the next chapter, we develop the basic framework for the scattering of waves by condensed matter by looking at how electromagnetic waves scatter from the electrons contained in the particles. Although this is strictly relevant only for the scattering of X-rays and visible light, much of the formalism that develops will apply equally to other waves, including particle waves (electrons or neutrons) that interact with things other than electrons. In the following chapter (Chapter 6), we look at how X-rays scatter from crystals. There we will find scattering that is reminiscent of how visible light is scattered by a diffraction grating in that the scattered radiation exits as a set of discrete beams. This discrete (Bragg) diffraction is contrasted in Chapter 7 by the continuous pattern of scattering produced by glasses or liquids.

In the final chapter on scattering (Chapter 8), we examine how waves of a longer wavelength can be used to study structures of a larger extent. These include liquid crystals, whose symmetry is intermediate between those of crystals and liquids, and notable self-similar (i.e. fractal) objects such as polymers and aggregates.

5

Scattering theory

Introduction

In this chapter we develop a general formalism to describe the scattering of waves by a large system of particles and show that the scattering pattern relates directly to the structural arrangement of the particles. We develop this formalism using the specific example of light waves, composed of oscillating electromagnetic fields. But, in principle, the waves could represent any wave-like entity including matter waves such as traveling electrons or neutrons. The characteristic scattering pattern is known as the *static structure factor*, and it results from the collective interference of waves scattered by particles in the system. This interference is sensitive to the relative separation between the particles, and the static structure factor is shown to be just a spatial Fourier transform of the particle structure as it is represented by the density–density correlation function.

5.1 The dipole field

All condensed matter is constructed of atoms that contain nuclei and electrons. The nuclei reside at the atom center and the electrons, while bound up in the atom, orbit about the nucleus at a relatively large distance under the attraction of a Coulomb force. In considering the interaction of an atom with an external electric field, we know that both the electron and the nucleus experience opposing forces owing to their opposite charge. However, because neutrons and protons are about two thousand times more massive than the electron, we can largely disregard any disturbances in the location of the nucleus and instead focus on the motion of electrons alone.

To a reasonably good approximation, we can treat an electron in an atom as a negatively charged particle with mass m attached by a spring to a rigid nucleus. The spring constant can be expressed in terms of the natural, or resonant, frequency, ω_o, as $m\omega_o^2$. For convenience, we place the electron at the origin of a coordinate system, as shown in Fig. 5.1, where it experiences the effects of an incoming electromagnetic wave,

$$\vec{E}_\mathrm{i} = E_o \hat{z} \mathrm{e}^{i\vec{k}_i \cdot \vec{r} - i\omega t}, \tag{5.1}$$

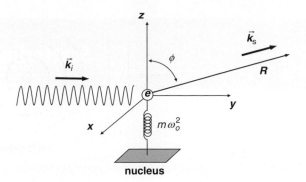

Figure 5.1 Classical model of the interaction of an electromagnetic wave with an electron orbiting a nucleus. The incident alternating electric field drives the electron to oscillate, thereby emitting dipole radiation to a detector (located at R).

where $k_i = 2\pi/\lambda$ is the wave vector and ω the angular frequency of the incident wave. We will take the wave to be polarized along the z-axis. At the origin, the electron experiences a periodic driving force

$$\vec{F}_{\text{driving}} = -eE_o e^{-i\omega t}\hat{z},\tag{5.2}$$

and if we assume that the incident electromagnetic wave is sufficiently weak that it produces only small displacements of the electron without ionizing the atom, then its motion is described by Newton's second law as

$$m\frac{d^2 z}{dt^2} = -m\omega_o^2 z - eE_o e^{-i\omega t}.\tag{5.3}$$

The solution to this equation of motion is of the form $z = z_o e^{-i\omega t}$, where

$$z_o = \frac{(eE_o/m)}{\left(\omega^2 - \omega_o^2\right)}.\tag{5.4}$$

As a result of the incident wave, the electron is driven into oscillatory motion at a frequency ω and amplitude z_o, and behaves as an *oscillating dipole* of dipole moment $p_o = ez_o$. From electrodynamics, this oscillating dipole radiates an additional electromagnetic wave outwards from its center, much like a radio antenna,

$$\begin{aligned}\vec{E}_S(\vec{R},t) &= \frac{-k_i^2}{4\pi\varepsilon_o}\left(p_o\sin\phi\,\hat{\phi}\right)\frac{e^{i\left(\vec{k}_S\cdot\vec{R}-\omega t\right)}}{R}\\[2mm] &= E_o\left\{\frac{k_i^2}{4\pi\varepsilon_o}\left(\frac{e^2/m}{\left(\omega_o^2 - \omega^2\right)}\sin\phi\,\hat{\phi}\right)\right\}\frac{e^{i\left(\vec{k}_S\cdot\vec{R}-\omega t\right)}}{R},\end{aligned}\tag{5.5}$$

where ε_o is the permeability of free space, and the remaining terms are defined in Fig. 5.1. Because the electron oscillates at the same frequency with which it

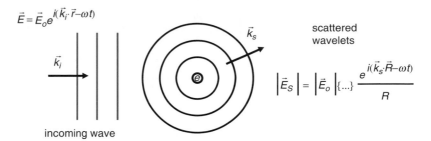

Figure 5.2 A view of the scattered radiation along the z-axis of Figure 5.1 showing the formation of spherical wavelets emanating from the scattering source.

is driven, the scattered field exhibits the same wavelength as the incident field. This sort of scattering event, in which the energy of the incoming and outgoing photons is the same, is known as *elastic scattering*.

The field produced by the electron has a donut-shaped profile as a result of the sine term in Eq. (5.5). Along the z-axis ($\phi = 0$), the field vanishes, while in the xy-plane the field is isotropic. Furthermore, the intensity, which is proportional to the square of the field, decreases as $1/R^2$, characteristic of a radiating source. From here on, we will restrict our discussion to observations made in the xy-plane where the field is isotropic. Here the scattered field from the electron appears much like the waves in a pond emanating from an oscillating bobber (see Fig. 5.2).

5.1.1 The scattering cross section

In this development, we have considered only the interaction of matter with an electromagnetic wave via its disturbance of the electron. The details of that interaction are contained in the bracketed term in Eq. (5.5), which can be viewed as a sort of "cross section" for the interaction: a measure of the strength of the interaction for a given magnitude of incident field (E_o). For most atoms, the resonant absorption frequencies, ω_o, reside in the UV range and hence for X-rays, the frequency of the incident wave is much larger than resonance ($\omega^2 >> \omega_o^2$). In this limit, the factor in brackets reduces to

$$\left\{ \frac{k_i^2}{4\pi\varepsilon_o} \left(\frac{e^2/m}{(\omega_o^2 - \omega^2)} \sin\phi \,\hat{\phi} \right) \right\} \approx \frac{-1}{4\pi\varepsilon_o} \left(\frac{k_i^2}{\omega^2} \frac{e^2}{m} \sin\phi \,\hat{\phi} \right) = \frac{-1}{4\pi\varepsilon_o} \left(\frac{e^2}{mc^2} \sin\phi \,\hat{\phi} \right),$$

$$(5.6)$$

which is independent of the wavelength and consistent with the *Thomson scattering* from a free electron. In contrast, for visible light where $\omega^2 << \omega_o^2$, the bracketed term reduces to

$$\left\{ \frac{k_i^2}{4\pi\varepsilon_o} \left(\frac{e^2/m}{(\omega_o^2 - \omega^2)} \sin\phi\,\hat{\phi} \right) \right\} \approx \frac{1}{4\pi\varepsilon_o} \left(\frac{k_i^2}{\omega_o^2} \frac{e^2}{m} \sin\phi\,\hat{\phi} \right) = \frac{\pi}{\varepsilon_o} \left(\frac{e^2}{m\omega_o^2} \sin\phi\,\hat{\phi} \right) \frac{1}{\lambda^2},$$

$$(5.7)$$

and is seen to vary inversely with the square of the wavelength of the incident light.

In the case of particle waves, a radiated field similar to that in Eq. (5.5) is generated. However, the corresponding bracketed term is quite different and involves elements that describe the cross section for the specific scattering process. For example, in neutron scattering this bracketed term would contain details associated with the quantum mechanical interaction of incoming neutrons with the nuclei of the matter, and the outgoing wave would contain information regarding the probability for a scattered neutron to be detected.

5.2 Interference

Let us now consider two electrons separated by a distance d, as depicted in Fig. 5.3. Each radiates a similar field, described by Eq. (5.5), but the separation necessarily introduces a relative phase shift due to differences in the optical path length taken to the detector. Thus, when the two fields combine at the detector, they may constructively or destructively interfere depending on the

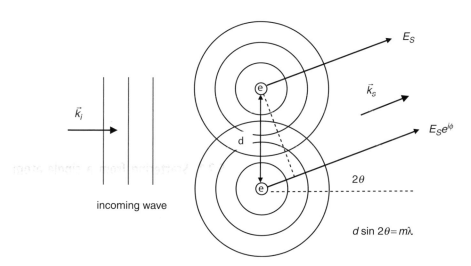

Figure 5.3   View of the interference of wavelets emanating from two scattering sources separated by a distance d. The condition for constructive interference is seen to be that for the optical two-slit interference pattern. Note the scattering angle (relative to the incident direction) is here defined to be 2θ.

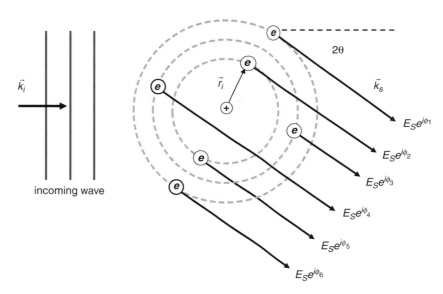

Figure 5.4 The cumulative interference of scattering from a collection of electrons contained in an arbitrary atom.

location of the detector and the separation of the two electrons. Consideration of Fig. 5.3 reveals that constructive interference will occur when:

$$d \sin(2\theta) = m\lambda, \tag{5.8}$$

a condition that should be familiar. It is merely the condition for a double-slit interference pattern! Here we now begin to see a glimpse of how the scattering of waves reveals structure in materials. We know when we change the spacing between these two scattering sources, the interference pattern will change in a corresponding fashion. If we move the sources together the interference fringes spread apart and if we move them apart the fringes pack together. This simple notion is the key to how scattering provides information about the particle arrangements in condensed matter.

5.2.1 Scattering from a single atom

Alright, now let us add more electrons. We next consider a single atom containing Z electrons in its orbital shells. We begin first with a crude atomic model of point electrons in orbit about the nucleus (see Fig. 5.4). Again we see that, because of the relative locations of the electrons, there will occur phase differences between the individual fields scattered by each electron such that the total field viewed at the detector is

$$E_S^{\text{tot}} = E_S \left\{ e^{i\phi_1} + e^{i\phi_2} + e^{i\phi_3} + e^{i\phi_4} + \cdots \right\} \tag{5.9}$$

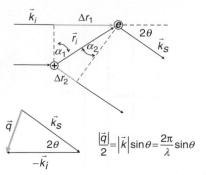

Figure 5.5 The scattering wave vector, $\vec{q} \equiv \vec{k}_s - \vec{k}_i$, is defined as the vector difference between the scattered wave vector and the incident wave vector. For the case of elastic scattering, its magnitude is simply related to the wavelength of the radiation and the scattering angle.

Scattering wave vector

We can reference the phase shifts (ϕ_i) relative to any fixed point and here we take that point to be the nucleus of the atom. As shown in Fig. 5.5, the path length difference of a given electron relative to the nucleus is

$$\Delta r_1 - \Delta r_2 = \frac{\vec{k}_i \cdot \vec{r}_i}{|k_i|} - \frac{\vec{k}_s \cdot \vec{r}_i}{|k_s|}. \tag{5.10}$$

But, because the scattering is elastic, $|k_s| = |k_i| = 2\pi/\lambda$ the phase difference can be expressed as

$$\phi_i \equiv \frac{2\pi(\Delta r_1 - \Delta r_2)}{\lambda} = (\vec{k}_i - \vec{k}_s) \cdot \vec{r}_i = -\vec{q} \cdot \vec{r}_i, \tag{5.11}$$

where $\vec{q} \equiv \vec{k}_s - \vec{k}_i$ is the (all important) *scattering wave vector*. For elastic scattering, the magnitude of $\vec{q}$ is related (see Fig. 5.5) to both the wavelength and scattering angle, 2θ, as

$$|\vec{q}| = \frac{4\pi}{\lambda} \sin \theta. \tag{5.12}$$

Then the net electric field generated by all the electrons in the atom and observed at the detector is given by

$$E_S^{\text{tot}} = E_S \sum_i e^{-i\vec{q} \cdot \vec{r}_i}. \tag{5.13}$$

Here we see the real utility of the scattering wave vector. It provides a simple means of keeping track of phase differences based upon relative positions of the electrons.

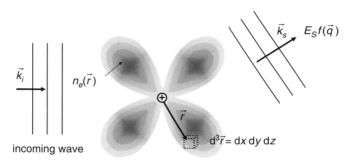

Figure 5.6 Unlike the illustration in Figure 5.4, electrons in an atom present themselves not as point scattering sources but rather as an electron number density, $n_e(\vec{r})$, distributed about the nucleus in a manner that depends upon the electronic configuration. The collective interference of waves scattered by the electron distribution is characterized by the atomic form factor obtained by integrating over the atom as in Eq. (5.15).

Atomic form factor

In reality, the electrons in atoms do not present themselves as point charges, but rather as a distribution in space whose net charge equals e. These electron clouds that surround the nucleus assume characteristic shapes depending on the specific electron configuration of the atom, as illustrated in Fig. 5.6. To accommodate the distributed nature of the electrons we merely need to convert Eq. (5.13) over to an integration of the electron density, where the fraction of an electron per unit volume at some position $\vec{r}$ from the nucleus is defined as $n_e(\vec{r})$. Then Eq. (5.13) becomes

$$E_S^{tot} = E_S \int_{atom} n_e(\vec{r}) e^{-i\vec{q}\cdot\vec{r}} d^3\vec{r}. \qquad (5.14)$$

Fortunately, the electron cloud distributions are the same for any atom of a given element and others have computed this atomic scattering for us. These are commonly provided as so-called *atomic form factors*, $f(\vec{q})$, defined as a ratio of the total scattered field relative to that of a single electron,

$$f(\vec{q}) \equiv \frac{E_S^{tot}}{E_S} = \int_{atom} n_e(\vec{r}) e^{-i\vec{q}\cdot\vec{r}} d^3\vec{r}. \qquad (5.15)$$

Examples of $f(\vec{q})$ for a few atoms are shown in Fig. 5.7. Note that in the limit that q approaches zero (forward scattering direction), $f(\vec{q})$ approaches the number of electrons, Z. This occurs because all the phases in Eq. (5.13) vanish in the forward direction ($\vec{q} = 0$), leading to a scattered field that is just Z times that of a single electron.

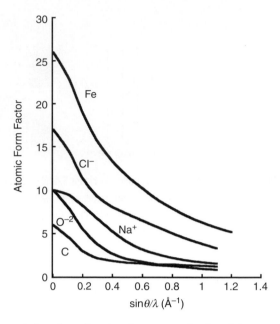

Figure 5.7 Variation of the atomic form factor for selected atoms as a function of the scattering wave vector $\vec{q}$. Overall, the form factor is proportional to the number, Z, of electrons in the atom and approaches Z as $\vec{q}$ approaches zero. (Data from Cullity (1978).)

5.3 Static structure factor

Finally, we proceed to the scattering from a collection of particles. Here, "particles" is a generic term for the objects that make up our condensed matter. The particles could be atoms, as depicted in Fig. 5.8, that each individually scatter a field given by the appropriate atomic form factor defined in Eq. (5.15). Alternatively, the particles could be any object containing a distribution of electrons described by some appropriate particle form factor that individually scatters as

$$E_{S,i} = E_S f_i(\vec{q}). \tag{5.16}$$

In any event, the particles in our collection have different relative positions and so interference again arises at the detector between the fields scattered by each particle. Once more, we can track the phase differences using the scattering wave vector and describe the detected field as

$$E_S^{\text{tot}} = E_S \sum_{\text{particles}} f_i(\vec{q}) e^{-i\vec{q}\cdot\vec{r}_i}, \tag{5.17}$$

where $\vec{r}_i$ is the position of the center of the ith particle.

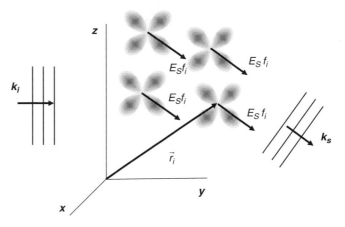

Figure 5.8   The collective interference from all of the electron density in a collection of particles can be reduced to the interference due to the relative positions of the particles themselves, each particle scattering as though it were a point particle. The scattering from each particle is however modulated in q by its internal electron structure, as conveyed by the particle form factor.

In scattering experiments, one generally detects not the electric field, but the intensity, which is proportional to the square modulus of the electric field. This is also true for neutron scattering. One does not detect the neutron wave function, but rather the probability for observing a neutron, which is proportional to the square of the wave function. For light waves, the intensity is

$$I_S = \left|E_S^{\text{tot}}\right|^2 = |E_S|^2 \left(\sum_i f_i(\vec{q})\mathrm{e}^{-i\vec{q}\cdot\vec{r}_i}\right)\left(\sum_j f_j(\vec{q})\mathrm{e}^{-i\vec{q}\cdot\vec{r}_j}\right)^{*}. \qquad (5.18)$$

For the time being, let us assume that all our particles are identical with identical form factors. Then,

$$I_S = \left|E_S^{\text{tot}}\right|^2 = |E_S|^2|f(\vec{q})|^2 N\left\{\frac{1}{N}\sum_i\sum_j \mathrm{e}^{-i\vec{q}\cdot(\vec{r}_i-\vec{r}_j)}\right\} = |E_S|^2|f(\vec{q})|^2 N S(\vec{q}),$$

$$(5.19)$$

where the *static structure factor*, is defined by

$$S(\vec{q}) = \frac{1}{N}\left\langle \sum_i\sum_j \mathrm{e}^{-i\vec{q}\cdot(\vec{r}_i-\vec{r}_j)}\right\rangle. \qquad (5.20)$$

Again, the angled brackets indicate an average taken over appropriate ensembles of the structure.

5.3.1 A relevant scattering length scale

What is $S(q)$? It is an overall measure of the effect of phase differences in the scattered field arising from the relative separation of the particles, *regardless of*

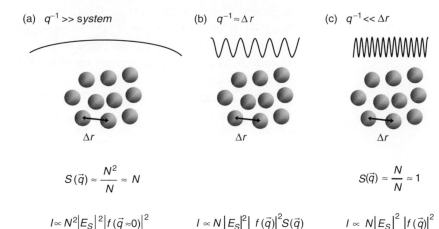

(a) $q^{-1} \gg$ system   (b) $q^{-1} \approx \Delta r$ (c) $q^{-1} \ll \Delta r$

$$S(\vec{q}) \approx \frac{N^2}{N} \approx N \qquad\qquad S(\vec{q}) \approx \frac{N}{N} \approx 1$$

$$I \propto N^2 |E_S|^2 |f(\vec{q} \approx 0)|^2 \qquad I \propto N |E_S|^2 |f(\vec{q})|^2 S(\vec{q}) \qquad I \propto N |E_S|^2 |f(\vec{q})|^2$$

Figure 5.9   Three significant regimes are illustrated for $S(q)$ depending on the particle spacing in relation to the scattering length. (a) When the scattering length is much larger than the collection of particles, interference appears as a sum of nearly equivalent phases for which $S(q)$ is proportional to the number of particles. (b) When the scattering length is comparable to the particle spacing there are significant angular variations to the scattered intensity. (c) When the scattering length is much smaller than the particle spacing, the phases arrive randomized and interference leads to $S(q) \approx 1$ with the q-dependence entering only from the particle form factor.

the sort of waves used in the scattering experiment. In this regard, the scattering wave vector plays a pivotal role. The inverse of the scattering wave vector represents a relevant scattering length scale, $l = 2\pi/q$, whose size in comparison to the mean particle spacing dictates the severity of the interference effects. This is illustrated in Fig. 5.9, where three different regimes are displayed.

When the scattering length scale is large compared to the size of the system of scatterers, such as when visible light ($q^{-1} \approx \lambda \approx 5000\,\text{Å}$) scatters from a smaller-sized collection of particles, the phase differences, $\vec{q} \cdot (\vec{r}_i - \vec{r}_j)$, between neighboring scattered waves are nearly identical and result in constructive interference:

$$S(\vec{q}) \equiv \frac{1}{N} \left\langle \sum_{i,j} e^{-i\vec{q} \cdot (\vec{r}_i - \vec{r}_j)} \right\rangle \approx \frac{N^2}{N} = N. \qquad (5.21)$$

Conversely, when the scattering length scale is small compared to the particle spacing, the phase differences, $\vec{q} \cdot (\vec{r}_i - \vec{r}_j)$, between waves scattered by neighboring particles are randomized and produce

$$S(\vec{q}) \equiv \frac{1}{N} \left\langle \sum_{i,j} e^{-i\vec{q} \cdot (\vec{r}_i - \vec{r}_j)} \right\rangle \approx \frac{N}{N} = 1. \qquad (5.22)$$

It is only when the scattering length scale is comparable to the particle spacing that large variations in the phase difference occur to produce significant angular variations in the scattered field, that are associated with the relative positions of the particles. Hence, the scattering wave vector (specifically its inverse) sets an important scale for examining the structure and must be chosen to correspond to the particle separations of interest. X-ray scattering operates on an Ångstrom scale and so probes particles roughly 1Å apart. Visible light operates on nearly a micron scale and so probes "particles" of a much larger size with correspondingly larger separation. Neutrons possess a range of scattering length scales that depend upon the temperature at which they are moderated (Ex. 3).

5.3.2 A Fourier relationship: the density–density correlation function

Before leaving the subject of scattering, we emphasize an important relationship between the spatial structure of matter and the angular pattern of the scattering that arises. As it turns out, the two are related by a Fourier transformation. To demonstrate this, we begin by exploiting the properties of Dirac delta functions, to replace the argument of the double summation in Eq. (5.20) by

$$e^{-i\vec{q}\cdot(\vec{r}_i-\vec{r}_j)} = \int d^3\vec{r}_1 \int d^3\vec{r}_2 \delta(\vec{r}_1 - \vec{r}_i)\delta(\vec{r}_2 - \vec{r}_j)e^{-i\vec{q}\cdot(\vec{r}_1-\vec{r}_2)}, \qquad (5.23)$$

allowing the structure factor to be expressed as,

$$S(\vec{q}) = \frac{1}{N}\int d^3\vec{r}_1 \int d^3\vec{r}_2 \left\langle \sum_i \sum_j \delta(\vec{r}_1 - \vec{r}_i)\delta(\vec{r}_2 - \vec{r}_j) \right\rangle e^{-i\vec{q}\cdot(\vec{r}_1-\vec{r}_2)}. \quad (5.24)$$

Since the density of particle centers can be expressed as a sum of Dirac delta functions,

$$n(\vec{r}) = \sum_{i=1,N} \delta(\vec{r} - \vec{r}_i), \qquad (5.25)$$

we can write,

$$S(\vec{q}) = \frac{1}{N}\int d^3\vec{r}_1 \int d^3\vec{r}_2 C(\vec{r}_1, \vec{r}_2)e^{-i\vec{q}\cdot(\vec{r}_1-\vec{r}_2)}, \qquad (5.26)$$

where

$$C(\vec{r}_1, \vec{r}_2) \equiv \langle n(\vec{r}_1)n(\vec{r}_2) \rangle = \left\langle \sum_{i=1,N} \sum_{j=1,N} \delta(\vec{r}_1 - \vec{r}_i)\delta(\vec{r}_2 - \vec{r}_j) \right\rangle \qquad (5.27)$$

is known as the *density–density correlation function*. Its description as a "correlation" function can be seen in Eq. (5.27) and Fig. 5.10, where there are two "search vectors" $\vec{r}_1$ and $\vec{r}_2$ that cause a correlation in $C(\vec{r}_1, \vec{r}_2)$ to light

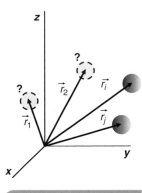

Figure 5.10

The density–density correlation function is defined by a set of search vectors ($\vec{r}_1$ and $\vec{r}_2$) searching for correlations as evidenced by the presence of particles located at both $\vec{r}_i$ and $\vec{r}_j$.

up only when each happens to point to the center of the same particle, or two different particles. Equation (5.26) thus conveys the following important mantra: *the angular pattern of scattered radiation* described by the static structure factor *is merely a (double) Fourier transform of the structure* of the material, as described by its density–density correlation function. In practical terms, this means that if the spacing between particles is frequently some value Δr, a corresponding peak will occur in a $S(q)$ near $q \approx 2\pi/\Delta r$.

Summary

- The structure of matter is assessed by the interference of the waves it scatters.

- The interference pattern caused by a collection of point scatterers is described by the *static structure factor*, $S(\vec{q})$.

- In scattering experiments, the inverse of the scattering wave vector, $\vec{q}$, sets a relevant length scale at which structural features are probed.

- The static structure factor is a Fourier transform of the density–density correlation function which describes the particle positions in matter.

Exercises

5.1. Estimate the range of q for both X-rays ($\lambda = 0.7$ Å) and visible light ($\lambda = 5000$ Å) for a typical range of scattering angles $2° < 2\theta < 180°$.

5.2. Returning to the oscillating dipole model for a single electron, consider an alternate case in which the electron experiences a damping force of the form $\vec{F}_{\text{damp}} = -\gamma \frac{dz}{dt}\hat{z}$. Show in this instance that the amplitude of oscillations is given by the complex quantity

$$z_o^* = \frac{eE_o/m}{\left(\omega^2 - \omega_o^2 + i\omega\gamma/m\right)},$$

whose real part is

$$\text{Re}\left[z_o^*\right] = \frac{eE_o/m\left(\omega^2 - \omega_o^2\right)}{\left(\left(\omega^2 - \omega_o^2\right)^2 + \left(\omega\gamma/m\right)^2\right)}.$$

5.3. In a neutron scattering experiment, an incident beam of neutrons is rendered nearly monochromatic by having the neutrons first achieve thermal equilibrium in a moderator gas maintained at a temperature T.

(a) What de Broglie wavelength would a room temperature (300 K) neutron have? (b) What temperature of moderator would be required to produce neutrons with a de Broglie wavelength near that of visible light (5000Å)?

5.4. Consider a 1D chain of atoms with spacing a that alternates between two atom types (A and B) with form factors f_A and f_B, respectively. An X-ray beam is incident with $\vec{k}_i$ perpendicular to the chain. (a) Show that the constructive interference condition is $m\lambda = 2a\cos\phi$, where ϕ is the angle between the diffracted beam and the line of atoms. (b) Show that for m odd the intensity of the diffracted beam is proportional to $|f_A - f_B|^2$, while for m even it is proportional to $|f_A + f_B|^2$. (c) Explain what happens to the intensity of the diffracted beam when all of the B atoms are replaced with A.

5.5. Three identical scattering particles are arranged in a straight line along the z-axis and separated by a common distance a. A plane wave is incident on the arrangement with $\vec{k}_i = (2\pi/\lambda)\hat{y}$. (a) Compute the static structure factor in the scattering plane and show that it can be expressed as: $S(q) = \frac{1}{3}\{1 + 4\cos\phi + 4\cos^2\phi\}$, where $\phi = \vec{q} \cdot a\hat{z}$. (b) If the spacing a equals the wavelength of the incident wave, list the scattering angles (2θ) for which $S(q)$ is maximum.

5.6. Show that, for a spherically symmetric electron distribution, the form factor given in Eq. (5.15) can be expressed as $f(\vec{q}) = 4\pi \int\limits_{\text{atom}} r^2 n_e(r) \left[\frac{\sin qr}{qr}\right] \, \mathrm{d}r$.

5.7. The electron number density for a hydrogen atom is given by the square of its ground state wave function as $n_e(r) = |\psi(r)|^2 = \frac{1}{\pi a_o^3} e^{-2r/a_o}$, where a_o is the Bohr radius, Show that the form factor is $f_H(q) = 16/(4 + (qa_o)^2)^2$.

5.8. Imagine a spherical scattering particle of radius R with a uniform density, n_o. Show that the particle's form factor is $f_{\text{sphere}}(q) = \frac{4\pi n_o}{q^3}[\sin qR - qR\cos qR]$.

Suggested reading

Griffiths is a good source for better understanding the origin of the dipole scattering which is presented rather abruptly as Eq. (5.5). The structure factor is developed in many of the other references. Beware though that it may appear differently depending on how it is defined and normalized.

D. J. Griffiths, *Introduction to Electrodynamics*, 3rd Ed. (Prentice Hall, New Jersey, 1999).

G. Strobl, *Condensed Matter Physics* (Springer-Verlag, Berlin, 2004).

P. M. Chaikin and T. C. Lubensky, *Principles of Condensed Matter Physics* (Cambridge University Press, New York, 2003).

H. C. van de Hulst, *Light Scattering by Small Particles* (John Wiley and Sons, New York, 1957).

6

Scattering by crystals

Introduction

In the previous chapter, we developed scattering theory and introduced the static structure factor as the Fourier counterpart to the density–density correlation function for particle arrangements in a solid. In that sense, the static structure factor measures how recurrent a given particle spacing is within the material. In this chapter, we consider the scattering of waves by an ordered crystal and establish the foundations of conventional crystallography, based upon Bragg diffraction from crystal planes defined by *Miller indices*. Here we show that the extremely periodic pattern of particle positions necessitates a discrete set of Fourier components and that scattering occurs only in a discrete set of scattering directions. Furthermore, the set of scattering wave vectors, q, associated with these limited allowed scattering directions form an important lattice in wave vector space, known as the *reciprocal lattice*. In later chapters we will repeatedly encounter this reciprocal space as an important consideration for the properties of any wave that travels within a crystal.

6.1 Scattering by a lattice

As discussed in Chapter 1, the atoms in a crystal are highly organized and the overall structure can be reduced to a small group of atoms (the basis set) that is repeatedly attached to an imaginary space lattice. The space lattice itself is defined by a set of translation vectors, $\vec{T} = h\vec{a}_1 + k\vec{a}_2 + l\vec{a}_3$, and possesses translational symmetry in addition to other symmetry properties due to its repetitive nature. The small group of atoms attached to each unit cell of the lattice is referenced by a set of basis vectors, $\vec{R}_i = x_i\vec{a}_1 + y_i\vec{a}_2 + z_i\vec{a}_3$, and these, together with the translation vectors, define the location of each atom in the crystal.

We now consider how incident waves, namely X-rays whose scattering length scale best matches the lattice spacing of common crystals, scatter from the orderly array of atoms found in a crystal. For the moment, we ignore the fact that these atoms are actually vibrating and pretend instead that they are fixed in space. Of course, we could treat the atoms as our scattering particles,

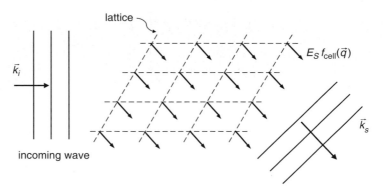

Figure 6.1 The collective interference of waves scattered by a crystal can be reduced to the interference arising from the lattice sites alone, each of which scatters as though it is a point particle scattering with a form factor, $f_{cell}(q)$, which contains all the interference effects due to contents of the basis set.

each scattering with an atomic form factor given in Eq. (5.15), and proceed to compute the static structure factor describing the interference arising from each and every atom in the crystal. However, the orderly nature of the crystal offers a practical short cut. Because each unit cell of the space lattice contains exactly the same small group of atoms, it is far more convenient to treat this small group of atoms in the unit cell as though they were a single scattering "particle" that scatters with a collective *cell* form factor:

$$f_{\text{cell}}(\vec{q}) = \sum_{\text{basis}} f_i(\vec{q}) e^{-i\vec{q}\cdot\vec{R}_i}, \qquad (6.1)$$

where $f_i(\vec{q})$ are the individual *atomic* form factors for each atom in the unit cell and $\vec{R}_i$ is the set of vectors describing their location. By adopting this choice for "particles", the remaining interference of scattered waves now arises from the relative locations of the lattice points of the space lattice, as is illustrated in Fig. 6.1.

How then does the orderly structure of a crystal affect the structure factor, $S(q)$ as given in Eq. (5.20)? First of all, the ensemble averaging inherent in Eq. (5.20) is unnecessary, because any reincarnation of the crystal will appear exactly identical. There is only one configuration that this ordered system can assume. Secondly, because the space lattice is ordered, the position of any lattice site (our "particles") at $\vec{r}_i$ can be accurately referenced to any other arbitrary lattice site at $\vec{r}_o$ by the set of translation vectors that define the lattice. Thus, we can express the structure factor for a crystal as:

$$S_{\text{crystal}}(\vec{q}) = \frac{1}{N}\sum_i\sum_j e^{-i\vec{q}\cdot(\vec{r}_i - \vec{r}_j)} = \frac{1}{N}\sum_i\sum_j e^{-i\vec{q}\cdot\left[(\vec{r}_o + \vec{T}_i) - (\vec{r}_o + \vec{T}_j)\right]}$$

$$= \frac{1}{N}\left|\sum_i e^{-i\vec{q}\cdot\vec{T}_i}\right|^2. \qquad (6.2)$$

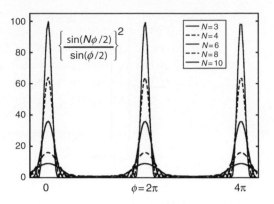

Figure 6.2   Interference pattern due to a one-dimensional chain of N particles. Note the intensity of the maxima increases as the square of the number of particles, while the total area under each peak remains constant. Consequently, for large N the peaks are extremely sharp and virtually all of the scattered intensity occurs only in discrete directions.

6.1.1 A set of allowed scattering wave vectors

Where (i.e. at what $\vec{q}$) does the most intense scattering occur? Intense scattering corresponds to constructive interference of the scattered waves and in Eq. (6.2) is seen to occur predominantly for a *discrete* set of $\vec{q} = \vec{G}_i$ for which

$$\vec{G}_i \cdot \vec{T}_i = 0, \ 2\pi, \ 4\pi, \ \cdots. \tag{6.3}$$

In fact, there is virtually *no scattering* present at any other scattering wave vectors! To demonstrate this, let us expand Eq. (6.2) by substituting in the complete set of translation vectors:

$$S_{\text{crystal}}(\vec{q}) = \frac{1}{N} \left| \sum_{h=0}^{N_1-1} e^{-ih(\vec{q}\cdot\vec{a}_1)} \sum_{k=0}^{N_2-1} e^{-ik(\vec{q}\cdot\vec{a}_2)} \sum_{l=0}^{N_3-1} e^{-il(\vec{q}\cdot\vec{a}_3)} \right|^2. \tag{6.4}$$

Each summation runs over the number of lattice points found along each of the three crystal axes, and the total number of lattice points equals $N = N_1 N_2 N_3$. As an exercise (Ex. 1), one can show that Eq. (6.4) can be restated in the following form:

$$S_{\text{crystal}}(\vec{q}) = \frac{1}{N} \left\{ \frac{\sin(N_1 \vec{q} \cdot \vec{a}_1/2)}{\sin(\vec{q} \cdot \vec{a}_1/2)} \right\}^2 \left\{ \frac{\sin(N_2 \vec{q} \cdot \vec{a}_2/2)}{\sin(\vec{q} \cdot \vec{a}_2/2)} \right\}^2 \left\{ \frac{\sin(N_3 \vec{q} \cdot \vec{a}_3/2)}{\sin(\vec{q} \cdot \vec{a}_3/2)} \right\}^2. \tag{6.5}$$

As an illustration of the sharpness of the scattering profile, a plot of the quantity in the first bracket of Eq. (6.5) is shown in Fig. 6.2 for increasing values of N_1. Notice how the maxima occurring at $\phi = \vec{q} \cdot \vec{a}_1 = h2\pi$ increase rapidly with increasing N_1, while the region in between the peaks vanishes.

The height of each maximum is proportional to the square of the number of lattice sites, N_1^2, and the pattern shown in Fig. 6.2 is identical to that of the intensity of light reflected from a diffraction grating in classical optics. This equivalence is not coincidental, because the bracketed term describes the interference of waves by a single line of equally spaced "particles" (unit cells in this instance), which, in fact, is how a diffraction grating operates.

The Laue conditions

Notice though that $S(q)$ in Eq. (6.5) is a *product* of three such bracketed terms. This means that intense scattering from a 3D crystal will require not just one bracketed quantity to be maximized, but all three bracketed quantities must *simultaneously* be maximized. Thus intense scattering is found only when the three *Laue conditions*:

$$\vec{q} \cdot \vec{a}_1 = 2\pi h$$
$$\vec{q} \cdot \vec{a}_2 = 2\pi k \tag{6.6}$$
$$\vec{q} \cdot \vec{a}_3 = 2\pi l,$$

are simultaneously satisfied. If any one of these three conditions is not met for a given scattering wave vector $\vec{q} = \vec{k}_s - \vec{k}_i$, then there is no scattering observed in the corresponding $\vec{k}_s$ direction.

6.2 Reciprocal lattice

From the discussion above, we conclude that scattering from a crystal will only occur for a discrete set of scattering wave vectors $\vec{q} = \vec{G}_i$. We define this discrete set as

$$\vec{G}_i = \vec{G}_{hkl} = h\vec{b}_1 + k\vec{b}_2 + l\vec{b}_3, \tag{6.7}$$

where h,k,l are the complete set of integers. In order for this set of scattering wave vectors to satisfy the Laue conditions in Eq. (6.6), we need

$$\vec{b}_i \cdot \vec{a}_j = 2\pi\delta_{ij}, \tag{6.8}$$

where $\delta_{ij} = 1$ if $i = j$ and $\delta_{ij} = 0$ if $i \neq j$. A suitable choice for the vectors, $\vec{b}_i$, is then

$$\vec{b}_1 = 2\pi\frac{\vec{a}_2 \times \vec{a}_3}{V_{\text{cell}}}, \quad \vec{b}_2 = 2\pi\frac{\vec{a}_3 \times \vec{a}_1}{V_{\text{cell}}}, \quad \vec{b}_3 = 2\pi\frac{\vec{a}_1 \times \vec{a}_2}{V_{\text{cell}}}. \tag{6.9}$$

Adopting this notation, the scattering from a crystal can be summarized as: "intense scattering from a crystal occurs only for those $\vec{q}$ that equal any one of the translation vectors, $\vec{G}_{hkl}$, that define the *reciprocal lattice*". What is the reciprocal lattice? Firstly, it is reciprocal because the vectors that define it in Eq. (6.9) have dimensions of reciprocal length. But, most importantly, the reciprocal lattice is a "road map" for the diffraction from the crystal. The reciprocal lattice is fundamentally related to the original space lattice and, in fact, is a spatial Fourier transform of it. It details all the allowed $\vec{q}$ for which scattering can possibly occur.

As an example, consider the reciprocal lattice associated with a simple cubic (SC) space lattice. The lattice vectors of the SC lattice are equal in length and mutually orthogonal:

$$\vec{a}_1 = a\hat{x}, \quad \vec{a}_2 = a\hat{y}, \quad \vec{a}_3 = a\hat{z}, \tag{6.10}$$

and so the lattice vectors of the reciprocal space are then given by

$$\vec{b}_1 = \frac{2\pi}{a}\hat{x}, \quad \vec{b}_2 = \frac{2\pi}{a}\hat{y}, \quad \vec{b}_3 = \frac{2\pi}{a}\hat{z}. \tag{6.11}$$

Thus the reciprocal lattice defined by Eq. (6.11) is seen to generate another SC lattice, but with lattice spacing equal to $2\pi/a$. In a similar manner, one can demonstrate (Ex. 2 and Ex. 3) that the BCC space lattice transforms into a reciprocal lattice with an FCC structure, while the FCC space lattice transforms into a reciprocal lattice with a BCC structure. Thus, if we measure all the discrete $\vec{q}\ (= \vec{G}_{hkl})$ for which intense scattering is observed from the crystal, we obtain the reciprocal lattice and, in turn, can identify the structure of the space lattice.

6.3 Crystal planes

6.3.1 Miller indices

One might think that a crystallographer's order of business then would be to examine the discrete scattering from a crystal, construct the reciprocal lattice and finally Fourier transform that to obtain the spatial structure of the crystal. But this is not the approach adopted by most crystallographers. Instead, crystallography emphasizes the role of *crystal planes* in the diffraction process.

Because of its orderly structure, one finds in crystals a number of well-defined planes formed by lattice points, as illustrated in Fig. 6.3, and crystallographers have developed a useful system for indexing them in terms of so-called *Miller indices*, which happen to be just the set of integers h, k and l introduced earlier. In this indexing scheme, a given crystal plane is labeled by its set of indices as "(hkl)". As an example, the (320) planes of a SC lattice are shown in Fig. 6.3 and the recipe for determining the indices is as follows. First locate where the plane crosses the three axes of the space lattice. These are often at some fraction of a lattice spacing. For the example in Fig. 6.3, one

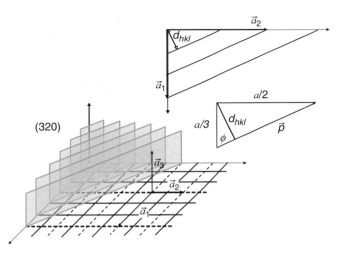

Figure 6.3 The (320) planes of a crystal lattice. Upper right-hand figures represent a view of the planes taken along the $\vec{a}_3$ axis.

of the planes shown crosses the $\vec{a}_1$ axis at $1\vec{a}_1$, the $\vec{a}_2$ axis at $(3/2)\vec{a}_2$ and the $\vec{a}_3$ axis at infinity. Next, invert these fractions (1/1, 2/3, 0) and multiply by the least common denominator to obtain a set of integers ($h = 3$, $k = 2$, and $l = 0$). A view taken down the z-axis shows sequential (320) planes separated by a fixed distance d_{hkl}, known as the plane spacing. By far, the most prominent planes for diffraction experiments are the (100), (110) and (111) planes. Examples of these for the SC lattice are illustrated in Fig. 6.4.

How do these planes play a role in the scattering? The proof is left as an exercise (Ex. 6), but one can show quite generally that: (1) $\vec{G}_{hkl}$ is directed *normal* to the (hkl) planes:

$$\hat{G}_{hkl} = \hat{n}_{hkl}, \tag{6.12}$$

and has a *magnitude* inversely related to the spacing between the (hkl) planes:

$$\left| \vec{G}_{hkl} \right| = \frac{2\pi}{d_{hkl}}. \tag{6.13}$$

To demonstrate the validity of these statements in the absence of a proof, consider again the (320) planes of the SC lattice discussed in Fig. 6.3. From Eq. (6.7), the magnitude of $\vec{G}_{hkl}$, for a SC lattice is given by

$$\left| \vec{G}_{hkl} \right|_{\text{cubic}} = \sqrt{\vec{G}_{hkl} \cdot \vec{G}_{hkl}} = \frac{2\pi}{a} \sqrt{h^2 + k^2 + l^2}, \tag{6.14}$$

and the direction is given by

$$\left(\hat{G}_{hkl} \right)_{\text{cubic}} = \frac{\vec{G}_{hkl}}{\left| \vec{G}_{hkl} \right|} = \frac{h\hat{x} + k\hat{y} + l\hat{z}}{\sqrt{h^2 + k^2 + l^2}}. \tag{6.15}$$

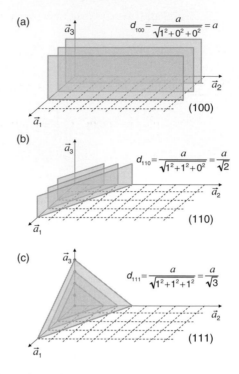

(a) $d_{100} = \frac{a}{\sqrt{1^2+0^2+0^2}} = a$

(100)

(b) $d_{110} = \frac{a}{\sqrt{1^2+1^2+0^2}} = \frac{a}{\sqrt{2}}$

(110)

(c) $d_{111} = \frac{a}{\sqrt{1^2+1^2+1^2}} = \frac{a}{\sqrt{3}}$

(111)

Figure 6.4   Prominent crystal planes of the cubic lattice.

For the (320) planes these then predict $d_{320} = \frac{a}{\sqrt{3^2+2^2+0^2}} = \frac{a}{\sqrt{13}}$ and $\hat{n}_{320} = \frac{3\hat{x}+2\hat{y}}{\sqrt{13}}$. From the right-angled triangle shown in Fig. 6.3, the angle $\phi = 56.3°$ and so $d_{320} = \frac{a\sin(56.3)}{3} = \frac{a}{\sqrt{13}}$. As for the normal vector, it can be obtained by a cross product of the vector $\vec{p}$ in Fig. 6.3 with the z-axis. This then results in $\hat{n}_{320} = \frac{\vec{p}\times\hat{z}}{|\vec{p}|} = \frac{3\hat{x}+2\hat{y}}{\sqrt{13}}$, in full agreement with the predictions of Eq. (6.15).

6.3.2 Bragg diffraction

Let us now look at the scattering of waves in relation to the (hkl) planes of a crystal. Imagine, as shown in Fig. 6.5a, an incident wave (incident upon a plane at some arbitrary angle θ) and a scattered wave (reflected at some arbitrary angle ϕ). Since the allowed scattering condition requires the scattered wave vector,

$$\vec{q} = \vec{k}_s - \vec{k}_i = \vec{G}_{hkl}, \tag{6.16}$$

to match one of the translation vectors of the reciprocal lattice, the *direction* of $\vec{q}$ must match that of $\vec{G}_{hkl}$. This only occurs if the incident and reflected angles are equal, such that the vector addition of Eq. (6.16) forms an isosceles triangle. In optics, this is simply known as the law of reflection.

(a)

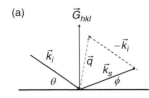

(b)

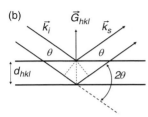

Figure 6.5

(a) The allowed scattering
condition of Eq. (6.16) requires
that $\vec{q}$ be parallel with $\vec{G}_{hkl}$.
Hence, for elastic scattering, the
law of reflection must be
maintained for scattering from
crystal planes. (b) Bragg
scattering of Eq. (6.19) is seen as
the constructive interference of
waves scattered by adjacent
planes, whenever a path
difference equal to any multiple
of the wavelength is introduced.

In addition, our scattering constraint (Eq. (6.16)) can be rearranged to read

$$\vec{k}_s \cdot \vec{k}_s = \left(\vec{k}_i + \vec{G}_{hkl}\right) \cdot \left(\vec{k}_i + \vec{G}_{hkl}\right) = \left|\vec{k}_i\right|^2 + 2\vec{k}_i \cdot \vec{G}_{hkl} + \left|\vec{G}_{hkl}\right|^2, \quad (6.17)$$

and for the case of elastic scattering $\left(\left|\vec{k}_s\right| = \left|\vec{k}_i\right| = 2\pi/\lambda\right)$ becomes,

$$\left|\vec{G}_{hkl}\right|^2 = -2\vec{k}_i \cdot \vec{G}_{hkl} = \frac{4\pi}{\lambda}\left|\vec{G}_{hkl}\right|\cos(90 - \theta) = \frac{4\pi}{\lambda}\left|\vec{G}_{hkl}\right|\sin\theta. \quad (6.18)$$

Introducing $\left|\vec{G}_{hkl}\right| = 2\pi/d_{hkl}$, we find our scattering condition is simply the familiar *Bragg diffraction* law:

$$\lambda = 2d_{hkl}\sin\theta, \quad (6.19)$$

illustrated in Fig. 6.5.

A common approach to determination of crystal structure using the Bragg law is the powder diffraction method. In this technique, a beam of monochromatic X-rays is directed at a small specimen of pulverized material, rotated about an axis. The pulverizing and rotating both serve to position small crystals at all possible orientations so that any Bragg diffraction that is allowed will be realized. The intensity of X-rays scattered as a function of the scattering angle, 2θ, is recorded to produce typical powder diffraction patterns, like those illustrated in Fig. 6.6 for both CsCl and NaCl. Given the incident wavelength and scattering angle of a Bragg peak, Eq. (6.19) can be used to determine the spacing of the planes, d_{hkl}, and so identify the corresponding Miller label (Ex. 7 and Ex. 8).

6.3.3 Missing reflections

We have seen that all $\vec{q} = \vec{G}_{hkl}$ will produce a constructive interference leading to a maximum in the static structure factor, $S_{\text{crystal}}(\vec{G}_{hkl}) = N$. However, this alone does not guarantee that the reflection will be observed. Recall that the scattered intensity is given by

$$I_S(\vec{G}_{hkl}) = \left|E_S^{\text{tot}}\right|^2 = N|E_S|^2\left|f_{\text{cell}}(\vec{G}_{hkl})\right|^2 S(\vec{G}_{hkl}), \quad (6.20)$$

and so the intensity also depends upon the specific arrangement of objects in the unit cell that contribute to the cell form factor, $f_{\text{cell}}(\vec{G}_{hkl})$. Consider, for example, CsCl in its SC structure. The basis set consists of a Cl atom at $\vec{R} = 0\vec{a}_1 + 0\vec{a}_2 + 0\vec{a}_3$ that scatters individually with an atomic form factor $f_{\text{Cl}}(\vec{G}_{hkl})$, and a Cs atom at $\vec{R} = (1/2)\vec{a}_1 + (1/2)\vec{a}_2 + (1/2)\vec{a}_3$ that scatters individually with an atomic form factor $f_{\text{Cs}}(\vec{G}_{hkl})$. Taken together, each cell of the SC lattice scatters as a "particle", with a cell form factor

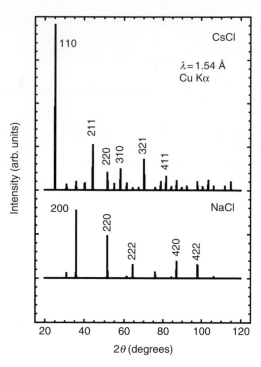

Figure 6.6 Computer-generated powder diffraction patterns for CsCl and NaCl crystals, with Miller indices listed for prominent reflections.

$$\left(f_{\text{cell}}(\vec{G}_{hkl})\right)_{\text{CsCl}} = \sum_{\text{cell}} f_i(\vec{G}_{hkl})e^{-i\vec{G}_{hkl}\cdot\vec{R}_i}$$

$$= \sum_i f_i(\vec{G}_{hkl})\exp\left\{-i(h\vec{b}_1 + k\vec{b}_2 + l\vec{b}_3)\cdot(x_i\vec{a}_1 + y_i\vec{a}_2 + z_i\vec{a}_3)\right\}$$

$$= f_{\text{Cl}}(\vec{G}_{hkl}) + f_{\text{Cs}}(\vec{G}_{hkl})\exp\{-i\pi(h + k + l)\}$$

$$(6.21)$$

If we study reflections from a plane in CsCl for which the sum of the Miller indices is odd (e.g. the (100) plane), we find that the cell form factor in Eq. (6.21) is reduced to the *difference* between the two atomic form factors, and would vanish completely if the two atomic form factors happened to be equivalent. The source of this extinguishing effect is readily understood from a consideration of the interference shown in Fig. 6.7a. Because the Cs atom is located midway between the (100) planes, its reflected waves suffer only half the optical path length difference of waves reflected by the Cl, and consequently interfere destructively. By comparison, reflections from the (110) planes (also shown in Fig. 6.7b), whose sum of indices in Eq. (6.21) is even, involve planes in which both the Cs and Cl constructively interfere to produce a maximal cell form factor that is the *sum* of the two atomic form factors.

By similar reasoning, one finds that certain reflections present in the SC structure are missing in both the BCC and FCC conventional lattices. The unit

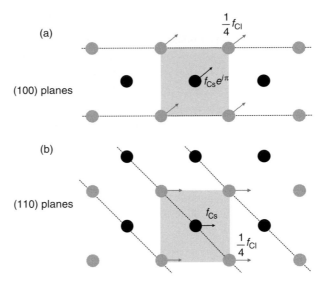

Figure 6.7 An analysis of the interference for (a) (100) and (b) (110) planes of CsCl. In the case of the (100) reflection, the Cs atom is located midway between the planes and thus suffers only a half-wavelength path difference relative to reflections by Cl atoms. This results in destructive interference and weakening of the scattered intensity. In the case of the (110) reflection, both Cs and Cl reside in the plane leading to constructive interference and a strengthening of the scattered intensity.

cell of the BCC lattice likewise consists of two *lattice* points: one at $\vec{R} = 0\vec{a}_1 + 0\vec{a}_2 + 0\vec{a}_3$ and the other at $\vec{R} = (1/2)\vec{a}_1 + (1/2)\vec{a}_2 + (1/2)\vec{a}_3$, and *regardless of the contents of the basis atoms*, it produces a cell form factor,

$$
\begin{aligned}
\left(f_{\text{cell}}(\vec{G}_{hkl})\right)_{\text{BCC}} &= f_{\text{basis}}(\vec{G}_{hkl})\{1 + \exp(-i\pi(h + k + l))\} \\
&= \begin{cases} 0 & \text{if } h + k + l \text{ is odd} \\ 2f_{\text{basis}}(\vec{G}_{hkl}) & \text{if } h + k + l \text{ is even,} \end{cases}
\end{aligned} \tag{6.22}
$$

which vanishes whenever the sum of Miller indices equals an odd number. One can show (Ex. 5) similarly, that for the FCC conventional lattice, the cell form factor vanishes for those reflections in which the Miller indices are partly even and partly odd, while constructive interference occurs when the indices are either all odd or all even integers:

$$
\left(f_{\text{cell}}(\vec{G}_{hkl})\right)_{\text{FCC}} = \begin{cases} 0 & \text{if } h, k, l \text{ are partly even/odd} \\ 4f_{\text{basis}}(\vec{G}_{hkl}) & \text{if } h, k, l \text{ are all even/all odd} \end{cases} \tag{6.23}
$$

An example of these missing reflections can be seen in Fig. 6.6 for the instance of NaCl. Note, in each case of constructive interference, the cell form factor is just the basis form factor multiplied by the number of lattice sites in the conventional unit cell.

Summary

- The translational symmetry of a crystal results in scattering only at a discrete set of scattering wave vectors, $\vec{q} = \vec{G}_{hkl}$.

- The set of allowed scattering wave vectors, $\vec{G}_i = \vec{G}_{hkl} = h\vec{b}_1 + k\vec{b}_2 + l\vec{b}_3$, forms the *reciprocal lattice* – a roadmap of allowed Bragg reflections.

- The integers (h, k and l) are also known as the *Miller indices* and define certain crystal planes with spacing $d_{hkl} = 2\pi/\left|\vec{G}_{hkl}\right|$, from which the allowed reflections are described by Bragg's law, $\lambda = 2d_{hkl}\sin\theta$.

- Scattering from a crystal depends not only on the $S(\vec{q})$ of the space lattice, but also on the contents of the conventional unit cell whose form factors may cancel. In some instances, an allowed reflection for $S(\vec{q})$ may be absent.

Exercises

6.1. Derive Eq. (6.5) starting from Eq. (6.4).

6.2. Show that the reciprocal space of the BCC lattice forms a FCC lattice.

6.3. Show that the reciprocal space of the FCC lattice forms a BCC lattice.

6.4. The primitive cell of the hexagonal lattice can be defined by the following lattice vectors:

$$\vec{a}_1 = \frac{\sqrt{3}a}{2}\hat{x} + \frac{a}{2}\hat{y}, \quad \vec{a}_2 = -\frac{\sqrt{3}a}{2}\hat{x} + \frac{a}{2}\hat{y}, \quad \vec{a}_3 = c\hat{z}.$$

  (a) Show that the volume of the primitive cell is $\frac{\sqrt{3}}{2}a^2c$.

  (b) Determine the corresponding lattice vectors describing a primitive cell of the reciprocal lattice and demonstrate that the hexagonal lattice is its own reciprocal (aside from a rotation).

6.5. Show that the cell form factor of the FCC lattice is given as in Eq. (6.23).

6.6. For any arbitrary (hkl) plane in a crystal lattice: (a) Prove that $\vec{G}_{hkl} = h\vec{b}_1 + k\vec{b}_2 + l\vec{b}_3$ is perpendicular to the plane. (b) Prove that the separation between any two adjacent planes is $d_{hkl} = 2\pi/\left|\vec{G}_{hkl}\right|$.

6.7. Students have conducted a powder diffraction experiment on iron in a phase known to be cubic. Their X-ray source was the CuKα radiation emitted from an X-ray tube at 1.542 Å and strong peaks were observed at the angles listed in the table below.

  (a) Compute the interplanar spacing corresponding to each reflection.

  (b) Determine the values of hkl for each reflection by seeking the (common) lattice parameter for this Fe crystal.

2θ	d_{hkl}	a	(hkl)
44			
64			
82			

(c) Specify whether the crystal structure is SC, FCC or BCC and justify your answer.

6.8. In Fig. 6.6 of the text, the powder diffraction patterns of CsCl and NaCl are presented with their major reflections labeled by corresponding Miller indices. Determine from the figures the lattice spacing for each crystal. Based on the reflections present, explain how the measurements confirm that CsCl is SC and NaCl is FCC.

6.9. Figure 6.8 shows two crystal planes within a unit cell near the origin. Determine the Miller indices for each plane.

6.10. Figure 6.9 shows two crystal planes within a unit cell near the origin. Determine the Miller indices for each plane.

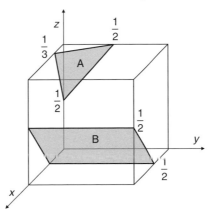

Figure 6.8

6.11. Sketch within a cubic unit cell the following planes:
(a) (011), (b) (112), (c) (122), (d) (131), (e) (013)

6.12. Consider the cubic lattices of SC, BCC and FCC with a monatomic basis of atoms. For each of these lattices, determine the number of atoms per unit conventional cell that would be found within the (a) (100), (b) (110) and (c) (111) planes. As an example, for the FCC structure the (110) planes are shown in Fig. 6.10 (top view) and are seen to contain a total of two atoms per cell.

6.13. Consider the scattering of X-rays by a diamond crystal. Although the FCC cell of which the diamond structure is composed permits reflections only when hkl are all even or all odd, the basis set of atoms in the diamond unit cell can interfere so as to still cancel a reflection. Determine and list (in order of increasing h, k and l) the first five sets of hkl for which this cancellation due to the basis contents will occur.

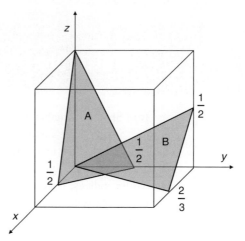

Figure 6.9

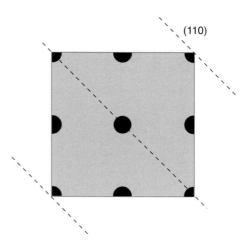

Figure 6.10

Suggested reading

Almost any optics textbook would have a development of the intensity pattern for a diffraction grating. I only list here the text by Hecht and Zajac with which I am most familiar. Cullity's textbook is an excellent source for experimental X-ray diffraction techniques, including powder diffraction and the analysis of diffraction patterns.

E. Hecht and A. Zajac, *Optics* (Addison-Wesley, Reading, Mass., 1974).
B. D. Cullity, *Elements of X-ray Diffraction*, 2nd Ed. (Addison-Wesley, Reading, Mass, 1978).
C. Kittel, *Introduction to Solid State Physics*, 8th Ed. (John Wiley and Sons, 2005).

7

Scattering by amorphous matter

Introduction

In contrast to the sharp, discrete scattering that occurs in crystals as a result of their perfect periodicity, amorphous materials possess a distribution of particle spacings and display scattering that is far more continuous as a function of the scattering wave vector. In this chapter, we explore in detail the relationship between the structure factor of a liquid or glass and the corresponding short-range order described by the pair distribution function introduced in Chapter 2. We demonstrate that $S(q)$ is (mostly) a Fourier transform of the pair distribution function. Thus again, prominent features of the static structure factor point to recurrent particle spacings present in the material, and provide vital experimental clues to the short-range order.

In this chapter, we also look at how visible light is scattered by liquids and glasses. Unlike X-rays that probe mainly the short-range order resident over just a few coordination layers, the larger wavelength of visible light makes it sensitive to larger-scaled density variations caused by thermal fluctuations. In this alternative scattering regime, the pattern of density fluctuations is described by the van Hove correlation function which, again, is related to $S(q)$ by a Fourier transform.

7.1 The amorphous structure factor

In the last chapter, we saw how discrete symmetry of the ordered state gave rise to a discrete set of scattering conditions. What happens when this discrete symmetry is absent as it is in the case of an amorphous solid? We begin again with our general statement of the structure factor in Eq. (5.26).

$$S(\vec{q}) = \frac{1}{N} \int d^3\vec{r}_1 \int d^3\vec{r}_2 C(\vec{r}_1, \vec{r}_2) e^{-i\vec{q}\cdot(\vec{r}_1 - \vec{r}_2)}, \tag{7.1}$$

involving the density–density correlation function,

$$C(\vec{r}_1, \vec{r}_2) \equiv \langle n(\vec{r}_1)n(\vec{r}_2)\rangle = \left\langle \sum_{i=1,N}\sum_{j=1,N} \delta(\vec{r}_1 - \vec{r}_i)\delta(\vec{r}_2 - \vec{r}_j) \right\rangle, \tag{7.2}$$

with its two sampling vectors $\vec{r}_1$ and $\vec{r}_2$, whose function is to search space for the locations of particle centers. Unlike the crystal situation, repeated reincarnations of the amorphous structure of a glass will not be exactly the same and so the ensemble averaging in Eq. (7.2) must be retained. Let us suppose $\vec{r}_1$ happens to find a particle. What conditions apply for which $\vec{r}_2$ will also locate a particle? Recall in Chapter 2 that the structure of amorphous matter is rotationally invariant and can be characterized by the pair distribution function $g(r)$ which describes the average local density a distance $\vec{r}$ from any arbitrary particle center. Because of this rotational invariance, the precise vectors $\vec{r}_1$ and $\vec{r}_2$ are not as relevant as is their difference, $\vec{r} = \vec{r}_1 - \vec{r}_2$. Making this change in Eq. (7.1), we can express the structure factor of an amorphous solid as

$$S_{\text{glass}}(q) = \frac{1}{N} \int d^3\vec{r}\, e^{-i\vec{q}\cdot\vec{r}} \int d^3\vec{r}_2 \left\langle \sum_i \sum_j \delta(\vec{r} + \vec{r}_2 - \vec{r}_i)\delta(\vec{r}_2 - \vec{r}_j) \right\rangle. \quad (7.3)$$

Next, we will separate out those terms in the double summation which refer to the same particle,

$$S_{\text{glass}}(q) = \frac{1}{N} \int d^3\vec{r} \left\langle \sum_{\substack{i,j \\ i \neq j}} \int d^3\vec{r}_2 \delta(\vec{r} + \vec{r}_2 - \vec{r}_i)\delta(\vec{r}_2 - \vec{r}_j) \right\rangle e^{-i\vec{q}\cdot\vec{r}}$$

$$+ \frac{1}{N} \int d^3\vec{r} \left\langle \sum_i \int d^3\vec{r}_2 \delta(\vec{r} + \vec{r}_2 - \vec{r}_i)\delta(\vec{r}_2 - \vec{r}_i) \right\rangle e^{-i\vec{q}\cdot\vec{r}}, \quad (7.4)$$

and apply the properties of the Dirac delta function to obtain,

$$S_{\text{glass}}(q) = \frac{1}{N} \left\{ \int d^3\vec{r} \left\langle \sum_j \sum_{\substack{i \\ i \neq j}} \delta(\vec{r} - (\vec{r}_i - \vec{r}_j)) \right\rangle e^{-i\vec{q}\cdot\vec{r}} + \sum_i \int d^3\vec{r}\,\delta(\vec{r})e^{-i\vec{q}\cdot\vec{r}} \right\}. \quad (7.5)$$

The second integral above is just the Fourier transform of a Dirac delta function, which equals unity. The first integral contains a double sum in which a sampling vector ($\vec{r}$) is compared against a separation between two particles. However, because the system is rotationally invariant, the double sum is unnecessary. Consider the second summation (running over i). If we pick some arbitrary particle at $\vec{r}_j = \vec{r}_o$, the sum over all other particles ($\vec{r}_i$) should on average appear the same, regardless of the particular particle chosen for $\vec{r}_o$. Hence the double sum is just N times the average quantity for a single sum in which the search vector $\vec{r}$ is compared to the distance from an arbitrarily chosen central particle at $\vec{r}_o$:

$$S_{\text{glass}}(q) = \frac{1}{N} \left\{ \int d^3\vec{r}\, N \left\langle \sum_{\substack{i \\ i \neq o}} \delta(\vec{r} - (\vec{r}_i - \vec{r}_o)) \right\rangle e^{-i\vec{q}\cdot\vec{r}} + N \right\}. \quad (7.6)$$

Consider this single summation. It looks like the expression of a density in terms of a summation of Dirac delta functions, but one in which distances are measured relative to an origin fixed on a central particle and one in which the density of the central particle itself is excluded. This ensemble-averaged quantity is in fact the average radial density defined by the pair distribution function, $g(r)$, that was introduced in Chapter 2:

$$\left\langle \sum_{\substack{i \\ i \neq o}} \delta(\vec{r} - (\vec{r}_i - \vec{r}_o)) \right\rangle = \langle n(\vec{r}) \rangle_{\text{excl}} = \langle n \rangle g(\vec{r}). \tag{7.7}$$

Thus the structure factor for an amorphous solid can be expressed as

$$S_{\text{glass}}(q) = 1 + \langle n \rangle \int d^3 \vec{r} g(r) e^{-i\vec{q} \cdot \vec{r}}, \tag{7.8}$$

where it is seen that, aside from an isotropic (i.e. q-independent) contribution, $S(q)$ in an amorphous system *is just a Fourier transform of the pair distribution function*. In practical terms, this means that prominent particle spacings present in $g(r)$, such as the nearest neighbor distance, $\Delta r = 2b$, will appear as a prominent peak in $S(q)$ near $q \approx 2\pi/\Delta r$.

7.1.1 Equivalence for liquids and glasses

Although Eq. (7.8) is developed for an amorphous *solid*, it applies equally well for the time-averaged scattering from a *liquid*. In a liquid, the particles are in motion and the instantaneous scattering varies in time. This time-dependent variation in the scattered intensity contains important information regarding the dynamics, which will be discussed in a later chapter concerning the so-called *dynamic* structure factor. However, when the scattered intensity is averaged over time, the reincarnations of the structure driven by the particle motions appear to be identical to ensemble averages, and the resulting angular dependence of the scattering from a liquid is given by a *static* structure factor:

$$S_{\text{liquid}}(q) = S_{\text{glass}}(q) = S_{\text{amorph}}(q) = 1 + \langle n \rangle \int d^3 \vec{r} g(r) e^{-i\vec{q} \cdot \vec{r}}, \tag{7.9}$$

identical to that for the glass.

7.1.2 Investigating short-range order

Suppose we are interested in understanding the short-range order (i.e. features inherent in the first few peaks of $g(r)$ that correspond to the first few coordination shells) of a glassy solid such as amorphous germanium. These features correspond to particle separations comparable in size ($r = b$) to the particles

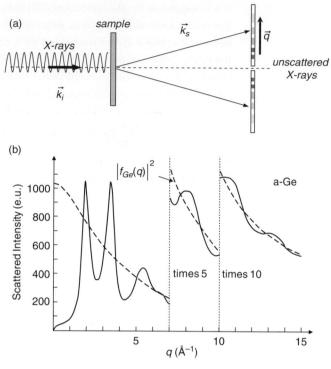

Figure 7.1 (a) X-ray scattering from thin film of amorphous germanium produces a series of ring fringes. (b) The fringe intensity, as a function of q, is seen to be comprised of undulations caused by relative atomic positions superimposed on the atomic form factor of Ge (dashed curve). Note the changes in vertical scale at large q. (Adapted from Temkin, Paul and Connell (1973).)

themselves and so an appropriate scattering wave vector would be $q^{-1} \approx b$, such as obtained with X-rays or neutrons. We could then direct a beam of, say, X-rays at the glass and study the scattering intensity as a function of the scattering angle. An example of the resulting spectrum for germanium is shown in Fig. 7.1, which displays a series of undulations superimposed on the atomic form factor of Ge.

To obtain $g(r)$, Eq. (7.8) instructs that we compute the inverse Fourier transform of $[S(q) - 1]/\langle n \rangle$. But before we do this, let us first recognize that Eq. (7.8) can be expressed alternatively as,

$$S_{\text{amorph}}(q) = 1 + \langle n \rangle \int d^3\vec{r}[g(r) - 1]e^{-i\vec{q}\cdot\vec{r}} + \langle n \rangle \int d^3\vec{r}e^{-i\vec{q}\cdot\vec{r}}, \qquad (7.10)$$

where we have both subtracted and added unity to the pair distribution function. The second integral involves the Fourier transform of unity that equals a Dirac delta function,

$$S_{\text{amorph}}(q) - 1 = \langle n \rangle \int d^3\vec{r}[g(r) - 1]e^{-i\vec{q}\cdot\vec{r}} + \langle n \rangle (2\pi)^3 \delta(\vec{q}), \qquad (7.11)$$

and corresponds to radiation "scattered" into the forward direction – a direction that is not accessible for our X-ray experiment because of the large amount of unscattered beam that also exits at that location. Aside from this forward scattering which is often blocked by a finite-sized beam stop, we can now express $g(r)$ in terms of the inverse Fourier transform,

$$g(r) - 1 = (2\pi)^{-3} \langle n \rangle^{-1} \int d^3\vec{q} \left[S_{\text{amorph}}(q) - 1 \right] e^{+i\vec{q}\cdot\vec{r}}. \qquad (7.12)$$

Since this is an integral over q-space, we are free to set $\vec{r}$ along any direction. For convenience, let us set it along the z-axis of a spherical coordinate system. Then one can show (Ex. 2) that

$$g(r) - 1 = (2\pi)^{-3} \langle n \rangle^{-1} \int\limits_0^{2\pi} d\phi \int\limits_0^\infty \int\limits_0^\pi e^{+iqr\cos\theta} \sin\theta \, d\theta \left[S_{\text{amorph}}(q) - 1 \right] q^2 \, dq$$

$$= (2\pi)^{-2} \langle n \rangle^{-1} \int\limits_0^\infty \left[S_{\text{amorph}}(q) - 1 \right] q^2 \left[\frac{2 \sin qr}{qr} \right] dq. \qquad (7.13)$$

The radial distribution function is then obtained as,

$$4\pi r^2 \langle n \rangle g(r) = 4\pi r^2 \langle n \rangle + \frac{2r}{\pi} \int\limits_0^\infty q \left[S_{\text{amorph}}(q) - 1 \right] \sin qr \, dq, \qquad (7.14)$$

of which an example for amorphous germanium is shown in Fig. 7.2. Note in this figure, that the most prominent spacing is the nearest neighbor separation given by the first peak near $2b \approx 2.5$Å. This prominent spacing

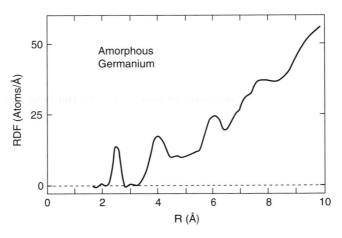

Figure 7.2   Radial distribution function of thin film of amorphous germanium obtained by Fourier transformation of Figure 7.1. (Adapted from Temkin, Paul and Connell (1973).)

is seen to correspond closely with the first peak in $S(q)$ in Fig. 7.1 occurring near $q \approx 2\pi/\Delta r \approx 2$ Å^{-1}.

7.1.3 Rayleigh scattering

There is in Eq. (7.8) a curious q-independent, or isotropic, contribution to the static structure factor in amorphous materials, known as the *Rayleigh scattering*. This contribution is seen to arise from our separation (in Eq. (7.5)) of the terms in $S(q)$ that involve interference (purely constructive interference) of a particle with itself. In Fig. 7.1, this Rayleigh contribution is indicated by the dashed line that falls off with increasing q. Why is it not q-independent here? Note that Fig. 7.1 represents the *intensity* of scattering, which if we recall from Chapter 5 is given as

$$I_S = \left|E_S^{\text{tot}}\right|^2 = |E_S|^2 |f(q)|^2 N S(q), \tag{7.15}$$

and includes any inherent q-dependence of the scattering cross section ($|E_S|^2$) or of the particle form factor, $f(q)$. For the q range of an X-ray experiment, $f(q)$ exhibits considerable q-dependence, and the dashed line in Fig. 7.1 is just seen to be proportional to $|f(q)|^2$ for a Ge atom.

But suppose we examined an alternative q range, like that of visible light, for which $q \approx 0$ in Fig. 7.1 and $f(q) \approx$ constant. Suppose we had an ideal gas of germanium (or any other small particles) and directed a visible laser beam into it. The particles of an ideal gas are in constant motion, zipping about making elastic collisions with other particles. If we take one particle as a central particle and compute the pair distribution function over many ensembles, we will find that the probability of finding a second particle at any distance from our central particle is just that associated with the average density of the gas. Thus, $g(r) \approx 1$ for all r outside the radius of our central particle, and from Eq. (7.8), the scattered intensity,

$$I_S = \left|E_S^{\text{tot}}\right|^2 = |E_S|^2 |f(q \approx 0)|^2 N \left\{1 + \langle n \rangle (2\pi)^3 \delta(\vec{q})\right\}, \tag{7.16}$$

reduces (aside from scattering in the forward direction) to only the Rayleigh contribution: an isotropic scattering with magnitude proportional to the number of scattering particles and the scattering cross section, $|E_S|^2$. Furthermore, we saw in Chapter 5 that the cross section for visible elastic scattering (Eq. (5.7)) differs from that for X-rays in that it varies as

$$|E_S|^2 \propto \lambda^{-4}, \tag{7.17}$$

and so preferentially scatters light with shorter wavelengths.

Blue skies and red sunsets

All of this leads up to an explanation for why the sky on Earth is blue while the lunar sky (that seen if you were on the moon) is black. The origin for this

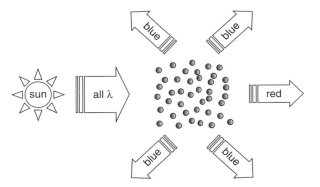

Figure 7.3 Rayleigh scattering explains the blueness of the sky and the redness of sunsets as the result of preferential scattering of shorter wavelength light by gas molecules in the atmosphere.

difference is the atmosphere that surrounds the Earth, but which is absent on the Moon. We can now understand the blueness of our sky as a consequence of Rayleigh scattering of sunlight by the atoms in our atmosphere, as illustrated in Fig. 7.3. Incident light from the sun contains a broad spectrum of wavelengths. As we see in Eq. (7.17), the gas molecules in the atmosphere scatter blue light more strongly than red. Viewed in a quantum mechanical framework, this means that the incident sunlight (which contains roughly equivalent numbers of all color photons) has its blue photons *redirected* by the gas molecules more often than it does red ones. As a consequence, when we view the sky above us (away from the sun), we see more blue photons arriving than red photons and hence observe a blueness in the light. Near sunset, a view of the light coming directly from the sun has had most of its blue photons removed by scattering, leaving behind an excess of red.

7.2 Light scattering by density fluctuations

In discussing short-range order in amorphous materials, we focused our attention mainly on those undulations in $g(r)$ that occur within the first few coordination shells. Any undulations at larger r could be safely ignored for X-ray scattering experiments that probe only the $q^{-1} \approx 0.1\text{--}5$ Å range, but cannot be ignored when visible light ($q^{-1} \approx 10^3$ Å) is employed. If we were to look more closely at $g(r)$ in this larger range of r, as illustrated in Fig. 7.4, we would discover that undulations do still exist, although they are much reduced in magnitude in comparison with those describing the short-range order.

Where do these small undulations come from? These density variations arise from fluctuations inherent in any disordered thermodynamic system. Consider a fluid of particles in thermal equilibrium, such as illustrated in Fig. 7.5. Suppose we partition this system into smaller regions and compare the density

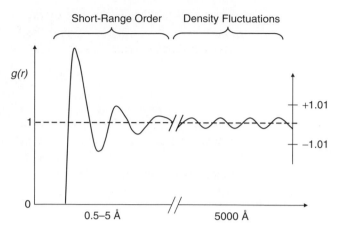

Figure 7.4   Beyond short-range order, corresponding to undulations in $g(r)$ in the Angstrom range, there exist much weaker features in $g(r)$ that arise from thermodynamic density fluctuations present to some degree in all disordered materials. These large length scale undulations contribute primarily to scattering by comparably long wavelength radiation (i.e. visible light).

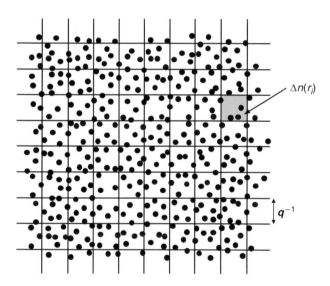

Figure 7.5   An amorphous system of scattering particles partitioned into boxes of arbitrary size. The number of particles found in each box varies from one box to the next, resulting in density fluctuations on length scales comparable to the size of the box. The magnitude of the density fluctuation is seen to diminish with increasing box size.

$n(r)$ in each box whose center is at $\vec{r}$. Because the particles are moving about, we will necessarily find instances in which some boxes contain greater and lesser density of particles than the overall density. We could characterize the mean squared variations in density by $\langle \Delta n^2 \rangle = \langle (n(r) - \langle n \rangle)^2 \rangle$, and would find that, regardless of the size of our partitions, some variation would always be observed, although the size of $\langle \Delta n^2 \rangle$ would necessarily decrease as the partitions grew larger. The magnitude of these inherent density fluctuations is given in thermodynamics as

$$\frac{\langle \Delta n^2 \rangle}{\langle n \rangle^2} = \frac{k_B T}{V} \chi_T, \tag{7.18}$$

where $\chi_T = -V^{-1}(\partial V / \partial P)$ is the isothermal compressibility. For an ideal gas whose compressibility is large, the density fluctuations are large in comparison with the average density. For liquids, whose compressibility is roughly 10^{-4} times smaller, the density fluctuations are less well resolved in comparison with the average density. Nevertheless, these fluctuations can result in detectable amounts of scattered light.

7.2.1 The van Hove space correlation function

Given that the undulations in *g(r)* at large *r* are spawned by thermodynamic fluctuations, it is customary to emphasize explicitly the role of these density fluctuations in the static structure factor. Returning to our starting point in Eq. (7.2), we begin by separating out the average density in the density–density correlation function,

$$C(\vec{r}_1, \vec{r}_2) \equiv \langle n(\vec{r}_1) n(\vec{r}_2) \rangle = \langle [\langle n \rangle + \Delta n(\vec{r}_1)][\langle n \rangle + \Delta n(\vec{r}_2)] \rangle$$
$$= \langle n \rangle^2 + \langle \Delta n(\vec{r}_1) \Delta n(\vec{r}_2) \rangle, \tag{7.19}$$

so that Eq. (7.1) now reads as

$$S(\vec{q}) = \frac{1}{N} \int d^3 \vec{r}_1 \int d^3 \vec{r}_2 \left\{ \langle n \rangle^2 + \langle \Delta n(\vec{r}_1) \Delta n(\vec{r}_2) \rangle \right\} e^{-i\vec{q} \cdot (\vec{r}_1 - \vec{r}_2)}. \tag{7.20}$$

Again, the rotational invariance of the amorphous structure implies that only the difference $\vec{r} = \vec{r}_1 - \vec{r}_2$ matters, so,

$$S_{\text{amorph}}(q) = \frac{1}{N} \left\{ \langle n \rangle^2 \int d^3 \vec{r} e^{-i\vec{q}\cdot\vec{r}} \int d^3 \vec{r}_2 + \int d^3 \vec{r} e^{-i\vec{q}\cdot\vec{r}} \int d^3 \vec{r}_2 \langle \Delta n(\vec{r} + \vec{r}_2) \Delta n(\vec{r}_2) \rangle \right\}. \tag{7.21}$$

The integrations over $d^3 \vec{r}_2$ correspond to sampling with the second search vector at different possible origins. Because of the invariance, all possible origins will appear the same (with ensemble averaging) and cannot depend upon the specific choice of $\vec{r}_2$. Thus the result of integrating $d^3 \vec{r}_2$ will be the volume of the scattering region, V, and so

$$S_{\text{amorph}}(q) = \langle n \rangle (2\pi)^3 \delta(\vec{q}) + \int d^3 \vec{r} [\langle \Delta n(\vec{r}) \Delta n(0) \rangle / \langle n \rangle] e^{-i\vec{q}\cdot\vec{r}}$$
$$= \langle n \rangle (2\pi)^3 \delta(\vec{q}) + \int d^3 \vec{r} G(r) e^{-i\vec{q}\cdot\vec{r}}, \tag{7.22}$$

where the *van Hove space correlation function* is defined here as,

$$G(r) \equiv \langle \Delta n(0) \Delta n(\vec{r}) \rangle / \langle n \rangle. \tag{7.23}$$

Physically, $G(r)$ represents the probability that if a density fluctuation (corresponding to one of the boxes in Fig. 7.5) is present at the origin, another will be found at a distance r away. Aside from the innocuous forward scattering, the scattering due to these random fluctuations appears as just a Fourier transform of $G(r)$.

A tale of two formalisms

In our development of scattering from amorphous materials, above, we have now developed two complementary perspectives on the relation between structure and scattering as conveyed by the static structure factor in Eq. (7.8) and Eq. (7.22). Both are equivalent, but each focuses attention on differing relevant scattering length scales. In Eq. (7.8), where $S(q)$ is related to the Fourier transform of $g(r)$, the emphasis is placed on small-scale structure ($q^{-1} \approx b$), where *individual particle positions* are highly influenced by local, pairwise *interactions* (i.e. bonds). This short-range order is best described by the strong undulations seen in $g(r)$, and is quite unrelated to that emphasized by $G(r)$ in Eq. (7.23) where *collections of particles* spanning sizes of order $q^{-1} \gg b$ are considered as a *continuum* fluid. By comparison, the structure at these larger length scales is not strongly influenced by individual interactions, but rather by thermodynamic potentials and *hydrodynamic motions* that cause small fluctuations in density about a global average value.

Fiber optic attenuation

Before leaving the subject of light scattering, we illustrate two significant consequences that follow from $G(r)$ in Eq. (7.23). The first concerns the application of fiber optics for telecommunication. During the late 1980s there was a major shift towards replacing conventional telephone wires with optical fibers along which data, encoded in the light, could be transferred at much higher speeds and sent together with other data sharing the same fiber. The most common light used is $\lambda \approx 1.5$ microns, for which silica fibers have a low absorption allowing the signal to travel the farthest before needing to be strengthened again. However, Eq. (7.22) reveals a remaining limitation that arises from the presence of inherent density variations in the glass in or around the 1.5 micron scale which scatter away light. Research is currently aimed at refining glass production and fiber drawing procedures in an effort to reduce these fluctuations and increase the transmission range of commercial fibers.

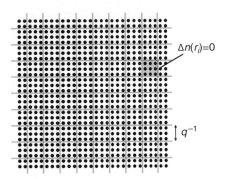

$\Delta n(r_i)=0$

$\updownarrow q^{-1}$

Figure 7.6   The Ewald–Oseen extinction theorem. An ordered system of particles is again partitioned into boxes of arbitrary size (but larger than the particle spacing). Because the particles are ordered, each box contains identical amounts of particles and thus produces none of the density fluctuations seen for a disordered system in Figure 7.5. Without these density fluctuations there is no scattering in other than the forward direction.

Ewald–Oseen extinction

As a second illustration of the consequences of Eq. (7.22), consider the scattering of visible light by an ideal crystal. If we again partition the crystal into many small boxes, as shown in Fig. 7.6, what will we find regarding the variations in the density between boxes? Well, as long as the boxes are larger than the unit cell of the crystal, there can be no variation in the density: *each box must contain exactly the same density as the entire crystal.* According to Eq. (7.23), this would mean that $G(r) = 0$ for all $r \gg b$, the size of a unit cell, and that visible light would have no scattering aside from that in the forward direction. This result illustrates the classic *Ewald–Oseen extinction theorem* of optics: for $\lambda \gg$ separation, the small differences in phase arising from *regularly spaced* scatters are seen to *perfectly cancel* in all directions, except the forward direction, where the net result is a refracted ray traveling at a reduced speed c/n in the medium of refractive index n. Thus, only when the scatters are *irregularly* spaced do they produce scattering in other than the forward direction.

7.2.2 Intermediate-range order: SAXS and SANS

As we have seen, there is a vast array of undulations in $g(r)$, which we have separated into those at short length scales that arise from the pairwise interparticle interactions, and those at long length scales that stem from random fluctuations. In some materials, there often appear *intermediate range* undulations in $g(r)$ ($r \approx 10$ to 100s Å) that are not derived purely from random fluctuations, but rather appear as a consequence of the nature of extended bonding configurations. By way of an illustration, consider again the structure of amorphous SiO_2 in which the short-range order (SRO) is characterized by a

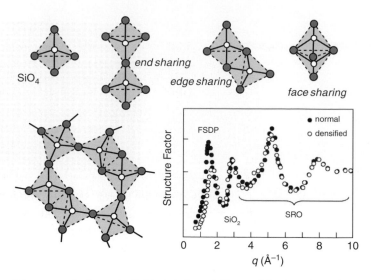

Figure 7.7   Intermediate-range order in covalently bonded glasses is illustrated by the variety of ways in
which SiO₄ tetrahedra can link together. Variations in the linkages often lead to voids in the glass
structure and the appearance of a first sharp diffraction peak (FSDP). This peak is seen in small
angle neutron scattering for both a normal (undensified) specimen and one that has been
densified to 20% less free volume. Note that the SRO of the densified sample is largely unaffected,
while the FSDP shifts to higher q consistent with the compaction of voids. (Adapted from
Susman *et al.* (1991).)

tetrahedral unit, SiO₄, shown in Fig. 7.7. In moving outwards along *g(r)*, we
are in a sense coarse-graining the structure into one that is no longer focused
on how one atom is coordinated with another, but rather how one tetrahedral
unit is coordinated with another. As illustrated in Fig. 7.7, there is a variety of
ways in which two tetrahedra could link together to influence the structural
features in *g(r)* beyond the SRO regime, and herein lie the combined effects of
pairwise bonding constraints, coupled with statistical variability, that make up
intermediate-range order (IRO).

To study this IRO requires a shortening of the scattering wave vector which
can be achieved by probing smaller scattering angles (see Eq. (5.12)). In such
small angle X-ray scattering (SAXS), scattered X-rays are detected at angles as
low as 0.1°, for which $q \approx 0.015$ Å^{-1} and $l \approx 2\pi/q$ probes particle separations
in the order of several hundred Ångstroms. Similarly, *small angle neutron
scattering* (SANS) can be employed to achieve scattering wave vectors in the
range between 0.006 Å^{-1} and 1 Å^{-1}, to probe structural features ranging from
about 6 Å to 1 micron. In many covalently bonded glasses, including SiO₂,
X-ray scattering studies have observed a distinct diffraction peak at a q below
that of the SRO. This "first sharp diffraction peak" is evidence of IRO in the
structure and is believed to be the result of correlations that develop between the
regions of empty space not occupied by the SiO₄ tetrahedra.

Summary

- For amorphous matter possessing rotational invariance, the static structure factor is related to the pair distribution function by a *Fourier transform*, $S_{\text{glass}}(q) = 1 + \langle n \rangle \int d^3 \vec{r} g(r) e^{-i\vec{q}\cdot\vec{r}}$. In X-ray scattering experiments, $g(r)$ can be obtained by an inverse Fourier transform of the observed, q-dependent scattered intensity.

- The amorphous structure factor contains an isotropic (q-independent) contribution known as *Rayleigh scattering* that dominates in the limit of small q.

- The amorphous structure factor can alternatively be expressed in terms of density fluctuations as $S_{\text{amorph}}(q) \approx \int d^3 \vec{r} G(r) e^{-i\vec{q}\cdot\vec{r}}$, where $G(r) \equiv \langle \Delta n(0) \Delta n(\vec{r}) \rangle / \langle n \rangle$ is the *van Hove space correlation function*.

- As a consequence of the Ewald–Oseen extinction theorem, light is not scattered by a perfect crystal because it contains no density fluctuations comparable to the wavelength of light.

- SAXS and SANS are techniques designed to probe intermediate-range order existing between about 5 Å and 1 micron.

Exercises

7.1. Estimate the range of q for both X-rays ($\lambda = 0.7$ Å) and visible light ($\lambda = 5000$ Å) for a *small angle* scattering experiment for which $0.1° < 2\theta < 20°$.

7.2. Derive Eq. (7.13) starting from Eq. (7.12).

7.3. Consider the thermal density fluctuations present in an ideal gas. Show that the isothermal compressibility equals the inverse of the pressure and that the average mean squared density fluctuation, $\langle \Delta n^2 \rangle$, is proportional to the number of gas particles.

7.4. Clouds are composed of a suspension of spherical water droplets with a range of sizes but with an average diameter roughly 50 times the wavelength of visible light. Since clouds appear white, the Rayleigh scattering must be such that all wavelengths are scattered equally. Explain how this occurs. (Hint: you may want to refer back to the result of Ex. 5.8).

7.5. The intensity of an optical signal sent along an optical fiber degrades over distance due to Rayleigh scattering from small density fluctuations as $I(x) = I_o \exp(-\alpha x)$, where the attenuation factor is obtained by integrating the scattered intensity in all directions, $\alpha = \frac{1}{V} \int \left(\frac{I_S}{I_o} \right) R^2 \mathrm{d}\Omega$, where V is the scattering volume element and $\mathrm{d}\Omega$ is the solid angle. Assume that the density fluctuations are randomly distributed with a mean separation 1/10th the wavelength of the light and that each has a common magnitude given by Eq. (7.18). Show that the Rayleigh contribution to the attenuation factor is given as $\alpha = \frac{8\pi^3}{3} \frac{1}{\lambda^4} |f|^2 C k_B T \chi_T$, where $C = \left(\frac{\langle n \rangle e^2}{\varepsilon_o m_e \omega_o^2} \right)^2$, and f is an unspecified form factor associated with the scattering profile of an individual density fluctuation.

Suggested reading

Greenler's book is full of nice pictures and explains a variety of naturally occurring optical phenomena (including the blueness of the sky) at a level accessible to even a non-scientist. Those interested in a proof of the Ewald–Oseen theorem should consult the textbook by Born and Wolf.

P. M. Chaikin and T. C. Lubensky, *Principles of Condensed Matter Physics* (Cambridge University Press, New York, 2003).

J. Zarzycki, *Glasses and the Vitreous State* (Cambridge University Press, New York, 1991).

R. Greenler, *Rainbows, Halos, and Glories* (Cambridge University Press, New York, 1980).

M. Born and E. Wolf, *Principles of Optics*, 6th Ed. (Cambridge University Press, Cambridge, 1997).

Berne, B. J. and Pecora, R. *Dynamic Light Scattering*, (John Wiley and Sons, New York, 1976).

8

Self-similar structures and liquid crystals

Introduction

In this final chapter on the subject of scattering, we examine the structure of extended, but finite-sized composite objects constructed of a very large number of individual particles. Examples include polymer molecules composed of many repeated individual chemical units, and aggregation clusters that form when many individual particles randomly assemble into a larger structure. In both instances, we will see that the amorphous structures of these macroscopic-sized objects display *self-similarity* – a continuous hierarchy of structures that appear identical on many alternative length scales. This self-similarity appears in the pair distribution function as a power law dependence on radial distance, much unlike the sort of *g(r)* curves we have examined thus far, and which transforms into Fourier space as a corresponding power law variation of *S(q)*.

Also in this chapter, we conclude our survey of structures and scattering with a brief look at liquid crystals and microemulsions, whose structures undergo a series of transitions with symmetries that are intermediate between that of crystals and liquids. In these materials the particles are able to spontaneously self-assemble into more ordered structures as a result of only weak, inter-particle forces.

8.1 Polymers

As the name implies, polymers are generally very large molecules constructed of a large number of repeated chemical units. Each unit is referred to as a "mono-mer" and thus a collection of these is a "poly-mer". By and large, polymers form long chains and such is the case in polystyrene, shown in Fig. 8.1. The chain structure typically consists of a grouping of atoms in the monomer that interlink with other monomers forming a "backbone", together with other groupings present in the monomer, known as side groups, that do not participate in the linkage as such. While chains are most common, cross-linking units can be introduced to create branching points on the chains which allow two or more chains to interconnect, much like the additions of As and Ge to selenium glasses discussed near the conclusion of Chapter 2.

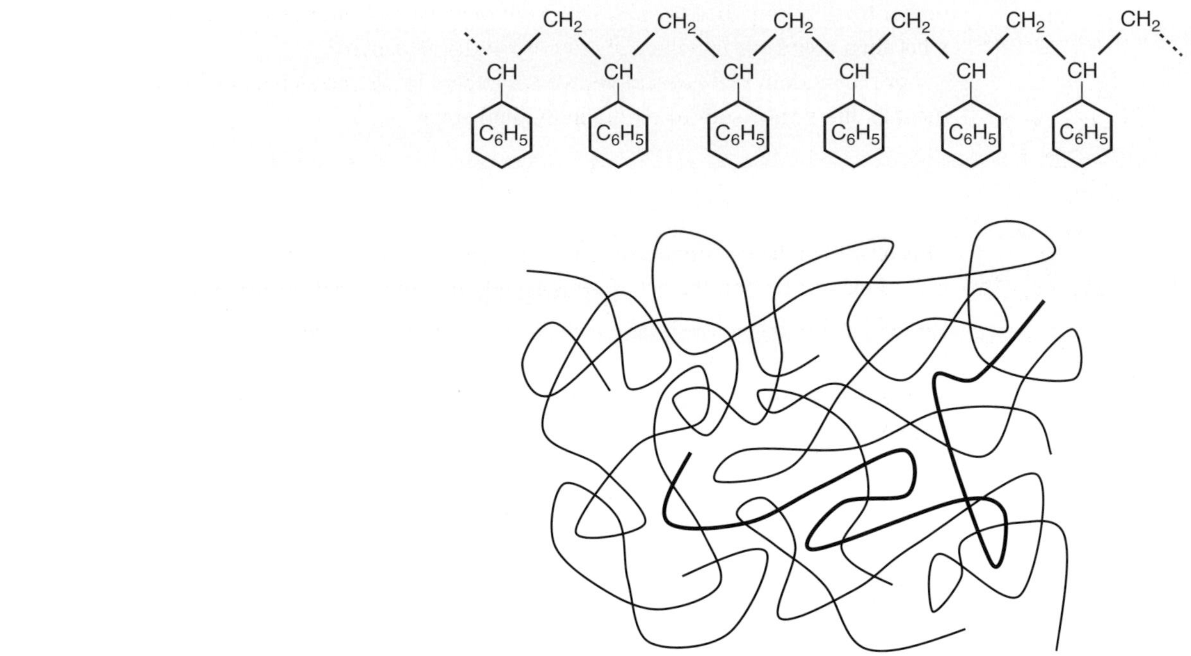

Figure 8.1 Molecular structure of poly(styrene) illustrating the hydrocarbon backbone and pendant side groups. Lower figure illustrates the coiled structure of a typical polymer chain in the melt, and entanglements occurring with other polymer chains.

Although the polymer structure requires linkages between monomer units, the specific orientation or angle made by each such linkage is largely unrestricted. This allows the chain to twist and bend about at each individual linkage so as to assume a variety of possible configurations. A condensed phase of many such polymer chains could aptly be compared to a bowl of cooked spaghetti noodles in which each polymer noodle twists and bends in a random manner to produce a highly entangled mixture, as illustrated in Fig. 8.1. How might we model the structure of a single such polymer chain that exhibits so much indifference about the shape it chooses to assume? The randomness of its shape would seem to be an impediment. But, as we shall see, randomness is key to understanding the structure because the random twists and bends of the chain can be adequately modeled using what is known as a *random walk*.

8.1.1 The random walk

Consider a random walk scenario in which a person (a walker) starts off from the origin and begins making a set of N sequential steps each of length b, as illustrated in Fig. 8.2. If the walker walks in the same direction each time, he/she would experience a net displacement $R = Nb$ after the N steps are completed. However, if the direction of each step were *randomly* chosen

(a) $R \propto Nb$

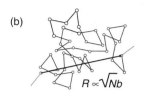

(b)

$R \propto \sqrt{N}b$

Figure 8.2

(a) A non-random walk produces a net displacement that is proportional to the number of steps taken by the walker.

(b) In the random walk, the net displacement of the walker varies as the square root of the number of steps.

(either by throwing dice or using the result of some random number generator), what then could we conclude about the net displacement?

For the random walk described, we could certainly express the net displacement as the vector sum of all the individual steps,

$$\vec{R} = \sum_{i=1,N} \vec{r}_i. \tag{8.1}$$

This result would, of course, be different each time we set the walker to his/her task, depending upon the actual set of random choices that occurred. Instead, we seek some reproducible and robust measure of the average net displacement of the walker such as might be obtained by performing an ensemble average of many repeated trials. This ensemble-averaged, root-mean-squared displacement is

$$\langle R \rangle = \sqrt{\langle \vec{R} \cdot \vec{R} \rangle} = \left\{ \left\langle \sum_{i=1,N} \sum_{j=1,N} \vec{r}_i \cdot \vec{r}_j \right\rangle \right\}^{1/2}, \tag{8.2}$$

and if we separate the double summation into self and pair terms, it becomes

$$
\begin{aligned}
\langle R \rangle &= \left\{ \sum_{i=1,N} \langle \vec{r}_i \cdot \vec{r}_i \rangle + \sum_{i=1,N} \sum_{\substack{j=1,N \\ j \neq i}} \langle \vec{r}_i \cdot \vec{r}_j \rangle \right\}^{1/2} \\
&= \left\{ Nb^2 + b^2 \sum_{i=1,N} \sum_{\substack{j=1,N \\ j \neq i}} \langle \cos \phi_{ij} \rangle \right\}^{1/2} = \sqrt{N}b.
\end{aligned}
\tag{8.3}
$$

The second term above, corresponding to the dot product of any two random pairs of displacements, vanishes because the $\cos \phi$, averaged over all possible values of ϕ, is zero. As a result, we find that the average net displacement made during the random walk is generally less than that of a walk taken in a straight line and increases only as the square root of the number of steps taken.

Self-similarity

An important feature of the random walk is its inherent self-similar nature. Although we started the walker off from the origin and observed his/her displacement to increase as the square root of the number of steps taken, we could have considered the walk to have "started" later at any arbitrary location of the walker and we would still find that the displacement from that point on also increases as the square root of the number of additional steps taken. In addition, we could "rescale" the walk without changing the basic square-root-of-the-number-of-steps feature. To see this, imagine we divided up the walk into M groups of N_o steps each. Each set of N_o steps will produce an average displacement $L_o = \sqrt{N_o}b$, which we could call a "leap". The net displacement following M such leaps is then,

$$\langle R \rangle = \sqrt{MN_o}b = \sqrt{M}\sqrt{N_o}b = \sqrt{M}L_o, \tag{8.4}$$

and appears the same when rescaled to the larger, leap-sized steps. This property, that a process or structure appears identically on differing scales, is known as *self-similarity* and is a hallmark of many objects and processes studied in condensed matter physics.

Pair distribution function

We can apply the random walk scenario discussed above to model the structure of a polymer coil by imagining that the polymer is constructed by the walker, who attaches a monomer to the one previous as he/she walks along. After the walk is completed, the resulting polymer will have a configuration that matches all the twists and bends that occurred in the path taken by the walker. To characterize this structure, we now seek an expression for the pair distribution function, $g(r)$. Recall from our discussion in Chapter 2, that this function is defined in relation to a central monomer and is determined by the number of monomers we find in a volume element $(4\pi r^2 dr)$ a distance r away. Because of the self-similarity, the structure around any arbitrary central monomer will mimic that of a random walk starting from that point, and thus the number of monomers contained inside a sphere of radius r about the central monomer is given by Eq. (8.3) as

$$N = \left(\frac{r}{b}\right)^{\alpha}, \tag{8.5}$$

where $\alpha = 2$. As r is increased by dr, the number of additional monomers included inside increases as

$$dN = \alpha \left(\frac{r}{b}\right)^{\alpha} \frac{1}{r} dr, \tag{8.6}$$

and the pair distribution function (defined in Eq. (2.3)) then becomes

$$g(r) = \langle n \rangle^{-1} \frac{dN}{4\pi r^2 dr} = \frac{\langle n \rangle^{-1}}{4\pi r^2} \frac{dN}{dr} = \frac{\langle n \rangle^{-1}}{4\pi r^2} \left(\alpha \left(\frac{r}{b}\right)^{\alpha} \frac{1}{r}\right). \tag{8.7}$$

We need to make one adjustment to this expression. The self-similarity upon which it is based is only valid for distances that remain within the polymer coil itself. Each polymer chain has a limited number of monomer units and assumes a finite size, $\xi = \sqrt{N}b$. For polystyrene with a mass number of $N \approx 10^4$ and monomer size of about $b \approx 5\text{Å}$, a typical polymer coil would extend no more than about 500Å in radius. Beyond this range, the number of additional monomers ceases to increase with increasing radius and to accommodate this limitation, we add to $g(r)$ an ad hoc, exponential cutoff,

$$g(r) = \frac{\langle n \rangle^{-1}}{4\pi r^2} \left(\alpha \left(\frac{r}{b} \right)^\alpha \frac{1}{r} \right) e^{-r/\xi}. \tag{8.8}$$

Because the structure of a polymer coil is isotropic, the waves scattered by the monomer particles in a polymer coil are described by the structure factor appropriate to amorphous matter, which was introduced in Chapter 7,

$$\begin{aligned} S_{\text{SelfSim}}(q) &= 1 + \langle n \rangle \int d^3\vec{r} \, g(r) e^{-i\vec{q}\cdot\vec{r}} \\ &= 1 + \langle n \rangle \int g(r) e^{-iqr\cos\theta} r^2 dr \sin\theta d\theta d\phi, \end{aligned} \tag{8.9}$$

which, for $g(r)$ in Eq. (8.8), becomes

$$S_{\text{SelfSim}}(q) = 1 + \alpha \int\limits_{r=0}^{r=\infty} \left(\frac{r}{b} \right)^\alpha \frac{1}{r} e^{-r/\xi} \frac{\sin qr}{qr} dr.$$

We next make the change of variables using $x = qr$, to obtain

$$S_{\text{SelfSim}}(q) = 1 + \alpha \left(\frac{1}{qb} \right)^\alpha \int\limits_0^\infty x^{\alpha-2} \sin x \, e^{-x/q\xi} dx. \tag{8.10}$$

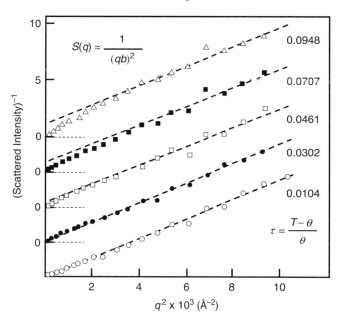

Figure 8.3 Small angle neutron scattering from protonated polystyrene in deuterated cyclohexane as a function of reduced temperature approaching the theta temperature θ. The scattering is seen to exhibit a q^2-dependence consistent with polymer coils generated by a random walk (see Eq. (8.11) with $\alpha = 2$. (Adapted from Farnoux et al. (1978).)

(a)   Random Walk

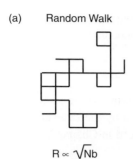

$R \propto \sqrt{N}b$

(b)   Self-avoiding Walk

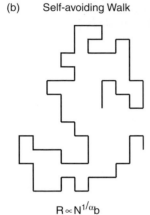

$R \propto N^{1/\alpha}b$

Figure 8.4

A comparison of the random walk structure with the more swollen structure obtained in the self-avoiding walk.

The definite integral differs significantly from unity only for $q^{-1} > \xi$ (Ex. 8) and, since our interest is only in the region for which $b < q^{-1} < \xi$, the structure factor may be approximated simply as a power law of the form,

$$S_{\text{SelfSim}}(q) \approx \left(\frac{1}{qb}\right)^{\alpha}. \tag{8.11}$$

However, if we scattered waves from a large mixture of polymer coils, such as depicted in Fig. 8.1, we would not observe the scattering suggested by Eq. (8.11). This expression describes the interference of scattered waves arising from the relative positions of monomers in the *same* chain. In a melt, this chain would be entangled with other similar chains, as illustrated in Fig. 8.1, and the resulting structure factor would contain interferences from these other monomer units as well. To view the scattering from a single coil (or a few coils at a low concentration) requires some of the polymers to be chemically altered (i.e. tagged), such that their monomer units scatter more strongly (via a larger monomer form factor, $f_{\text{m}}(q)$) than other surrounding polymer coils. An example of the scattering in such a situation is shown in Fig. 8.3, where it can be seen that the structure of the polymer coil in a melt-like, "theta" solvent is well described by the random walk model of Eq. (8.11) with $\alpha = 2$.

8.1.2 Swollen polymers: self-avoiding walks

It is a little surprising that the random walk model in which a walker is allowed to retrace portions of his/her previous path would be so successful for describing a polymer whose monomers are certainly unable to occupy the same location. However, in the melt the coil is severely restricted by entanglements with neighboring coils, and these entanglements can force the monomers into close proximity in such a fashion that the random walk manages to capture the compact structure. When polymers are dissolved in a solvent at dilute concentrations, the above polymer entanglements are absent and the coils expand (or swell) slightly. To model the structure of the swollen polymers suspended in a good solvent will require a slight modification of the random walk scenario, known as a *self-avoiding walk* (SAW).

In the self-avoiding walk we again allow a walker to make steps of length b in randomly chosen directions and we have the walker assemble the units of the chain as he/she proceeds. However, if the randomly chosen step should at any point require the newly added monomer to reside at a location coincident with a previously assembled unit, the choice is rejected and another random choice is made. In this way, the polymer is forced to avoid itself much as it does in the presence of a good solvent. A comparison of the structures generated by both a random walk and a SAW, shown in Fig. 8.4, illustrate the more swollen character of the resulting polymer coil.

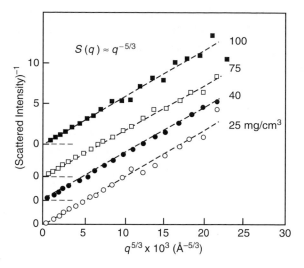

Figure 8.5 Neutron scattering from polymer in a good solvent as a function of polymer concentration, showing good agreement with the self-avoiding walk structure. (Adapted from Farnoux *et al.* (1978).)

Computer simulations of the SAW scenario conducted in various dimensions (d) can be summarized by the evolution of the mean displacement of the walker with the number of steps taken, which now varies as

$$\langle R \rangle = N^{1/\alpha} b, \tag{8.12}$$

where $\alpha = \frac{d+2}{3}$.

This variation closely resembles that of the random walk (differing only in terms of the exponent) and again is seen to exhibit self-similarity. For $d = 3$, $\alpha = 5/3$ and is slightly smaller than that ($\alpha = 2$) of the random walk, consistent with the expectation that the self-avoidance will enlarge the overall size of the polymer coil. Again we can obtain the corresponding structure factor via Eq. (8.11) as

$$S_{\text{SAW}}(q) \approx \left(\frac{1}{qb} \right)^{5/3}, \tag{8.13}$$

which is seen to agree well with the neutron scattering from polymer coils in a good solvent, shown in Fig. 8.5.

8.2 Aggregates

Consider constructing a "polymer" by an alternative route. This time we begin with a solution containing individual monomers. The solution is specially designed to initially keep the monomers dissolved, but can be chemically altered to induce the monomers to stick together whenever they come into

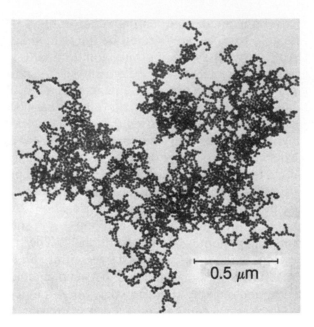

Tunneling electron micrograph of a gold colloidal cluster containing 4739 gold particles, obtained by DLCA. (Reproduced from Weitz and Oliveria (1984).)

contact with one another. As these monomers diffuse about in the solution, they occasionally encounter a second monomer and combine to form a dimer. Additional encounters lead to the formation of larger-sized polymer molecules in a process known as *aggregation*.

However, unlike the chain polymers formed by the random walk, these polymer aggregates will form clumpy structures with a dendritic (tree-like) quality, containing multiple levels of branching, as illustrated in Fig. 8.6. A careful examination of this figure would reveal that the dendritic pattern is reproduced when viewed at multiple levels of magnification, suggesting that these aggregated clusters contain a self-similar structure much like the polymer coils discussed previously. Note also, that as the cluster grows during the aggregation process, incoming monomers generally do not travel far into the interior of the cluster before making contact and sticking. Rather, they tend to stick to monomers on the outer edge of the cluster. As a result, one sees that the cluster develops considerable amounts of empty space within its interior and the total mass (total number of monomers sticking together) does not increase as the cube of the cluster size, as do most other forms of condensed matter. Instead the mass of the cluster is seen to increase as

$$N \approx \left(\frac{r}{b}\right)^{D_f}, \tag{8.14}$$

where the exponent, D_f, is a *fractal dimension*, less than three.

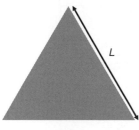

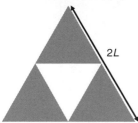

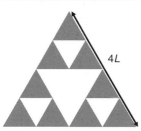

Figure 8.7

Self-similarity of the Sierpinski gasket. In each new generation, the size is doubled while the mass is tripled. At any point in the generation the overall structure appears similar to any earlier or later generation.

In the literature on aggregation one finds two important variants of the aggregation process. The first of these is *diffusion limited aggregation* (DLA), in which clusters form primarily by the addition of monomer particles, as described above. In DLA, the aggregate typically develops a fractal dimension of about $D_f = 2.5$, which is consistent with a reasonably small amount of empty space. Alternatively, a second aggregation process, known as *diffusion limited cluster aggregation* (DLCA), involves the aggregation of pre-existing clusters into much larger aggregates. In this case, the resulting aggregates tend to contain a larger amount of empty space and exhibit $D_f \approx 1.8$.

8.2.1 Fractals

Our aggregating cluster is an example of self-similarity, known as a *fractal*. Although fractals are found extensively in nature, the self-similarity can be best understood by examining certain geometrically engineered fractals. An example is the so-called Sierpinski gasket, shown in Fig. 8.7, which is generated by the sequential "aggregation" of equilateral triangles of side L. In the first step, three triangles are assembled to form a larger triangle (of side $2L$) with a triangular void in the center. In the next step, three of these larger triangles are combined to form a yet larger triangle of side $4L$. The process is repeated indefinitely, leading to a large-sized structure containing a considerable amount of empty space.

To determine the fractal dimension for this object, we consider how its mass increases with its increasing size. From Eq. (8.14),

$$\ln N \approx D_f \ln\left(\frac{r}{b}\right), \tag{8.15}$$

which is the equation of a line on a double-logarithmic scale whose slope is the fractal dimension,

$$D_f = \frac{\Delta \ln N}{\Delta \ln(r/b)} = \frac{\ln(N_{i+1}/N_i)}{\ln(r_{i+1}/r_i)}. \tag{8.16}$$

Returning to the Sierpinski gasket, we see that in any iteration (i) of the aggregation process, the mass increases by a factor of three, while the size doubles. Thus the fractal dimension for the Sierpinski gasket is

$$D_f = \frac{\ln(3)}{\ln(2)} = 1.5849\ldots,$$

and is seen to be much less than 3, indicating that a large amount of empty space is present.

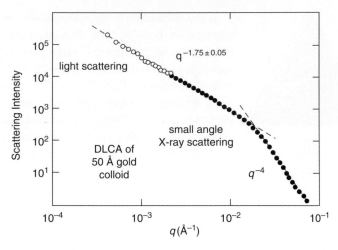

Figure 8.8 Combined light and SAXS from DLCA gold colloid clusters formed in solution (an example of which is found in Figure 8.6). Over a range of scattering wave vectors the scattered intensity decreases as a power law with slope 1.75, equal to the fractal dimension. (Adapted from Dimon *et al.* (1986) and Weitz *et al.* (1985).)

Because the aggregates are self-similar, we can determine the pair distribution function, as we did for the chain polymers, by counting the number of monomers in a sphere of radius r,

$$N \approx \left(\frac{r}{b}\right)^{\alpha} = \left(\frac{r}{b}\right)^{D_f},$$
(8.17)

where the exponent α is in this instance identified as the fractal dimension D_f. As before, we find the structure factor via Eq. (8.11) as

$$S_{\text{fractal}}(q) \approx \left(\frac{1}{qb}\right)^{\alpha} \approx (qb)^{-D_f}.$$
(8.18)

An example of $S(q)$ obtained by combined laser light scattering and small angle X-ray scattering for the DLCA of 50Å gold particles in a solvent is shown in Fig. 8.8. In this double-logarithmic presentation, the slope of $S(q)$ provides a direct measure of the fractal dimension.

8.2.2 Example: soot formation

As another good example of scattering from self-similar objects, we look now at scattering studies on soot particles that form in the combustion zone above a fuel-enriched flame. Anyone who has done welding with an acetylene torch is familiar with the black fluff that forms when the torch is initially ignited. This soot is primarily composed of carbon spherules that stick together as they cool vertically above the flame. As these particles continue to rise in the flame, they

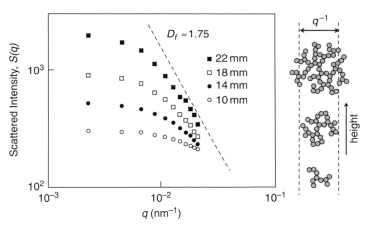

Figure 8.9 Light scattering from soot particles obtained at various distances above the burner. At the greatest height, one sees evidence of a power law with the same DLCA fractal dimension observed for gold colloidal clusters formed in solutions (Figure 8.8). At small wave vectors, and lower heights, a q-independent Rayleigh scattering is observed as a result of the reduced size of the soot particles (here smaller than the wavelength of the light). (Adapted from Sorensen *et al.* (1998).)

aggregate in a manner similar to the DLCA of gold particles in a solvent and form dendritic clusters with sizes comparable to the wavelength of visible light.

Figure 8.9 shows results of light scattering conducted at differing heights above the burner. Near the burner, the aggregation process is in its infancy and the particle sizes remain small in comparison with the scattering length scale (q^{-1}) of the incident light. Here, the structure factor mimics the isotropic (q-independent) scattering seen in Rayleigh scattering from a random collection of small clusters. Higher in the flame, where aggregation has been more extensive and the average particle size is larger, the scattering length scale begins to penetrate the interior structure of the clusters where self-similarity is present. At the highest point in the flame, $S(q)$ is seen to vary roughly as $q^{-1.75}$, indicating that these soot particles have the same fractal dimension as that for clusters of gold produced by aggregation carried out in a solution.

Guinier regime

In Fig. 8.9 one sees that $S(q)$ "bends over" from its Rayleigh (q-independent) dependence towards q^{-D_f} dependence as the scattering length scale decreases from large lengths that probe collections of several clusters, to smaller lengths that probe the interior structure of the clusters themselves. Hence the value of q^{-1} near the transition region, known as the *Guinier regime*, should provide a measure of the average size of the clusters themselves. Indeed, in Fig. 8.9 one senses that the location of this transition point shifts to smaller q (larger q^{-1}) as the distance above the burner increases in accord with our expectation that the clusters are growing in size.

The transition is not sharply defined, and to obtain a more accurate measure of the cluster size we examine the behavior of $S(q)$ in the limit that q^{-1} approaches the cluster size. Again, we can express the structure factor in Eq. (8.9) as

$$S(q) = 1 + 4\pi \langle n \rangle \int g(r) \left\{ \frac{\sin qr}{qr} \right\} r^2 dr. \qquad (8.19)$$

Since our q^{-1} of interest is near $q^{-1} \approx \xi$, the quantity qr remains less than unity throughout this integration and we are justified in expanding the term in brackets as,

$$\frac{\sin x}{x} \approx 1 - \frac{x^2}{3!},$$

to obtain

$$S_{\text{Guinier}}(q) \approx 1 + \langle n \rangle \int g(r) 4\pi r^2 dr - \langle n \rangle \int g(r) \frac{(qr)^2}{6} 4\pi r^2 dr. \qquad (8.20)$$

The first integral in Eq. (8.20) is just a computation of the coordination number (see Eq. (2.4)) carried out over the entire cluster. This results in counting all the N_m monomers that make up the cluster with the exception of the central monomer about which $g(r)$ is defined to exclude,

$$S_{\text{Guinier}}(q) \approx 1 + (N_m - 1) - \frac{q^2}{6} \int \langle n \rangle r^2 g(r) d^3 \vec{r}. \qquad (8.21)$$

The second integral is seen to be similar to the definition of a *radius of gyration* for the cluster,

$$R_G^2 \equiv \frac{\int (\vec{r} - \vec{r}_{CM})^2 n(\vec{r}) d^3 \vec{r}}{\int n(\vec{r}) d^3 \vec{r}}, \qquad (8.22)$$

where $\vec{r}_{CM}$ is the position of the cluster's center of mass. Indeed, one can show (Ex. 6) that

$$2R_G^2 \approx \frac{\int r^2 \langle n \rangle g(r) d^3 \vec{r}}{N_m}, \qquad (8.23)$$

and that $S(q)$ near the transition region is given approximately as

$$S_{\text{Guinier}}(q) \approx N_m \left\{ 1 - \frac{q^2 R_G^2}{3} \right\}. \qquad (8.24)$$

The utility of Eq. (8.24) lies in its universal applicability since the result is independent of $g(r)$ and therefore independent of the actual internal structure of

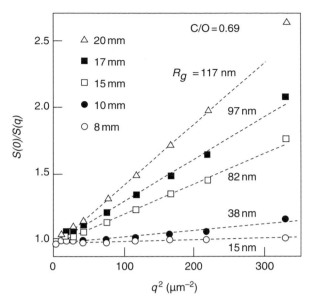

Figure 8.10 Guinier plot of the scattering from soot particles shown previously in Figure 8.9. The radius of gyration of the particles is determined from the slope of the dashed lines using the inverse of Eq. (8.24) under a small q limit. Note that the size of the soot particles increases with increasing distance above the burner due to increased aggregation time. (Adapted from Gangopadhyay *et al.* (1991).)

the objects themselves. Furthermore, as illustrated in Fig. 8.10, where the measurements presented in Fig. 8.9 are plotted in a linearized manner, the radius of gyration of the object can be determined directly from the experiment.

Delving deeper: the Porod law

One reason why we have focused on aggregates is because they possess a wide range of structural regimes and offer an excellent example of how q^{-1} functions as a relevant scattering length scale – a telescoping measuring stick – that probes structure at different levels as we change q. With this in mind, let us pause to reflect on the evolution of the scattering from a system of soot clusters as the scattering length scale (q^{-1}) decreases. This evolution is illustrated in Fig. 8.11. At very large q^{-1}, larger than the average separation $(\Delta r_{\text{cluster}})$ between clusters, our scattering records mainly the interference of scattered waves arising from the relative locations of a gaseous *system* of macro-sized particles (i.e. the clusters). The location of these large gas particles is random and is described by a pair distribution function $g_{\text{system}}(r > \Delta r_{\text{cluster}}) \approx 1$ which produces an isotropic (Rayleigh) scattering with $S_{\text{system}}(q > \Delta r_{\text{cluster}}^{-1}) \approx 1$. The normalized scattered intensity, given by Eq. (5.19), would be

$$I'_S \equiv \frac{I_S}{|E_S|^2} = N_c |f_c(q \approx 0)|^2, \qquad (8.25)$$

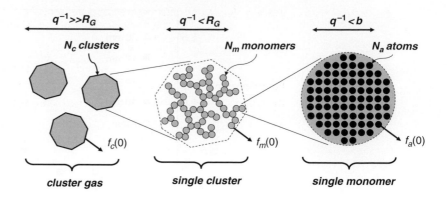

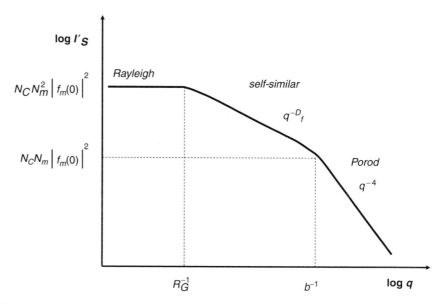

Figure 8.11   Illustration of the various scattering regimes of soot aerosols as the scattering length scale decreases. Each regime appears as a distinct power law in the static structure factor.

where N_c is the number of clusters in the system and $f_c(q)$ is the form factor that describes the characteristic scattering from a *single* cluster.

As the scattering length decreases further, it approaches that of the size (R_G or ξ) of the individual clusters and the scattering begins to explore the internal structure of a cluster itself. In this instance, we now replace the cluster form factor $f_c(q)$ by the corresponding structure factor of a cluster, $S_c(q)$, containing N_m monomers each of which scatters with a *monomer* form factor, $f_m(q)$:

$$I'_S = N_c|f_c(q)|^2 = N_c\left\{N_m|f_m(q \approx 0)|^2 S_c(q)\right\}. \tag{8.26}$$

Just at $q^{-1} \approx R_G$, the cluster structure is described by the Guinier law in Eq. (8.24),

$$I'_S = N_c N_m^2 |f_m(q \approx 0)|^2 \left\{ 1 - \frac{q^2 R_G^2}{3} \right\}, \tag{8.27}$$

but proceeding deeper into $q^{-1} < R_G$, we begin to probe the self-similar arrangement of monomer positions that comprise the internal structure of the cluster,

$$I'_S = N_c N_m |f_m(q \approx 0)|^2 \left\{ 1 + \left(\frac{1}{qb} \right)^{D_f} \right\}, \tag{8.28}$$

until we eventually shrink q^{-1} down to the level of the monomer size, b, where

$$I'_S = N_c N_m |f_m(q)|^2. \tag{8.29}$$

Now what happens? At this level we can again replace the monomer form factor by a corresponding structure factor of a monomer, $S_m(q)$,

$$I'_S = N_c N_m \left\{ N_a |f_a(q \approx 0)|^2 S_m(q) \right\}, \tag{8.30}$$

where N_a is the number of atoms in a monomer and $f_a(q)$ the atomic form factor. But what is the internal structure of the atoms in a monomer? Let us assume a crystalline structure with atoms arranged on a lattice with lattice spacing a. We have seen (Eq. (7.20)) that $S_m(q)$ can be alternatively expressed in terms of density fluctuations as

$$S_m(q) = \frac{1}{N_a} \int d^3 \vec{r}_1 \int d^3 \vec{r}_2 \left\{ \langle n \rangle^2 + \langle \Delta n(\vec{r}_1) \Delta n(\vec{r}_2) \rangle \right\} e^{-i\vec{q} \cdot (\vec{r}_1 - \vec{r}_2)}, \tag{8.31}$$

which, aside from a forward scattering contribution, reduces to

$$S_m(\vec{q}) = \frac{1}{N_a} \int d^3 \vec{r}_1 \int d^3 \vec{r}_2 \langle \Delta n(\vec{r}_1) \Delta n(\vec{r}_2) \rangle e^{-i\vec{q} \cdot (\vec{r}_1 - \vec{r}_2)}. \tag{8.32}$$

Consider now a regime of scattering lengths where $b > q^{-1} \gg a$, as illustrated in Fig. 8.12a. For this regime, the fluctuations in atom density vanish everywhere in the *interior* of the monomer because of its ordered structure. The only contribution occurs on the *surface* of the monomer where there is a *discontinuity* in the atom density, and so

$$S_m(\vec{q}) \approx \frac{1}{N_a} \langle \Delta n_{surface}^2 \rangle \sum_{\substack{i,j \\ surface}} e^{-i\vec{q} \cdot (\vec{r}_i - \vec{r}_j)}, \tag{8.33}$$

where we have coarse-grained our remaining surface integration into a summation of boxes of size q^{-1} as illustrated in Fig. 8.12a. For the regime of $q^{-1} < \Delta r$ that we are considering, the double summation of interference terms will randomize to become proportional to the number of boxes on the surface, $N_s \approx b^2/(q^{-1})^2$. Meanwhile, the magnitude of the surface density fluctuations decreases in

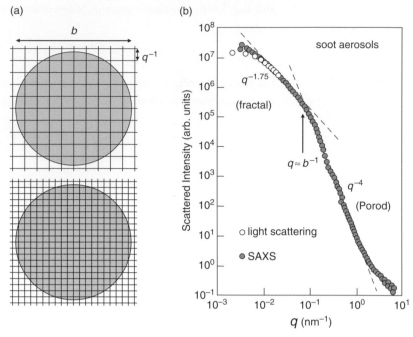

Figure 8.12 (a) Figure illustrating the partitioning of a monomer particle into cells of fixed density (interior) that contribute no scattering by virtue of the Ewald–Oseen extinction, and cells of variable density (density fluctuations) located only on the surface of the particle whose mean density fluctuation scales with the cell size. The Porod scattering law arises solely from the presence of an interface. (b) Combined light and SAXS measurements conducted on soot aerosols. A fractal regime is indicated at small q, which crosses over into the Porod regime where the intensity decreases as the fourth power of the scattering wave vector. (Adapted from Sorensen *et al.* (1998).)

proportion to the decreasing box size, $\sqrt{\langle \Delta n^2 \rangle} \approx (q^{-1})^3 / b^3$. Thus, we obtain in this scattering regime the so-called *Porod scattering law,*

$$I_S' = N_c N_m |f_a(q \approx 0)|^2 N_s \langle \Delta n^2 \rangle = N_c N_m |f_a(q \approx 0)|^2 \left(\frac{1}{qb}\right)^4. \qquad (8.34)$$

An example of the Porod law is illustrated in Fig. 8.12b, where SAXS has been used to extend the q-range of studies of soot particles to scales within the size of the monomers.

8.3 Liquid crystals

Thus far we have explored two extremes of structure: the well-ordered structure present in crystalline materials for which only a limited set of symmetry operations exist, and the reasonably random structure of

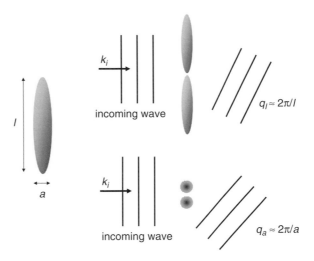

Figure 8.13 Illustration of the oblong shape of a typical liquid crystal molecule and the two prominent particle spacings relevant for scattering. Note the small spacing corresponds to a larger scattering angle.

amorphous materials for which an unlimited number of symmetry operations are possible, owing to their average translational and rotational invariance. In this final section regarding scattering from matter, we examine so-called *liquid crystals* which possess both the orderliness of a crystal and the randomness of a liquid, simultaneously. These materials consist of liquids whose constituent particles undergo partial ordering transitions, either with respect to temperature in the case of *thermotropic* liquid crystals, or with respect to composition in the case of *lyotropic* liquid crystals. As a result of these transitions, the infinite symmetry of the liquid is systematically decreased or "broken" and mesoscopic range ordering appears spontaneously through a process of *self-assembly*.

8.3.1 Thermotropic liquid crystals

We begin by considering the thermotropic liquid crystals whose particles consist of single molecules with an asymmetric profile. These particles generally have a rod-like shape with a length (l) significantly greater than their waist (a), as illustrated in Fig. 8.13. In the *isotropic phase*, found at high temperatures, thermal agitation serves to effectively randomize both the positions and orientations of these oblong particles, resulting in the rotational and translational invariance common to normal liquids. Because of the asymmetry of the particles, the short-range order of the isotropic phase, as described by a pair distribution function, exhibits two distinct correlations associated with two prominent particle–particle spacings. As illustrated in Fig. 8.13, the particles in this randomized liquid are predominantly separated either by the distance l,

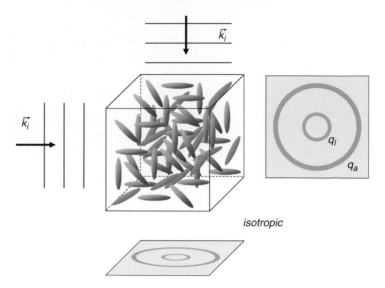

Figure 8.14 Scattering patterns from the isotropic phase are identical in any direction of incidence because of the rotational and translational invariance. Two diffuse halos occur: one at a small scattering angle corresponding to the length (l) of the molecules and one at a larger scattering angle corresponding to the width (a) of the molecules.

corresponding to the length of the particles, or by the distance a, corresponding to the width of the particles. In this instance $S(q)$, as measured by the scattering of waves with an appropriate q^{-1}, would display two broadened halos at scattering angles in the forward direction where $q_l \approx 2\pi/l$ and $q_a \approx 2\pi/a$, as shown in Fig. 8.14.

Nematic phase

As the temperature is lowered, reduced thermal activity allows the system to enter into a partially ordered state known as the *nematic phase*. In this phase the rod-like particles retain their random positions, but assume a roughly common orientation in space, specified by a unit vector known as the *director*, $\hat{n}$, as illustrated in Fig. 8.15. The phase is thermodynamically favored because of the improved packing efficiency and increased cohesion that result when the particles are aligned relative to one another.

In the nematic phase, translational invariance remains but the rotational symmetry is broken as the system no longer appears the same (on average) when rotated in certain ways. For example, if we rotate about any axis perpendicular to the director we will sense a roughly 2-fold symmetry. However, if we rotate about the director, the system will appear to be rotationally invariant. These simultaneous features of both order and disorder are what give liquid crystals their name.

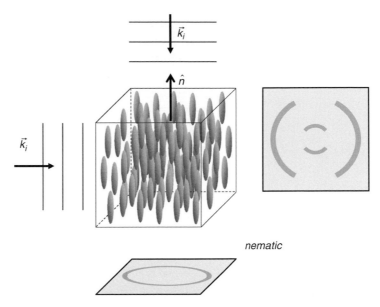

nematic

Figure 8.15   Scattering patterns from the nematic phase are sensitive to the direction of incidence.
Radiation incident from the top observes a random pattern of objects with spacing of
approximately a and produces a single diffuse halo at the larger scattering angle, q_a. Radiation
incident from the sides observes a collection of scatterers with discrete rotational symmetry.
In the vertical direction, the prominent spacing is l and interference leads to two lobes at a
small scattering angle. In the horizontal direction, the prominent spacing (a) leads to two lobes at
a large scattering angle.

What happens if we scatter from the nematic phase? The answer now
depends upon the orientation of the scattering experiment. As shown in
Fig. 8.15, if our incident waves are directed in parallel with the director,
interference will arise only from the spacing a between the oriented particles
and a single halo will appear corresponding to $q_a \approx 2\pi/a$. However, if our
incident waves are directed from the sides (perpendicular to the director) two
partial halos will appear. The outer halo arises from short spacing (a), which,
because of the orientation of the particles, is emphasized only in the horizontal
directions. By contrast, the inner halo caused by the long spacing (l) is
emphasized only in the vertical direction.

Liquid crystal displays

Liquid crystals find practical application in liquid crystal displays or LCDs.
These displays consist of a nematic phase sandwiched between two crossed
polarized plates, as shown in Fig. 8.16. The inner surfaces of each plate are
treated in a way that promotes the alignment of the oblong particles along a
direction in parallel with the polarizer and leads to a continuous rotation of the

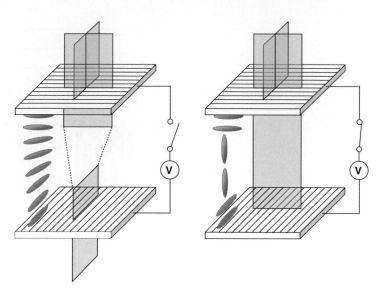

Figure 8.16   Operating principles of a liquid crystal display.

director by 90° as one travels through the display. Unpolarized light incident on the display is polarized upon entering the device and its polarization is preferentially rotated in passing through the twisted nematic phase, allowing considerable amounts of light to exit through the second polarizer. However, when an electric field is applied, an induced dipole moment develops on the particles causing those in the interior of the liquid to align vertically with the field. This alignment destroys the twisted phase. Incident light polarized by the first polarizing filter is no longer rotated and is blocked by the second filter causing a darkened appearance.

Smectic phases

When the nematic phase is cooled further it enters into the *smectic (A)* and *smectic (C)* phases, illustrated in Fig. 8.17. In these phases, the oriented particles of the nematic phase separate into a set of vertical planes with a plane spacing roughly equal to the particle length, *l*. Symmetry is broken further as the translational invariance is now partially lost. If we translate vertically along the director, we find a repeating pattern in the average particle positions owing to their ordering into planes. However, if we translate horizontally, we find no pattern and the translational invariance remains intact.

Again, scattering of waves by the smectic (*A*) will continue to display a partial halo in the horizontal plane associated with the random particle separations of roughly *a*. In the vertical direction, the vertical ordering into planes will result in reinforcement of the constructive interference (quasi-Bragg diffraction) in that direction, forcing the partial halo to shrink to a smeared

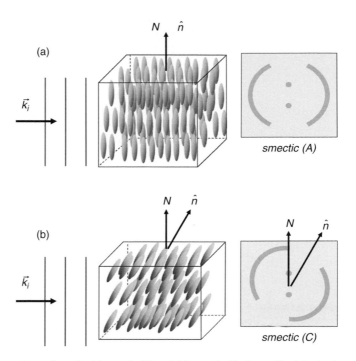

Figure 8.17 Scattering patterns from the (a) smectic (*A*) and (b) smectic (*C*) phases. The introduction of translation symmetry along the vertical direction results in narrowing of the vertical lobes (discussed with regard to the nematic phase in Figure 8.14) into two diffuse spots. In the horizontal direction, translational symmetry remains absent and the two lobes at large scattering angle persist. In the case of the smectic (*C*), these lobes are rotated by an angle equal to the angle between the normal of the planes, *N*, and the director, $\hat{n}$.

point at $q_l \approx 2\pi/l$. A similar pattern emerges for the smectic (C) except for the rotation of the $q_a \approx 2\pi/a$ partial halo by the relative angle, f, between the director and the plane normal, and an increase in the location of the Bragg spots to $q_C \approx 2\pi/l\cos\phi$ (see Fig. 8.17b).

8.3.2 Lyotropic liquid crystals: micelles and microemulsions

Unlike thermotropic liquid crystals that are composed of a single asymmetric molecule species and whose ordering arises from decreasing temperature, *lyotropic liquid crystals* consist of mixtures of particles that enter into a variety of partially ordered phases as a result of the relative concentrations. Key to these liquid crystals is the incorporation of certain *amphiphilic* particles that interact with other particles in an asymmetric fashion. Most common among the amphiphilic molecules are the *lipids* that consist of short hydrocarbon chains terminated at one end by a head group. The head group contains a charge or dipole moment that favors the presence of water (hydrophilic), while

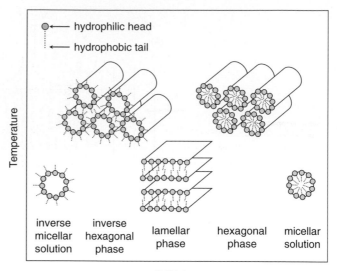

Figure 8.18 Schematic phase diagram for lyotropic liquid crystals as a function of water content (not to scale), illustrating the variety of macroscale structures that result from self-assembly. (Adapted from Seddon (1990).)

the hydrocarbon tail avoids water (hydrophobic) and prefers oils. If you have ever washed dishes by hand, you have already encountered these amphiphilic molecules. Dish soap contains lipid-based surfactants that, when added to oily water, surround the oil by sinking their hydrocarbon tails into the oil while leaving their hydrophilic heads exposed to the water. This removes oil from the surfaces of dishes and allows the oil to be suspended in the water and eventually directed down the drain.

In the absence of oil, mixtures of water and lipid assume a variety of stable phases depending on the relative concentrations, as shown in Fig. 8.18. At very low concentrations of lipid, the amphiphilic molecules find no benefit in joining together unless there are sufficient numbers to create an enclosed structure, known as a *micelle*, in which the hydrocarbon tails are safely concealed away from the water with the hydrophilic heads forming a spherical shell. As the concentration of lipids increases, these micelles coalesce in an effort to better shield their hydrophilic tails and, for a range of concentrations, the micelles form into long tubes that pack into an ordered hexagonal structure. With increasing concentration, these tubes coalesce to form sheets and the system enters the *lamellar* phase in which stacks of lipid bilayers form separated by layers of water. At even higher concentrations, the phases form in reverse order as the hydrophilic heads now begin to encase ever decreasing amounts of water.

Ternary mixtures of water, lipid (surfactant) and oil are referred to as *microemulsions* and like the binary mixtures discussed above, these systems

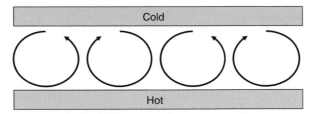

Figure 8.19 Rayleigh–Benard convection. At small temperature differences between the upper and lower plates, heat merely conducts through the fluid. However, at some critical temperature difference, thermal convection sets in and produces alternating regions of ascending and descending fluid that results in circular convection patterns.

undergo a variety of phase transitions depending upon the relative concentrations of the three components.

Self-assembly

Lyotropic liquid crystals provide an excellent example of the *static self-assembly* that is often encountered in condensed matter physics wherein a system, interacting only by weak forces (van der Waals or hydrogen bonding), spontaneously assumes an ordered configuration on length scales much larger than the average particle separation. This sort of self-assembly is quite unlike the molecular-scale ordering that occurs with crystallization which is driven by much stronger interactions occurring at shorter distances, and the spontaneity of the ordering is evident in a simple experiment. Imagine we had a tank containing a lamellar phase of lipid bilayers, which we stirred vigorously so as to break up the ordered structure. Without any changes to the composition, temperature or pressure, the molecules in the system would, over time, reassemble themselves into the lamellar structure.

Microemulsions, including the ordered phases of the lipid/water system, are an example of static self-assembly which occurs at equilibrium. However, there are a number of situations in which self-assembly arises out of equilibrium when the system is dissipating energy. A prime example of this sort of *dynamic self-assembly* is found in the phenomenon of Rayleigh–Benard convection. In this process, a thin layer of liquid is sandwiched between two flat surfaces of differing temperature, with the hotter surface at the bottom. When the temperature difference between the two surfaces is small, heat from the lower surface migrates in a uniform manner to the upper surface by thermal conduction. But when the temperature difference is increased sufficiently, a rather dramatic transition to convection occurs which drives the development of a series of convection cells, as illustrated in Fig. 8.19. Again, spontaneous ordering develops on a macroscopic level. Provided the temperature difference is maintained, the rotation of the cells (alternating from clockwise to counter-clockwise) remains stable.

Self-assembly is currently an important emerging theme in the development of nanoscience and technology. The past approach to miniaturization of devices (e.g. the miniaturization of computer chips) has been a "top-down" approach in which existing production technologies are merely scaled to shorter and shorter length scales. Self-assembly offers a radically different, "bottom-up" approach, in which nanoscale objects are manufactured by the building up of the object from molecular components through the process of self-assembly.

Summary

- Polymer coils are self-similar objects that can be adequately modeled by either a random walk or a self-avoiding random walk.

- Aggregates produced by diffusion-limited aggregation processes (DLA or DLCA) are also self-similar and characterized by a fractal dimension, $D_f < 3$.

- The structure factor of a self-similar object quite generally follows a power law of the form $S_{\text{SelfSim}}(q) \approx (qb)^{-D_f}$.

- The size of an aggregated cluster can be estimated by a Guinier analysis of the structure factor, $S_{\text{Guinier}}(q) \approx N_m \left\{1 - \frac{q^2 R_G^2}{3}\right\}$, where R_G is the radius of gyration of the cluster.

- Liquid crystals undergo phase transitions in which symmetries intermediate between liquid and crystal occur.

- Self-assembly refers to the spontaneous development of macroscopic order induced by a weak interaction.

Exercises

8.1. Estimate the radius of gyration of a $N_m = 10^7$ molecular weight linear polymer in a good solvent.

8.2. Using either a random number generator or dice, execute a 2D random walk on a piece of graph paper for a minimum of 100 steps. Record the net displacement after each step and plot this against the step number. Evaluate the result in the light of the theoretical prediction of Eq. (8.3).

(More ambitious students are encouraged to use a spreadsheet or write a computer algorithm to complete this exercise.)

8.3. Using either a random number generator or dice, execute a 2D self-avoiding walk on a piece of graph paper for a minimum of 50 steps. (If you get trapped before 50 steps have occurred, back up several steps and try again.) Record the net displacement after each step and plot this against the step number. Evaluate the result in the light of the theoretical prediction of Eq. (8.12). (More ambitious students are encouraged to use a spreadsheet or write a computer algorithm to complete this exercise.)

8.4. Shown in Fig. 8.20 are two geometrical fractals embedded in a 2D space. Both of these are commonly used as broadband antennae in modern cell phones. Determine the fractal dimension of each.

(a)   (b)

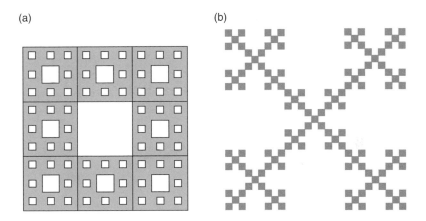

Figure 8.20

8.5. Shown in Fig. 8.21 are two geometrical fractals embedded in a 3D space. Determine the fractal dimension of each.

(a)   (b)

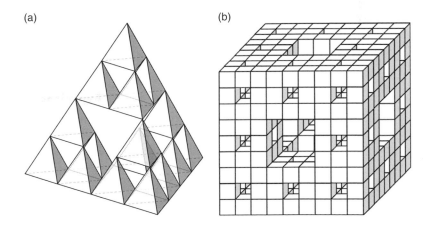

Figure 8.21

8.6. Show that the radius of gyration of a fractal cluster containing N_m monomers is given by Eq. (8.23).

8.7. Shown in Fig. 8.22 is the static structure factor for some system of aggregated particles. (a) Identify the q-ranges associated with the Rayleigh, Guinier, self-similar, and Porod regimes. (b) Estimate the mean cluster size (and explain your answer). (c) Estimate the mean monomer size (and explain your answer). (d) Determine the fractal dimension of the particles.

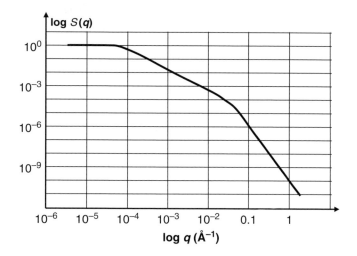

Figure 8.22

8.8. In arriving at Eq. (8.11) from Eq. (8.10), it was stated that the definite integral is roughly unity over the q-range of interest. (a) For the case of the random walk ($\alpha = 2$), compute this integral and verify that it is indeed nearly unity for $q^{-1} < \xi$. (b) Also, verify that the isotropic (Rayleigh) term may be ignored in the present q-range of interest.

8.9. In the light scattering study of soot aerosols, researchers determined the radius of gyration of the soot clusters in Fig. 8.10 by using the Guinier law of Eq. (8.24). (a) First show that for small q, the Guinier expression can be inverted to obtain $S(0)/S(q) \approx \left[1 + (qR_G)^2/3\right]$. (b) Use this result to verify the values of the radius of gyration given in Fig. 8.10 for the three highest positions (15, 17 and 20 mm) in the flame.

8.10. In the lab, students measured the intensity of light scattered by a suspension of polystyrene spheres in water at several angles and produced a Guinier plot of the normalized inverse scattering intensity, as reproduced in Fig. 8.23. (a) Use this to determine the mean radius of gyration of the spheres. (b) The manufacturer specifies the spheres to have a diameter of 109 nm. Assuming the spheres are of uniform

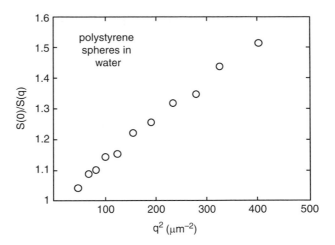

Figure 8.23

density, determine from the radius of gyration the corresponding mean sphere diameter. (Caution: this diameter is not simply $2 R_G$. You need to first determine how the radius of gyration is related to the radius of a solid sphere.)

Suggested reading

Mandelbrot's text is a classic and is filled with many figures that really develop the theme of fractals and self-similarity at an introductory level. The review by Sorensen provides a comprehensive overview of scattering by fractal materials. The article by Whitesides and Grzybowski gives several additional examples of self-assembly and its technological significance.

G. Strobl, *Condensed Matter Physics* (Springer-Verlag, Berlin, 2004).

P. M. Chaikin and T. C. Lubensky, *Principles of Condensed Matter Physics* (Cambridge University Press, New York, 2003).

R. Zallen, *The Physics of Amorphous Materials* (John Wiley and Sons, New York, 1983).

B. B. Mandelbrot, *The Fractal Geometry of Nature* (W. H. Freeman and Co., New York, 1983).

C. M. Sorensen, "Light Scattering by Fractal Aggregates: A Review," Aerosol Sci. Tech. **35**(2), 648–687 (2001).

G. M. Whitesides and B. Grzybowski, "Self-Assembly at All Scales," *Science* **295**, 2418 (2002).

the sorption, diffusion and transport properties of these materials.

PART III

DYNAMICS

A steaming cup of coffee is sitting on my desk. Aside from the steam, there is little else that would suggest any other activity is present. The cup and it contents appear to be "at rest". But a little closer examination reveals a small ripple of waves on the surface of the liquid caused by a mechanical pump in the room next door. Indeed, if I rest my finger gently on the lip of the cup, I can feel the vibration. I can also feel the heat that has developed in the cup, now several minutes since I poured the coffee from the pot and added some creamer.

If I could examine this even more closely I would actually see that nothing is truly "at rest". The particles of the liquid are jostling about incessantly. The particles of the creamer that I added have clearly taken flight and diffused rapidly out into all regions of the coffee. The cup itself is also in motion. Its particles are undergoing incessant vibrations that are ultimately responsible for the heat I feel when I hold it.

Perhaps most interesting is that all this microscopic motion appears to be driven entirely by thermodynamics. The coffee and cup are sitting in a climate-controlled office and there is a constant flux of thermal energy (heat) entering and exiting both the coffee and the cup to keep things moving.

In discussing dynamics, we focus mostly on those microscopic motions present in liquids and solids that are driven by thermodynamic forces. We begin in Chapter 9 with the motion of particles in a liquid, and commence by adapting our concept of a structure factor to accommodate the time-dependent scattering from density fluctuations of a liquid caused by the incessant jostling. Following this, in a pair of chapters, we examine the vibrational motions of particles in crystals and glasses to demonstrate how these motions in turn determine thermal and transport properties of these materials.

In Chapters 12 and 13 we consider the inherent motions of conduction electrons in metallic crystals. Not only will we see how electron motions contribute to the thermal properties of a crystal, we will also discover why certain crystals are better or worse conductors of electricity.

In a final chapter on the subject of dynamics, we contrast all of this microscopic motion with the comparatively macroscopic deformations of a material that result when an external force is applied. Interestingly, we will discover that the response of a material to such bulk forces shares much in common with those microscopic dynamics present in the absence of the force.

9

Liquid dynamics

Introduction

In Chapter 5 we introduced the structure factor, $S(q)$, as the Fourier representation of the positions of a collection of fixed, elastically scattering, particles. In reality, these particles are rarely fixed. In a solid (crystal or glass), the particles are bound together by bonds and, while unable to wander about, are able to oscillate or vibrate about a fixed center of motion. In a liquid, the particles are even less constrained and are free to wander around over considerable distances. In this chapter we develop the *dynamic* structure factor as a straightforward extension of the static structure factor introduced previously, and apply it to examine the dynamics of liquid-like systems. In one instance, we consider the *Brownian diffusion* of macromolecules in a solvent, where the motion mimics that of the random walk we discussed in the previous chapter. In another instance, we show how thermodynamically driven density fluctuations present in a simple liquid are responsible for the characteristic *Rayleigh–Brillouin spectrum* of light scattering. We also take this opportunity to consider the special case of slow dynamics in polymer liquids and to briefly consider the nature of the liquid-to-glass transition that separates amorphous solids from their liquid counterparts.

9.1 Dynamic structure factor

We can think of a liquid as a time-dependent amorphous structure. In many ways, the structure of a liquid resembles the structure of a glass in that, at any instant in time, a "snapshot" of its $S(q)$ resembles that of the glass. Indeed, the only real difference between a liquid and a glass is the presence or absence, respectively, of long-range translational motion. In the liquid, the translational motion results from the incessant jostling of the particles allowing them to wander about. By virtue of this motion, particles of the liquid are able to rearrange on some characteristic time scale (related to the viscosity of the liquid) into different, but thermodynamically equivalent, amorphous configurations whose instantaneous structure resembles that of a glass. For certain glass forming liquids near their glass transition point, the characteristic time scale for

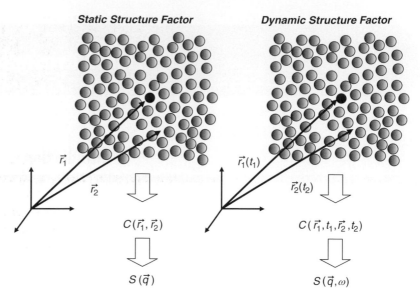

Static Structure Factor   **Dynamic Structure Factor**

$C(\vec{r}_1, \vec{r}_2)$ $C(\vec{r}_1, t_1, \vec{r}_2, t_2)$

$S(\vec{q})$ $S(\vec{q}, \omega)$

Figure 9.1 Comparison of the static and dynamic structure factors. (a) The static structure factor, $S(q)$, is derived from a spatial Fourier transform of a time-independent density–density correlation function. (b) The dynamic structure factor, $S(q, \omega)$, is derived from both a spatial and a temporal Fourier transform of a time-dependent density–density correlation function.

these rearrangements can become exceedingly long with some unusual consequences, as we will discuss later.

For time-independent structures, we found in Chapter 5 (Eq. (5.26)) that the structure factor was fundamentally related (by a double Fourier transform) to the density–density correlation function:

$$S(\vec{q}) = \frac{1}{N} \int d^3\vec{r}_1 \int d^3\vec{r}_2 \, C(\vec{r}_1, \vec{r}_2) e^{-i\vec{q}\cdot(\vec{r}_1 - \vec{r}_2)}, \tag{9.1}$$

where the density–density correlation function was defined as

$$C(\vec{r}_1, \vec{r}_2) \equiv \langle n(\vec{r}_1) n(\vec{r}_2) \rangle. \tag{9.2}$$

As illustrated again in Fig. 9.1, $C(\vec{r}_1, \vec{r}_2)$ contains two search vectors and represents a conditional probability that if a particle is located at $\vec{r}_1$, another will be found at $\vec{r}_2$.

This statement would still be completely true for a liquid if we were considering a discrete instant in time (i.e. a snapshot). However, in the liquid the structure is evolving as a result of the incessant jostling of the particles and we need to accommodate this by a time-dependent generalization of the density–density correlation function:

$$C(\vec{r}_1, \vec{r}_2, t_1, t_2) \equiv \langle n(\vec{r}_1, t_1) n(\vec{r}_2, t_2) \rangle. \tag{9.3}$$

Here the meaning of $C(\vec{r}_1, \vec{r}_2, t_1, t_2)$, illustrated in Fig. 9.1, now represents the probability that if a particle is located at $\vec{r}_1$ at time t_1, it or another particle will be found at $\vec{r}_2$ at another time t_2.

9.1.1 The van Hove correlation function

For a liquid at equilibrium, the average particle density, $\langle n \rangle$, is time-independent and so it is convenient to express the local density as the average density plus a *fluctuation*, $n(\vec{r}, t) = \langle n \rangle + \Delta n(\vec{r}, t)$, where $\Delta n(\vec{r}, t)$ is the local density fluctuation, which can be either positive or negative. The time-dependent density–density correlation function can then be restated in terms of density fluctuations as,

$$C(\vec{r}_1, \vec{r}_2, t_1, t_2) \equiv \langle [\langle n \rangle + \Delta n(\vec{r}_1, t_1)][\langle n \rangle + \Delta n(\vec{r}_2, t_2)] \rangle$$
$$= \langle n \rangle^2 + \langle \Delta n(\vec{r}_1, t_1)\Delta n(\vec{r}_2, t_2) \rangle, \quad (9.4)$$

and the resulting time-dependent structure factor given as,

$$S(\vec{q}, t) = \frac{1}{N} \int d^3\vec{r}_1 \int d^3\vec{r}_2 \left\{ \langle n \rangle^2 + \langle \Delta n(\vec{r}_1, t_1)\Delta n(\vec{r}_2, t_2) \rangle \right\} e^{-i\vec{q}\cdot(\vec{r}_1 - \vec{r}_2)}. \quad (9.5)$$

The quantity $S(\vec{q}, t)$, or its Fourier transform counterpart, $S(\vec{q}, \omega) = \int S(\vec{q}, t)e^{i\omega t}dt$, are both referred to as the *dynamic structure factor*.

Since the liquid remains disordered, it possesses the same rotational and translational invariance as that of an amorphous solid and so, again, only the relative positions of two density fluctuations, $\vec{r} = \vec{r}_1 - \vec{r}_2$, matter. Likewise, the absolute time origin is irrelevant and what matters is only the difference $t = t_1 - t_2$. Taking these into account,

$$S_{\text{liq}}(q, t) = \frac{1}{N} \int d^3\vec{r} \int d^3\vec{r}_2 \left\{ \langle n \rangle^2 + \langle \Delta n(\vec{r} + \vec{r}_2, t)\Delta n(\vec{r}_2, 0) \rangle \right\} e^{-i\vec{q}\cdot\vec{r}}$$
$$= \frac{1}{N} \left\{ \langle n \rangle^2 \int d^3\vec{r} e^{-i\vec{q}\cdot\vec{r}} \int d^3\vec{r}_2 + \int d^3\vec{r} e^{-i\vec{q}\cdot\vec{r}} \int d^3\vec{r}_2 \langle \Delta n(\vec{r} + \vec{r}_2, t)\Delta n(\vec{r}_2, 0) \rangle \right\}. \quad (9.6)$$

The integrations over $d^3\vec{r}_2$ correspond to sampling with the second search vector at different possible origins. But because of the invariance, the result cannot depend upon the specific choice of $\vec{r}_2$. Consequently, the result of integrating over $d^3\vec{r}_2$ will be the volume of the scattering region, V, and so

$$S_{\text{liq}}(q, t) = \frac{1}{N} \left\{ \langle n \rangle^2 V(2\pi)^3 \delta(\vec{q}) + V \int d^3\vec{r} e^{-i\vec{q}\cdot\vec{r}} \langle \Delta n(\vec{r}, t)\Delta n(0, 0) \rangle \right\}$$
$$= \langle n \rangle (2\pi)^3 \delta(\vec{q}) + \int d^3\vec{r} \langle \Delta n(\vec{r}, t)\Delta n(0, 0) \rangle / \langle n \rangle e^{-i\vec{q}\cdot\vec{r}} \quad (9.7)$$
$$= \langle n \rangle (2\pi)^3 \delta(\vec{q}) + \int d^3\vec{r} G(\vec{r}, t) e^{-i\vec{q}\cdot\vec{r}},$$

where the *van Hove space and time correlation function* is introduced here as

$$G(\vec{r}, t) \equiv \langle \Delta n(\vec{r}, t) \Delta n(0, 0) \rangle / \langle n \rangle. \qquad (9.8)$$

Aside from the innocuous forward scattering, the characteristic space and time dependence of the constantly evolving structure of a liquid is conveyed by a Fourier transform of the van Hove correlation function. Formally, this function represents the probability that if a density fluctuation is present at the origin at time $t = 0$, another will be found at $\vec{r}$ at a later time t.

9.1.2 Brownian motion: the random walk revisited

As an example of liquid-like dynamics, we first consider the dynamics of reasonably large-sized particles suspended in a solvent. In this instance, we ignore the scattering of the solvent and instead only consider the "density fluctuations" caused by movement of the macroparticles themselves. How do these particles move about? As a result of the incessant jostling of the solvent particles, the suspended particles experience a series of randomly directed kicks and execute what is known as *Brownian motion*. The finite viscosity of the solvent limits how far the macroparticle will move, and the result of these random kicks gives rise to a *random walk* performed by the macroparticle.

In principle, the van Hove correlation function for diffusing particles contains two contributions,

$$G(\vec{r}, t) = G_{\text{self}}(\vec{r}, t) + G_{\text{distinct}}(\vec{r}, t). \qquad (9.9)$$

The self part is associated with motions in which the fluctuation (i.e. macroparticle) present at $(\vec{r}, t)$ is the result of the *same* particle starting from the origin ($\vec{r} = 0$, $t = 0$) and *moving* to that location at a later time. The distinct part describes those situations in which the fluctuation at $(\vec{r}, t)$ results from the arrival of an altogether different particle at that location. Provided that our sample of diffusing macroparticles is sufficiently *dilute*, we can safely ignore $G_{\text{distinct}}(r, t)$ and associate the probability distribution of the random walk with that of $G_{\text{self}}(r, t)$ alone.

Our interest then is in determining the probability that a random walker will have arrived at $\vec{r}$ some time t after starting off from the origin. Consider then a 1D random walk whose walker takes steps of length b. In any arbitrary step, the probability that the walker will arrive at (x, t) is

$$G(x, t) = \frac{1}{2} G(x - b, t - \tau) + \frac{1}{2} G(x + b, t - \tau). \qquad (9.10)$$

The factor of one-half represents the equal odds that the walker will enter the site in question from an adjacent site either to the left or right of the site. If we apply a twofold Taylor expansion,

$$G(x \pm b, t - \tau) = G(x, t) - \tau \frac{\partial G}{\partial t} \pm b \frac{\partial G}{\partial x} + \frac{b^2}{2} \frac{\partial^2 G}{\partial x^2} + O(\tau^2). \qquad (9.11)$$

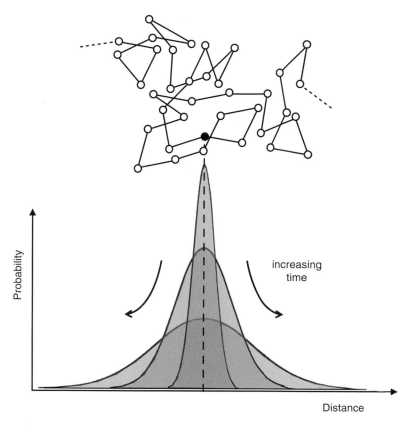

Figure 9.2   Probability distribution for the displacement of a diffusive particle. At $t = 0$, the particle has not yet moved and the probability is zero everywhere except at its starting location (i.e. a Dirac delta function). Over time, the many possible random walk trajectories of the particle lead to spreading of the probability distribution as described by Eq. (9.13).

Then Eq. (9.10) can be rearranged (Ex. 1) to read,

$$\frac{\partial G}{\partial t} = \frac{b^2}{2\tau}\frac{\partial^2 G}{\partial x^2} = D\frac{\partial^2 G}{\partial x^2}, \tag{9.12}$$

where $D = b^2/2\tau$ is known as the diffusion coefficient or *diffusivity*. Furthermore, if we recall that the mean-squared displacement of the random walk is proportional to the number of steps executed, $\langle x^2 \rangle = Nb^2 = (t/\tau)b^2$, then we see that this mean-squared displacement is also given as $\langle x^2 \rangle = 2Dt$.

Equation (9.12) is known as the *diffusion equation* and its solution (for an initial walker at the origin) is the Gaussian,

$$G(x,t) = \frac{1}{\sqrt{4\pi Dt}}\,\mathrm{e}^{-x^2/4Dt}, \tag{9.13}$$

which is illustrated together with a random walk trajectory in Fig. 9.2. Generalizing this to 3D, we have

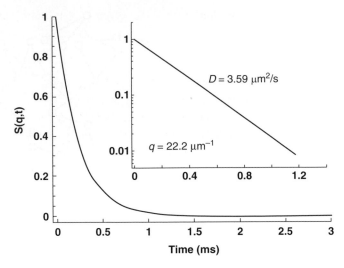

Figure 9.3   Dynamic structure factor observed for the diffusion of 0.1 μm polystyrene spheres in water. Inset figure is the same curve presented on a semi-log scale where the slope (see Eq. (9.17)) provides a measure of the diffusivity.

$$G(r,t) = G(x,t)G(y,t)G(z,t) = \frac{1}{(4\pi Dt)^{3/2}} e^{-r^2/4Dt}, \qquad (9.14)$$

which is the solution to

$$\frac{\partial G(r,t)}{\partial t} = D\nabla^2 G(r,t). \qquad (9.15)$$

Applying the Fourier transform to Eq. (9.15) then yields (Ex. 2) the following condition for the dynamic structure factor:

$$\frac{\partial S(q,t)}{\partial t} = -q^2 DS(q,t), \qquad (9.16)$$

whose solution is

$$S(q,t) = S(q,0)e^{-Dq^2 t}. \qquad (9.17)$$

One experimental technique for measuring $S(q,t)$ for particles diffusing in a solvent is *photon correlation spectroscopy*. An example for $S(q,t)$ thus obtained is shown in Fig. 9.3 for the scattering from a dilute suspension of small polystyrene balls of diameter $2b \approx 0.1$ micron in water. Analysis of the decay provides a direct measure of the diffusion coefficient. In many instances, the diffusion coefficient for diffusing macroparticles is related to the solution viscosity by the *Stokes–Einstein relation*,

$$D = \frac{kT}{6\pi\eta b}, \qquad (9.18)$$

resulting from a combination of Einstein's expression of the diffusivity,

$$D = \frac{kT}{\zeta} \tag{9.19}$$

in terms of the drag coefficient, ζ, and Stokes' relation, $\zeta = 6\pi\eta b$, for the drag experienced by a spherical particle of radius b moving in a medium of viscosity η. In these situations, scattering from a solution of unknown macroparticles together with knowledge of the solvent viscosity and temperature can be used to determine the particle size (Ex. 5).

9.1.3 Hydrodynamic modes in liquids

In the previous example, we ignored the scattering from solvent in which the macroparticles were dispersed, and considered only the scattering from the macroparticles themselves. Imagine now that these macroparticles are removed, leaving behind only the solvent, a molecular liquid, to scatter light. How do these molecular particles move about? As we have alluded to earlier, they execute some incessant jostling, continually bumping into one another. Naively, we might think that they experience a similar Brownian motion to that of the macroparticles discussed above. But, there is a difficulty here in that the particles being kicked are also the ones doing the kicking! Furthermore, the particles are now much smaller than the wavelength of visible light and the density fluctuations being probed in the van Hove correlation function correspond, not to the individual liquid molecules themselves, but to larger collections of molecules. In this case, the motion of this collection of particles is described by the motion of a *continuous* density distribution wherein thermo dynamic forces drive the formation of density fluctuations and ultimately determine their decay back to an equilibrium condition. Thus to describe the motion, we need to consider the role of thermodynamic fluctuations inherent in a liquid.

In a liquid at equilibrium, one requires only two independent thermodynamic fields to determine the remaining properties. For our purposes, we will use the entropy, S, and the pressure, P, such that any fluctuation in density can be expressed as

$$\Delta n(\vec{r}, t) = \left(\frac{\partial n}{\partial S}\right)_P \Delta S(\vec{r}, t) + \left(\frac{\partial n}{\partial P}\right)_S \Delta P(\vec{r}, t), \tag{9.20}$$

and the van Hove correlation function for these thermodynamic fluctuations, expressed as

$$G(\vec{r}, t) = \frac{1}{\langle n \rangle} \left(\frac{\partial n}{\partial S}\right)_P^2 \langle \Delta S(\vec{r}, t) \Delta S(0, 0) \rangle + \frac{1}{\langle n \rangle} \left(\frac{\partial n}{\partial P}\right)_S^2 \langle \Delta P(\vec{r}, t) \Delta P(0, 0) \rangle.$$

$$\tag{9.21}$$

Note that, because S and P are independent fields, their fluctuations are uncorrelated and the two cross terms (involving $\langle \Delta S(\vec{r}, t) \Delta P(0, 0)\rangle$ and $\langle \Delta P(\vec{r}, t) \Delta S(0, 0)\rangle$) necessarily vanish.

For a continuum fluid, the local density is governed by a set of conservation laws known as the *hydrodynamic equations*. In addition to S and P, they involve the density (n), the temperature (T) and the local velocity ($\vec{u}(\vec{r}, t)$). In their linearized form (appropriate to a fluid system with $\langle \vec{u}\rangle = 0$ that is not moving as a whole), these equations, expressed in terms of fluctuations, are:

$$\frac{\partial \Delta n(\vec{r}, t)}{\partial t} + \langle n\rangle \vec{\nabla} \cdot \Delta \vec{u}(\vec{r}, t) = 0, \tag{9.22}$$

where fluctuation in speed is $\Delta \vec{u}(\vec{r}, t) \equiv \vec{u}(\vec{r}, t) - \langle \vec{u}\rangle = \vec{u}(\vec{r}, t)$,

$$m\langle n\rangle \frac{\partial \Delta \vec{u}(\vec{r}, t)}{\partial t} = -\vec{\nabla}(\Delta P(\vec{r}, t)) + \eta_s \nabla^2 (\Delta \vec{u}(\vec{r}, t))$$
$$+ \left(\eta_V + \frac{1}{3}\eta_s\right) \vec{\nabla}\left(\vec{\nabla} \cdot \Delta \vec{u}(\vec{r}, t)\right), \tag{9.23}$$

where m is the particle mass, η_S the shear viscosity and η_V the bulk viscosity, and

$$\langle T\rangle \frac{\partial \Delta S(\vec{r}, t)}{\partial t} = \kappa_{\text{th}} \nabla^2 (\Delta T(\vec{r}, t)), \tag{9.24}$$

where κ_{th} is the thermal conductivity and $\Delta T(\vec{r}, t) \equiv T(\vec{r}, t) - \langle T\rangle$ is the temperature fluctuation.

Whew! Where do these equations come from? The first (Eq. 9.22) is a statement of the conservation of mass (per volume) and implies that, when the density increases at some point, it does so by the influx of new particles. The second equation (Eq. 9.23) is a statement of Newton's second law for the momentum per volume. The left-hand side contains the time derivative of the momentum caused by (on the right-hand side) the applied force ($-\vec{\nabla}P$) and a complicated drag force associated with the viscous properties of the liquid. The last equation (Eq. 9.24) is the heat equation. It is a combination of both conservation of energy $dQ = T dS = -\vec{\nabla} \cdot \vec{j}_Q$, and heat diffusion $\vec{j}_Q = -\kappa_{\text{th}} \vec{\nabla} T$.

The first and second equations can be combined by taking the divergence of Eq. (9.23) and substituting in Eq. (9.22) to eliminate $\Delta \vec{u}(\vec{r}, t)$,

$$m\frac{\partial^2 \Delta n}{\partial t^2} = \vec{\nabla}^2(\Delta P) + \frac{1}{\langle n\rangle}\left(\eta_V + \frac{4}{3}\eta_s\right)\frac{\partial}{\partial t}\left[\vec{\nabla}^2(\Delta n)\right]. \tag{9.25}$$

Next, we need to eliminate Δn and ΔT to obtain equations relating only ΔS and ΔP. Most generally,

$$\Delta n = \left(\frac{\partial n}{\partial S}\right)_P \Delta S + \left(\frac{\partial n}{\partial P}\right)_S \Delta P$$
$$\Delta T = \left(\frac{\partial T}{\partial S}\right)_P \Delta S + \left(\frac{\partial T}{\partial P}\right)_S \Delta P. \tag{9.26}$$

Although it would be *the proper thing to do*, substitution of these into Eq. (9.24) and Eq. (9.25) will lead to a set of coupled differential equations whose solution is more involved than I wish to go into here. If the reader will allow, I would like to assume *for the moment*, that

$$\left(\frac{\partial n}{\partial P}\right)_S \gg \left(\frac{\partial n}{\partial S}\right)_P \quad \text{and} \quad \left(\frac{\partial T}{\partial S}\right)_P \gg \left(\frac{\partial T}{\partial P}\right)_S. \tag{9.27}$$

Some reasoning for this assumption is the impression that density is more strongly altered by pressure than by entropy, and that temperature is more strongly altered by entropy (i.e. heat) than by pressure. Under this assumption, one can show that

$$\Delta n \approx \left(\frac{\partial n}{\partial P}\right)_S \Delta P = \left[\frac{-\langle n \rangle}{V}\frac{\partial V}{\partial P}\right]_S \Delta P = \langle n \rangle \chi_S \Delta P = \frac{1}{mv^2}\Delta P, \tag{9.28}$$

and

$$\Delta T \approx \left(\frac{\partial T}{\partial S}\right)_P \Delta S = \left[\frac{\langle T \rangle}{\rho C_p}\right]\Delta S, \tag{9.29}$$

where $\chi_S = 1/\rho v^2$ is the bulk modulus, v is the adiabatic speed of sound, C_p is the specific heat (per mass) at constant pressure and $\rho = m\langle n \rangle$ is the mass density. Again, the assumption made in Eq. (9.27) is NOT really correct, but if we indulge it for the time being, substitution of Eq. (9.28) into Eq. (9.25) and Eq. (9.29) into Eq. (9.24) then produces the following set of uncoupled equations:

$$\frac{\partial^2 \Delta P}{\partial t^2} = v^2 \vec{\nabla}^2(\Delta P) + D_V \frac{\partial}{\partial t}\left[\vec{\nabla}^2(\Delta P)\right], \tag{9.30}$$

where,

$$D_V = \frac{1}{\rho}\left(\eta_V + \frac{4}{3}\eta_s\right), \tag{9.31}$$

and

$$\frac{\partial \Delta S}{\partial t} = \frac{\kappa_{\text{th}}}{\rho C_p}\nabla^2(\Delta S) = D_T \nabla^2(\Delta S). \tag{9.32}$$

The astute reader may recognize in Eq. (9.30), the wave equation together with some sort of damping contribution, and in Eq. (9.32), a diffusion equation much like that of Eq. (9.15). We can now apply a Fourier transform to obtain,

$$\frac{\partial^2 \Delta P}{\partial t^2} = -q^2 v^2 \Delta P - q^2 D_V \frac{\partial \Delta P}{\partial t}, \tag{9.33}$$

$$\frac{\partial \Delta S}{\partial t} = -q^2 D_T \Delta S, \tag{9.34}$$

whose solutions are

$$\Delta P(\vec{q}, t) = \Delta P(\vec{q}, 0) \exp(-q^2 D_V t / 2) \exp(\pm i \omega_B t), \qquad (9.35)$$

$$\Delta S(\vec{q}, t) = \Delta S(\vec{q}, 0) \exp(-q^2 D_T t), \qquad (9.36)$$

where $\omega_B = q\upsilon \sqrt{1 - (q D_V / 2\upsilon)^2}$ is approximately $\omega_B \approx q\upsilon$ in situations of weak viscous damping. Again, the above results were obtained under the assumption of Eq. (9.27). The *true result*, is slightly modified such that

$$\langle \Delta P(\vec{q}, t) \Delta P(\vec{q}, 0) \rangle = \frac{1}{\gamma} \langle \Delta P^2(\vec{q}, 0) \rangle \exp(-q^2 \Gamma t) \exp(\pm i \omega_B t), \qquad (9.37)$$

$$\langle \Delta S(\vec{q}, t) \Delta S(\vec{q}, 0) \rangle = \frac{\gamma - 1}{\gamma} \langle \Delta S^2(\vec{q}, 0) \rangle \exp(-q^2 D_T t), \qquad (9.38)$$

where $\gamma = C_p / C_V$, $\Gamma = (1/2)[(\gamma - 1) D_T + D_V]$. The dynamic structure factor is then given by Eq. (9.7) and Eq. (9.21) as

$$
\begin{aligned}
S_{\text{liq}}(q, t) = {} & \frac{1}{\langle n \rangle} \left(\frac{\partial n}{\partial S} \right)_P^2 \langle \Delta S^2(\vec{q}, 0) \rangle \frac{\gamma - 1}{\gamma} \exp(-q^2 D_T t) \\
& + \frac{1}{\langle n \rangle} \left(\frac{\partial n}{\partial P} \right)_S^2 \langle \Delta P^2(\vec{q}, 0) \rangle \frac{1}{\gamma} \exp(-q^2 \Gamma t) \exp(\pm i \omega_B t),
\end{aligned}
\qquad (9.39)
$$

which, when Fourier transformed into frequency space, appears as a spectrum of scattered light consisting of two contributions,

$$S_{\text{Rayleigh}}(\vec{q}, \omega) = FT\left[S_{\text{entropy}}(\vec{q}, t) \right] \approx \frac{\langle \Delta n^2(\vec{q}, 0) \rangle}{\langle n \rangle} \left(1 - \frac{1}{\gamma} \right) \frac{D_T q^2}{\omega^2 + (D_T q^2)^2}, \qquad (9.40)$$

and

$$
\begin{aligned}
S_{\text{Brillouin}}(\vec{q}, \omega) = {} & FT\left[S_{\text{pressure}}(\vec{q}, t) \right] \\
\approx {} & \frac{\langle \Delta n^2(\vec{q}, 0) \rangle}{\langle n \rangle} \frac{1}{\gamma} \left\{ \frac{\Gamma q^2}{(\omega - \omega_B)^2 + (\Gamma q^2)^2} + \frac{\Gamma q^2}{(\omega + \omega_B)^2 + (\Gamma q^2)^2} \right\}.
\end{aligned}
\qquad (9.41)
$$

The resulting spectrum of scattered light is known as the *Rayleigh–Brillouin* spectrum, and is illustrated in Fig. 9.4. It is composed of three Lorentzian curves. One is centered at the frequency of the incident wave and two others are symmetrically shifted by $\omega_B = q\upsilon$. The central curve (known as the Rayleigh line) involves the thermal diffusivity, D_T, and is associated with the *non-propagating* decay of spontaneous density fluctuations through the diffusion of heat. The other two lines (known as the Brillouin doublet) involve the adiabatic speed of sound, υ, and both the viscous damping, D_V, as well as a smaller contribution due to thermal diffusivity. The Brillouin lines

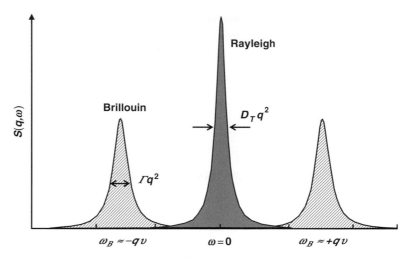

Figure 9.4
An illustration of the features in the Rayleigh–Brillouin spectrum of a simple liquid. A central Rayleigh line arises solely from the decay of density fluctuations driven by thermal diffusivity, and is flanked on either side by two Brillouin lines that are shifted in frequency by $\omega_B \approx qv$. The two Brillouin lines arise primarily from the propagation of density fluctuations in the form of acoustic waves (with both viscous and thermal damping characterized by the linewidth, Γ).

represent the *propagation* of density fluctuations in the form of (damped) *sound waves* traveling through the medium with speed v, and their frequency shift can be viewed much like that of a *Doppler shift*. Measurement of the Rayleigh–Brillouin spectrum thus allows for the determination of both the thermal diffusivity and the (longitudinal) speed of sound in a liquid or solid.

Lastly, the total integrated intensity of the Rayleigh–Brillouin spectrum is given as

$$I_{TOT} \propto \int_{-\infty}^{+\infty} S(q,\omega)\mathrm{d}\omega = I_R + 2I_B$$

$$= \frac{\left\langle \Delta n^2(\vec{q},0) \right\rangle}{\langle n \rangle} \left\{ \left(1 - \frac{1}{\gamma} \right) + \frac{1}{\gamma} \right\} = \frac{\left\langle \Delta n^2(\vec{q},0) \right\rangle}{\langle n \rangle}, \quad (9.42)$$

and is seen to be proportional to the average square of the density fluctuations which, as we observed in Chapter 7 (see Eq. (7.18)), can be expressed as

$$\frac{\left\langle \Delta n^2(\vec{q},0) \right\rangle}{\langle n \rangle^2} = \frac{kT}{V} \chi_T,$$

where

$$\chi_T \equiv \frac{1}{\langle n \rangle} \left(\frac{\partial \langle n \rangle}{\partial p} \right)_T \quad (9.43)$$

is the isothermal compressibility. Additionally, one sees that the ratio of the Rayleigh intensity to that of the Brillouin doublet, known as the *Landau–Placek* ratio,

$$R_{LP} = \frac{I_R}{2I_B} = \frac{(1 - 1/\gamma)}{1/\gamma} = \gamma - 1 = \frac{C_P}{C_V} - 1, \qquad (9.44)$$

provides a direct measure of the heat capacity ratio, $\gamma = C_P/C_V$.

9.2 Glass transition

Earlier in this chapter, we commented on the key difference between a liquid and a glass: a liquid possesses translational motion in which the particles are free to wander around, while the glass does not. A glass is merely the solid that forms when a liquid is sufficiently cooled without crystallization. Structurally, both the liquid and the glass appear the same in terms of average particle arrangements. Each is similarly disordered and described by a similar radial distribution function or static structure factor. The difference is that, in the liquid, the structure is continuously evolving on some characteristic time scale, τ, and by a series of *structural relaxations*, it traverses through multiple (thermodynamically equivalent) configurations.

Since the static structure is not changing, what then signals the transition from a liquid to a solid? As it happens there are several measurable properties of the liquid that exhibit changes near the glass transition temperature, T_g. One of these is the specific volume, illustrated in Fig. 9.5, whose temperature-

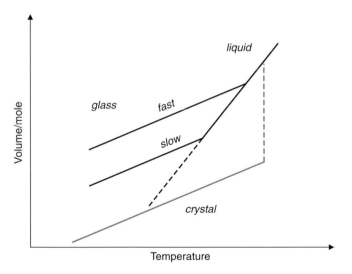

Figure 9.5 The temperature dependence of the specific volume for a glass-forming liquid. Depending on the cooling rate, a temperature is reached (the glass transition temperature) at which the thermal expansivity changes abruptly, signaling the transition from a liquid to a solid.

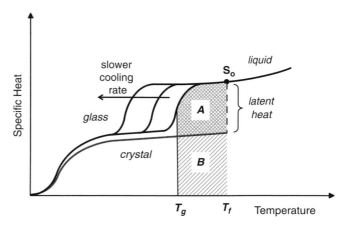

Figure 9.6
An illustration of the temperature variation of the specific heat for both glass formation and crystallization. For vitrification, the entropy loss corresponds to area $A+B$. For crystallization, the entropy loss corresponds to area B plus a contribution, L_f/T_f, due to latent heat of fusion. The Kauzmann temperature is reached when the entropy corresponding to area A equals that due to latent heat, L_f/T_f.

dependent slope (proportional to the thermal expansivity) is steeper above T_g than below. Above T_g, a portion of the expansivity is associated with the translational motions of the fluid. But below T_g, this translational motion is lost as a consequence of solidification and the expansivity mimics that of a crystalline solid for which only vibrational motions remain.

Another feature of the transition is its dependence on the rate at which the liquid is being cooled. As illustrated in Fig. 9.5, the transition occurs at a higher temperature when cooled rapidly, but at a lower temperature when cooled more slowly. This cooling rate feature is referred to by some rheologists as a *Deborah effect*, and is what makes the glass transition notoriously ambiguous. Because the transition involves no spontaneous change in structure, it inherently marks a temperature at which translational motions cease. However, there are many processes in nature, like the flow of glaciers in geology, which actually flow despite our human limitations to observe that flow. Who is to say when flow actually ceases? In the Deborah effect, the structural relaxation time (τ) of the fluid itself establishes a condition for observational time scales. If properties of the system are being observed on time scales shorter than τ, the properties will resemble that of a solid. Conversely, if the properties are observed over time scales longer than τ, the liquid behavior will appear.

9.2.1 Kauzmann paradox

The rate at which we cool the liquid thus establishes an observational time scale for the transfer of heat into and out of the liquid. This then has important consequences for the specific heat of the system. The temperature-dependent variation of the specific heat, C_p, is illustrated in Fig. 9.6 and, unsurprisingly, shows a cooling rate dependence. Above T_g, C_p exceeds that of the corresponding ordered solid as a result of the accessibility of translational degrees of freedom, in addition to vibrational degrees of freedom. Near the transition, the specific heat decreases rather abruptly to values comparable to (but slightly higher than) that of the crystal, as these translational degrees of freedom become inaccessible for the given observational time scale employed.

Consider now the evolution of the entropy of both the liquid and crystal phases as they are cooled below the melting point, T_f. As each is cooled, it loses entropy in amounts that are proportional to the area under each specific heat curve in Fig. 9.6. For example, the entropy of the liquid at any temperature $T < T_f$ can be expressed as

$$S_{\text{liq}}(T) = S_o - \int_T^{T_f} \frac{C_{p,\text{liq}}}{T} \, dT, \tag{9.45}$$

where S_o is the entropy of the liquid just above T_f. Because the specific heat of the liquid is larger than that of the crystal, it loses entropy more rapidly with cooling. However, the crystal loses a large chunk of entropy at T_f before cooling proceeds. This abrupt entropy loss is due to the ordering that occurs when the crystal state forms, and is given by $S_f = L_f/T_f$, where L_f is the latent heat of fusion. Thus for the crystal, the entropy at any temperature $T < T_f$ is

$$S_{\text{cryst}}(T) = S_o - \frac{L_f}{T_f} - \int_T^{T_f} \frac{C_{p,\text{cryst}}}{T} \, dT. \tag{9.46}$$

Kauzmann (1948) was the first to point out a predicament associated with the glass transition. He considered the evolution, with cooling, of the difference between the entropy of the liquid and that of the crystal,

$$\Delta S_{\text{excess}}(T) \equiv S_{\text{liq}}(T) - S_{\text{cryst}}(T) = \frac{L_f}{T_f} - \int_T^{T_f} \frac{[C_{p,\text{liq}} - C_{p,\text{cryst}}]}{T} \, dT, \tag{9.47}$$

and recognized that the value of the integral, corresponding to the cross-hatched area labeled A in Fig. 9.6, could be made arbitrarily large by extending the glass transition to lower temperatures by more gradual cooling rates. In fact, at some temperature T_K where

$$\frac{L_f}{T_f} = \int_{T_K}^{T_f} \frac{[C_{p,\text{liq}} - C_{p,\text{cryst}}]}{T} \, dT, \tag{9.48}$$

the entropy of the liquid would decrease below that of the crystal, in contradiction to traditional interpretations of entropy as a measure of disorder, embodied in the second law of thermodynamics.

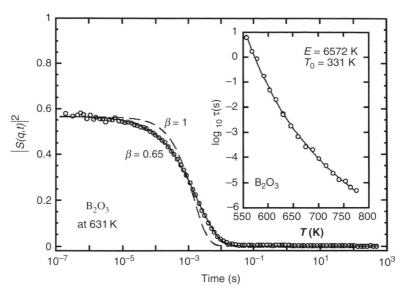

Figure 9.7 The dynamic structure factor for the structural relaxation in a molten oxide glass, B_2O_3, at 631 K. The solid curve visible beneath the data points is a fit using the stretched exponential (Eq. (9.50)) and the dashed curve is a regular exponential ($\beta = 1$) for comparison. The inset figure shows how the relaxation time increases dramatically with decreasing temperature and is well represented by Eq. (9.49) with $E = 6572$ K and $T_o = 331$ K.

The resolution to this apparent paradox lies in how the structural relaxation time of the liquid increases rapidly with cooling. In many instances, the relaxation time is proportional to the viscosity and near T_g, increases approximately as

$$\tau \propto \eta \approx \eta_o \exp\{E/(T - T_o)\}. \tag{9.49}$$

Numerous experimental studies have demonstrated a strong correlation between the divergence point T_o and the Kauzmann temperature T_K. Thus, it appears that the entropy paradox outlined above is narrowly avoided as no observational time scale exists for this entropy catastrophe to be observed.

9.2.2 Structural relaxation

What happens to $S(q, \omega)$ near the glass transition? With cooling, the Brillouin lines first broaden (due to the increasing viscosity of the liquid) but then narrow near T_g and shift to higher frequency as the medium begins to appear elastic to propagating sound waves. At these temperatures it is also possible to observe the emergence of an additional pair of Brillouin lines corresponding to the propagation of transverse waves. As for the central Rayleigh line, it is gradually encroached upon by a new, *non-hydrodynamic* (independent of q for $q^{-1} \gg$ particle spacings) mode whose linewidth narrows as the liquid

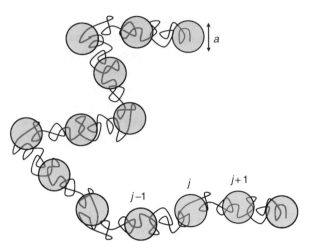

Figure 9.9 The Rouse "bead and spring" model for a polymer coil. A single coil is partitioned into alternating Rouse segments of size a. One segment (spring) represents the entropic forces that result when the coil is stretched. The other segment (bead) represents the viscous drag experienced by the coil when displaced in a fluid of other coils.

forces discussed above, along with the random Brownian forces needed to disturb it in the first place:

$$
\begin{aligned}
m\frac{d^2\Delta\vec{r}_j}{dt^2} &= \vec{f}_{j+1} + \vec{f}_{j-1} - \zeta\frac{d\Delta\vec{r}_j}{dt} + \vec{f}_{\text{random}} \\
&= C(\Delta\vec{r}_{j+1} - \Delta\vec{r}_j) - C(\Delta\vec{r}_j - \Delta\vec{r}_{j-1}) - \zeta\frac{d\Delta\vec{r}_j}{dt} + \vec{f}_{\text{random}}.
\end{aligned}
\tag{9.56}
$$

The beads of the chain are sufficiently damped that the inertial term in Eq. (9.56) can be neglected. Also, because the Brownian forces are random, their time averaged effect vanishes and so the displacement of the jth bead is given by,

$$
\frac{d\Delta\vec{r}_j}{dt} = \frac{C}{\zeta}(\Delta\vec{r}_{j+1} + \Delta\vec{r}_{j-1} - 2\Delta\vec{r}_j).
\tag{9.57}
$$

This vector equation corresponds to three independent equations, one for each direction (x, y and z) in space, that have a similar form,

$$
\frac{d\Delta x_j}{dt} = \frac{C}{\zeta}(\Delta x_{j+1} + \Delta x_{j-1} - 2\Delta x_j),
\tag{9.58}
$$

and similar wave-like solutions of the form,

$$
\Delta x_j(t) = \sum_{\text{modes},\,K} \left[x_{o,K}\cos(j\phi_K) \right] \exp(-t/\tau_K),
\tag{9.59}
$$

where

$$
\tau_K^{-1} = \frac{2C}{\zeta}(1 - \cos\phi_K) = \frac{4C}{\zeta}\sin^2(\phi_K/2).
\tag{9.60}
$$

(a)   (b)

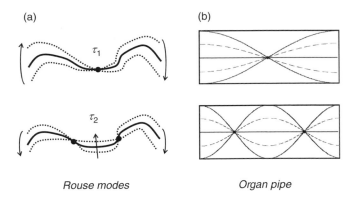

Rouse modes   Organ pipe

Figure 9.10   (a) An illustration of the first two lowest frequency Rouse modes for a polymer coil and (b) the corresponding standing wave patterns for sound in an open organ pipe, illustrating the equivalent number of nodes.

What does this solution mean? In Eq. (9.59) we see that the initial displacement of a bead is given by the bracketed term summed over a limited set of allowed *Rouse modes* that are consistent with the boundary conditions of the chain. Because the ends of the chain are unrestricted, these allowed modes (labeled by K) assume the same set of wavelengths as the standing wave patterns in an organ pipe (see Fig. 9.10), such that the relative phase angle between adjacent beads is

$$\phi_K = \frac{\pi}{(M-1)}K, \quad K = 1, 2, 3, \ldots, (M-1). \tag{9.61}$$

In Eq. (9.59) we see that the initial deformation of the chain is some linear combination of these allowed modes and that the deformation associated with a given mode decays back to equilibrium with a characteristic relaxation rate, given by Eq. (9.60). The overall relaxation of a Rouse chain is then ultimately limited by the relaxation of its slowest, fundamental mode for which $\phi_1 = \pi/(M-1)$. The relaxation time of this fundamental mode is

$$\tau_1^{-1} = \frac{4C}{\zeta}\sin^2(\pi/2(M-1)) \approx \frac{C}{\zeta}\left(\frac{\pi}{M-1}\right)^2 \quad \Rightarrow \quad \tau_1 \propto N^2, \tag{9.62}$$

and is seen to be proportional to the square of the number of monomers that make up the polymer molecule.

9.3.2 Reptation

The fundamental Rouse mode suggests a quadratic dependence of the viscoelastic relaxation time on the chain length or molecular weight of the polymer that is indeed seen for measurements conducted on polymers of varying molecular weight, as shown in Fig. 9.11. However, this success is restricted

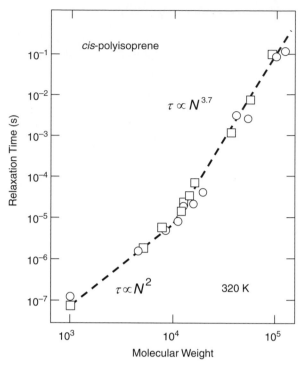

Figure 9.11 Structural relaxation time for a polymer melt of *cis*-polyisoprene obtained by dielectric spectroscopy (adapted from Boese and Kremer (1990).) At low molecular weight, the relaxation time exhibits a quadratic dependence consistent with the excitation of Rouse modes (see Eq. (9.62).) However, for long polymer chains with a molecular weight greater than 10^4, Rouse modes are suppressed and the relaxation is controlled by reptation.

to low molecular weights. Above some critical chain length, the relaxation time increases more dramatically and measurements favor

$$\tau \propto N^{\nu}, \quad 3.3 < \nu < 3.7. \tag{9.63}$$

It appears then that the simple Rouse model breaks down at larger molecular weights. Why is this? One thing that is missing from the Rouse model is the influence of other chains in the melt whose entanglements place additional constraints on the possible configurations allowed for any given polymer coil. At sufficiently high molecular weight, these entanglements will dominate and restrict the motion of a coil to that along a tube, as illustrated in Fig. 9.12. The contour of this tube matches the contour of the coil itself and represents the numerous entanglements produced by other polymer strands in the melt. The tube limits the lateral motions of the chain thus suppressing the wave-like Rouse modes. Instead, the polymer chain is forced to move along the contour of the tube by slithering, much in the fashion of a snake or other reptilian creature. For this reason, this new mode

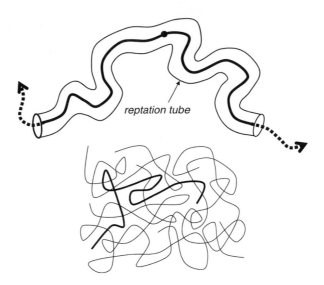

Figure 9.12 In the reptation model, motion of the polymer coil appears as a one-dimensional random walk restricted to the contour of the polymer coil itself. The restriction arises from entanglements of other neighboring polymer chains (illustrated in lower figure) which successfully suppress Rouse modes at high molecular weights.

of polymer motion is commonly referred to as *reptation* and is described by the *reptation model*.

In the reptation model, we inquire as to the characteristic time it takes a polymer chain to diffuse the entire length of its tube and thereby escape. This motion is effectively one-dimensional as it occurs along the contour of a tube of length $l = Ma$. Consider the motion of the central bead of the chain. We can think of its motion as that of a one-dimensional random walk along the tube whose mean-squared displacement after a time t was given earlier as $\langle x^2 \rangle = 2Dt$. The diffusion coefficient for this single bead is however significantly larger as it must pull and push all the other beads along with it. From the Einstein relation (Eq. (9.19)), the diffusion coefficient is

$$D = \frac{k_B T}{\zeta_{\text{chain}}} \propto \frac{k_B T}{M\zeta}, \tag{9.64}$$

and so the characteristic escape time – the time required for the mean-squared displacement to match the tube length – is

$$\tau_{\text{escape}} \approx \frac{\langle x^2 \rangle}{2D} \approx \frac{l^2_{\text{tube}}}{2D} \approx \frac{M^2 a^2}{2k_B T/M\zeta} \approx M^3 \left(\frac{a^2}{2k_B T/\zeta} \right) \propto N^3. \tag{9.65}$$

While the reptation model fails to predict the precise molecular weight dependence seen experimentally, it does provide approximately the correct dependence and, given its simplicity, provides a starting point for more advanced approaches.

Summary

- The dynamic structure factor is a time-dependent corollary to the static structure factor and is related by a Fourier transform to the van Hove space and time correlation function.

- Macromolecules in a solvent execute a diffusional motion known as Brownian motion which can be modeled by a random walk.

- The Rayleigh–Brillouin spectrum of a liquid results from density fluctuations. The central Rayleigh line is caused by non-propagating thermal diffusion. The two Brillouin lines are caused by propagating sound waves.

- The glass transition occurs when translational motion in a liquid is arrested. The transition point occurs when the structural relaxation time exceeds that of the observer.

- Low molecular weight polymers exhibit dynamics that are adequately described by the Rouse model. At higher molecular weights entanglements begin to dominate the dynamics resulting in reptational motion.

Exercises

9.1. Show how Eq. (9.12) arises from Eq. (9.11).

9.2. Show how Eq. (9.16) follows from Eq. (9.15). (Hint: the dot product $\vec{q} \cdot \vec{r}$ can be an impediment so you may want to work this out in Cartesian coordinates. Be prepared to integrate by parts a couple of times.)

9.3. Verify that Eq. (9.35) is indeed the solution to Eq. (9.33).

9.4. Discuss what would happen if the polymer coils of the Rouse model were grafted onto a substrate. That is, one end of each coil was attached to a rigid surface. How might this affect the Rouse modes? Determine the new fundamental Rouse mode for this situation.

9.5. Consider the results of the dynamic light scattering experiment shown in Fig. 9.3, in which the diffusivity of spherical particles in water was obtained at room temperature (20 °C). The manufacturer specifies that the spheres have a diameter of 109 nm. What diameter is obtained from the measurement? (Note: you will need to locate the viscosity of water elsewhere.)

9.6. Shown in Fig. 9.13 are measurements of the longitudinal Brillouin lines for a glassforming liquid of refractive index $n \approx 1.6$ at several temperatures (in K). The data were collected using 514 nm light collected in a

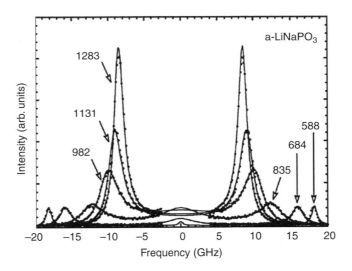

backscattered direction. Determine the longitudinal speed of sound for each set of Brillouin doublets and plot the result against temperature. Is there a discontinuity and, if so, why?

9.7. Consider a "biased" version of the 1D random walk problem for which Eq. (9.10) would be replaced by $G(x,t) = G(x-b, t-\tau) \times R + G(x+b, t-\tau) \times L$, where R and L are the (unequal) probabilities that a walker will select to move right or left, respectively. For the biased walk, we assume bias is in the forward direction, so that $R > L$. (a) Show that to the same order as in Eq. (9.11), the biased walk obeys the following relation: $\frac{\partial G}{\partial t} = -u\frac{\partial G}{\partial x} + D\frac{\partial^2 G}{\partial x^2}$, where $u = (R-L)b/\tau$ is the drift velocity. (b) Show also that the solution to this relation (for a particle starting at the origin) is given by $G(x,t) = \frac{1}{\sqrt{4\pi Dt}}e^{-(x-ut)^2/4Dt}$, and represents the same spreading as illustrated in Fig. (9.2), only in which the center of the distribution is displaced uniformly in time.

9.8. The pieces of glass found in windows of medieval cathedrals are typically thicker at the bottom than at the top and many have speculated that this is a consequence of the gradual flow of the glass (as an equilibrium liquid) over the past 800 years. Modern experiments with glasses produced using medieval compositions show temperature-dependent viscosity as in Eq. (9.49) with $\eta_o = 6.3 \times 10^{-5}$ Pas, $E = 12574$ K, and $T_o = 470$ K. Given that the structural relaxation time is approximately $\langle \tau \rangle = \eta/G_\infty$, where the high frequency shear modulus, G_∞, is typically 30 GPa, estimate the temperature at which a medieval glass would need to be maintained in order for it to exhibit any significant flow in 800 years. What does this tell you regarding the speculation that medieval glass has been flowing over the centuries? Explain your answer.

9.9. Derive the Einstein relation of Eq. (9.19) by considering the ensemble-averaged dynamics of a Brownian particle of mass m whose equation of motion is given by $m\frac{d^2\vec{r}}{dt^2} = \vec{f}_{\text{rand}} - \zeta\frac{d\vec{r}}{dt}$, where $\vec{f}_{\text{rand}}$ is a random force provided by the solvent molecules. Hint: you might begin by verifying the following two identities: $\frac{dr^2}{dt} = \frac{d}{dt}(\vec{r}\cdot\vec{r}) = 2(\vec{r}\cdot\frac{d\vec{r}}{dt})$ and $\vec{r}\cdot\frac{d^2\vec{r}}{dt^2} = \frac{d}{dt}(\vec{r}\cdot\frac{d\vec{r}}{dt}) - (\frac{d\vec{r}}{dt})^2$.

Suggested reading

Those interested in the true derivation of the Rayleigh–Brillouin spectrum are directed to the first three texts.

J. P. Hansen and I. R. McDonald, *Theory of Simple Liquids*, 2nd Ed. (Academic Press, New York, 1986).

P. M. Chaikin and T. C. Lubensky, *Principles of Condensed Matter Physics* (Cambridge University Press, New York, 2003).

B. J. Berne and R. Pecora, *Dynamic Light Scattering* (John Wiley and Sons, New York, 1976).

J. Zarzycki, *Glasses and the Vitreous State* (Cambridge University Press, New York, 1991).

G. Strobl, *Condensed Matter Physics* (Springer-Verlag, Berlin, 2004).

W. Kauzmann, Chem. *Rev.* **43**(2), 219 (1948).

B. J. Berne and R. Pecora. *Dynamic Light Scattering* (John Wiley and Sons, New York, 1976).

10

Crystal vibrations

Introduction

In the last chapter, we investigated the dynamics of liquids whose particles are free to wander about due to the reasonably weak level of inter-particle bonding. In a solid (crystal or glass), bonding between particles is stronger and the translational motion of the particles is arrested. Nevertheless, these "solid" particles continue to move and execute small, localized vibrations about a fixed point in space. In this chapter and the next, we investigate the nature of this vibrational motion and its impact on the thermal properties of a solid. Here we begin by considering a simple model of masses connected by ideal springs to demonstrate how vibrations of individual atoms are, in reality, a consequence of propagating waves traveling through the crystal lattice. In order to connect these waves with the quantum mechanical perspective of each atom behaving as a quantized harmonic oscillator, we find ourselves introducing the concept of a quantum of elastic wave, known as a *phonon*.

An important outcome of our development of quantized elastic waves is a growing appreciation for a special region of reciprocal space known as the *Brillouin zone*, which is populated by all the wave vectors, K, corresponding to allowed phonon waves in the crystal. For phonons whose K matches the edge of this zone, significant Bragg scattering results, to produce two equivalent standing wave patterns separated by an energy gap. We will revisit the Brillouin zone often in the chapters to come, and we will begin to appreciate the significance of this boundary for the motion of all waves that attempt to travel within a crystal.

10.1 Monatomic basis

A crystal is classified as a solid, but this does not imply that it is devoid of motion. At a microscopic level the atoms of the crystal lattice vibrate, and these vibrations ultimately control the thermodynamic properties of the solid. As discussed in Chapter 3, the bonding between any two particles is described by a pairwise interaction, $u(r_{ij})$, that contains

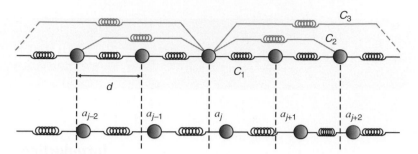

Figure 10.1   Elastic model of a one-dimensional monatomic crystal consisting of particles of mass m separated by springs of spring constant C_1. The upper string shows masses at their equilibrium positions. The lower string shows masses when they are displaced from equilibrium. Also, shown as weaker springs, are possible higher-order interactions between next-nearest neighbors and beyond.

both attractive and repulsive interactions. As illustrated for a 1D lattice in Fig. 10.1, each particle favors a point of stable equilibrium where no net force is experienced. When displaced from this point of equilibrium by $\vec{a}$, the particle experiences a restoring force, which for sufficiently small displacements is well-approximated by *Hooke's law* for ideal springs,

$$\vec{F}(\vec{a}) \approx -C_1\vec{a}, \qquad (10.1)$$

with a spring constant C_1 that generally depends upon the direction of displacement (i.e. longitudinal or transverse). In principle, there may be spring-like interactions between other than nearest-neighbor particles and these are then accommodated by higher-order spring connections ($C_2, C_3, \ldots$ etc.). Because these higher-order springs are presumably much weaker, we can ignore them for the time being.

We begin our analysis of vibrational motions by considering the 1D crystal, shown in Fig. 10.1, of identical particles each of mass m interacting via interconnecting springs. Although the figure suggests the case of longitudinal displacements, the analysis presented below applies equally well to transverse displacements. Since the 1D crystal contains an extended collection of coupled oscillators, it should come as no surprise that any given particle will not oscillate completely independent of its neighbors, but that the collection of particles will oscillate in some concerted manner giving rise to *waves*, such that the displacement of a particle at x is

$$a(x,t) = A \exp[i(Kx - \omega t)]. \qquad (10.2)$$

The motion of a particle at x is thus governed by the angular frequency ω, which in turn depends upon the wavelength (or wave vector $K = 2\pi/\lambda$) of the wave. Waves in a continuum travel with a fixed speed v for which $\omega = vK$. However, in the case of a discrete chain of particles, this relationship between ω and K will eventually fail as the wavelength shortens towards that of the spacing between the particles, where the continuum picture breaks down. In the following, we seek to develop the relationship between ω and K, known as a *dispersion relation*, which governs these waves in a discrete lattice.

Suppose a plane wave (see Eq. (10.2)) is propagating through the crystal along the x-axis. How is this traveling wave constrained by the elastic forces at a given location? Because the equilibrium positions of the particles are arranged in an orderly manner, we can label each by the index j and express the spatial phase in terms of the lattice spacing, d, such that the displacement of the jth particle from its equilibrium is:

$$a(x = jd, t) = a_j(t) = A \exp[i(Kjd - \omega t)]. \tag{10.3}$$

To obtain an equation of motion for the jth particle, we need to determine the forces acting upon it. There are two forces: one arising from the spring to the left (which connects the particle to the $(j-1)$th particle) and the spring to the right (which connects the particle to the $(j+1)$th particle). In each, the force is proportional to the net extension, Δl, of the spring. To the left, the force is given (see Fig. 10.1) by

$$F_j^{j-1} = -C_1 \Delta l_- = -C_1 \big(a_j(t) - a_{j-1}(t) \big), \tag{10.4}$$

and to the right, the force is given by

$$F_j^{j+1} = C_1 \Delta l_+ = C_1 \big(a_{j+1}(t) - a_j(t) \big). \tag{10.5}$$

Hence the net force on the jth particle is given by

$$F_j^{\text{net}} = C_1 \big[a_{j+1}(t) + a_{j-1}(t) - 2a_j(t) \big] = m \frac{d^2 a_j(t)}{dt^2} = -m\omega^2 a_j(t), \tag{10.6}$$

where Newton's second law has been included. We now choose solutions of the form:

$$\begin{aligned} a_{j-1}(t) &= A \exp[i(K(j-1)d - \omega t)] = a_j(t) e^{-iKd} \\ a_{j+1}(t) &= A \exp[i(K(j+1)d - \omega t)] = a_j(t) e^{+iKd}, \end{aligned} \tag{10.7}$$

and obtain from Eq. (10.6),

$$C_1 \big[a_j(t) e^{+iKd} + a_j(t) e^{-iKd} - 2a_j(t) \big] = -m\omega^2 a_j(t). \tag{10.8}$$

This can be rearranged to obtain

$$\omega^2 = -\frac{C_1}{m} \big[e^{+iKd} + e^{-iKd} - 2 \big] = \frac{4C_1}{m} \sin^2(Kd/2), \tag{10.9}$$

which relates the frequency of the wave to its wavelength and is thus the dispersion relation we desire.

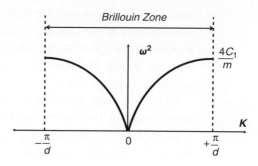

Figure 10.2 Dispersion curve (Eq. (10.9)) for the one-dimensional monatomic crystal with the limits imposed by the first Brillouin zone.

10.1.1 Dispersion relation

Let us now examine some of the features of this dispersion relation (plotted in Fig. 10.2) in closer detail. We begin in the long wavelength limit (near $K = 0$), where Eq. (10.9) can be expanded for small argument as:

$$\omega \approx 2\sqrt{\frac{C_1}{m}}\frac{Kd}{2} = \left[\sqrt{\frac{C_1 d}{m/d}}\right] K. \tag{10.10}$$

The quantity in brackets has been arranged on purpose to illuminate the similarity to that for the speed of a wave on a string,

$$\upsilon_o = \sqrt{\frac{\text{Tension}}{\text{mass/length}}} = \sqrt{\frac{T}{\mu}} = \sqrt{\frac{C_1 d}{m/d}}, \tag{10.11}$$

where $T = C_1 d$ is the tension, and $\mu = m/d$ is the mass per unit length. Thus we find, in the long wavelength limit, the waves in a crystal propagate like acoustical waves in a continuum, for which

$$\upsilon_o = \frac{\omega}{K} = f\lambda. \tag{10.12}$$

In the short wavelength limit, we see from Fig. 10.2 that the frequency approaches a limiting constant value, $\omega_{\text{max}} = 2\sqrt{\frac{C_1}{m}}$, as K approaches values of $\pm\pi/d$. Clearly there is an upper limit to the frequency associated with this special value of K. To appreciate this, consider the *group velocity* of the wave given by the slope of Fig. 10.2, or from a derivative of Eq. (10.9):

$$\upsilon_{\text{group}} = \frac{d\omega}{dK} = \upsilon_o \cos(Kd/2). \tag{10.13}$$

The group velocity is a measure of energy propagation and is seen to vanish at $K = \pm\pi/d$. What is happening here? Apparently, the energy carried by elastic waves is no longer moving from one location to another, but rather staying in place. Indeed, $K = \pm\pi/d$ corresponds to $\lambda = 2d$, so adjacent particles are moving exactly out of phase with one another to produce the appearance of a non-propagating, standing wave.

10.1.2 Brillouin zone

Extension of the above analysis to higher dimensions comes with several complications. Bonds along different crystal directions generally have different spring constants, and waves traveling along any direction will not necessarily conform to the 1D treatment above. However, for waves traveling along the [100], [110] and [111] directions in the simple cubic system, the particles in the corresponding planes move in unison and can be treated one-dimensionally with d now corresponding to the spacing between the planes.

Additionally, in the three-dimensional situation, the unique wave number $K = \pm\pi/d$, where the elastic wave assumes a standing wave behavior, corresponds to the edge of what is referred to as the first *Brillouin zone*. For elastic waves, this edge condition corresponds to the smallest *meaningful* wavelength for an elastic wave; i.e. a wavelength in which each adjacent atom moves exactly out of phase with its neighbor. Since the lattice consists of discrete atoms, there is no possibility for any elastic wave of shorter wavelength, as that would imply that there is some mythical mass present, *in between* the particles, that is oscillating. In reality, no such mass exists. This is quite different from electromagnetic waves traveling in the crystal, whose wavelength is not limited by these discrete pieces of mass.

Bragg diffraction and the Brillouin zone

There is, nevertheless, an interesting significance about the edge of the first Brillouin zone which plays a role for both elastic waves and electromagnetic waves (as well as other waves) traveling through the lattice. The edge marks a sort of "resonance" between the wave and the periodic structure of the crystal, wherein the conditions exist for strong constructive interference to occur. To see this, let us return to the subject of Bragg scattering for a moment. In Chapter 6, we observed that one of the expressions (see Eq. (6.18)) for the Bragg scattering condition was that

$$-2\vec{k}_i \cdot \vec{G}_{hkl} = \left|\vec{G}_{hkl}\right|^2, \qquad (10.14)$$

or

$$-\vec{k}_i \cdot \left(\frac{\vec{G}_{hkl}}{2}\right) = \left|\frac{\vec{G}_{hkl}}{2}\right|^2. \qquad (10.15)$$

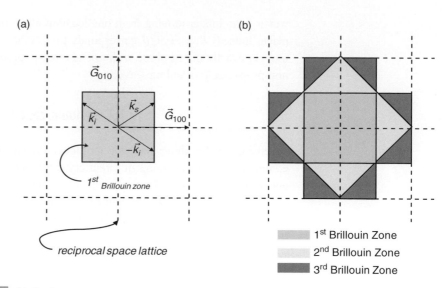

Figure 10.3 (a) The Bragg scattering condition shown vectorially for the first Brillouin zone. Bragg diffraction occurs only for wave vectors terminating on a Brillouin zone boundary, for which the projection of $-\vec{k}_i$ onto $\vec{G}_{hkl}$ equals half the reciprocal lattice vector. In these instances the scattered wave ($\vec{k}_s$) appears also to terminate at the Brillouin zone boundary. (b) Higher-order Brillouin zones obtained by successive Wigner–Seitz construction.

Consider the implication of this statement as it appears vectorially in the reciprocal space ($\vec{G}_{hkl}$), as illustrated in Fig. 10.3a. The condition for Bragg reflection, given by Eq. (10.15), indicates that the incident wave vector must have both a magnitude and direction such that its vector *terminates* on the edge of a Brillouin zone (so that the projection of $-\vec{k}_i$ with any $\vec{G}_{hkl}$ equals half the size of G_{hkl}). These special k "resonate" with the crystal structure in the sense that the corresponding waves will constructively interfere to produce strong Bragg scattering with a scattered wave vector $\vec{k}_s$ also terminating on the same Brillouin zone edge. Herein lies the resonant feature of the Brillouin zone. Multiple scattering occurs in which the scattered wave ($\vec{k}_s$) is itself scattered by the Brillouin zone back into the original incident wave $\vec{k}_i$ (which is scattered back into $\vec{k}_s$, ad infinitum). Moreover, you may recognize that the cell formed by a Brillouin zone boundary is a certain primitive cell that we have already encountered: it is in fact the *Wigner–Seitz* primitive cell (constructed not in real space, but in the reciprocal space defined by the reciprocal lattice). In addition to the first Brillouin zone, there exist also higher-order Brillouin zones (2nd, 3rd, etc.) shown in Fig. 10.3b, each with equal volume. These higher-order zones pertain to wave numbers in excess of those that are meaningful for elastic waves, but which are pertinent to other waves (e.g. electromagnetic) for which wavelengths shorter than *2d* remain meaningful.

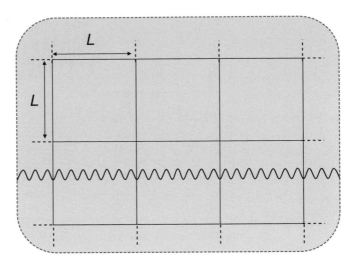

Figure 10.4   Periodic boundary conditions. An infinite crystal is partitioned into equivalent crystals of realistic size L. For each crystal to contain an equivalent set of discrete modes, the elastic waves must match at the partition boundary.

10.1.3 Boundary conditions and allowed modes

In our development of the dispersion relation for a 1D monatomic crystal, we have assumed that the crystal extended left and right indefinitely. But in real crystals, this is not the case and the presence of a finite boundary places additional constraints on wavelengths allowed to propagate. For example, if the particles at either end of the chain were rigidly fixed, the waves would be limited to the same discrete set of wavelengths, corresponding to the familiar standing waves of a string, $\lambda_n = 2L/n$, where L is the length. More generally, these end atoms are not fixed, but the finite size of the crystal nevertheless induces restrictions in the allowed wavelengths, or allowed "modes", of wave vector K. To establish the restricted set of allowed modes in a finite crystal of size $L=Nd$, imagine firstly an infinite-sized crystal that is partitioned into (laboratory-sized) pieces each of size L, as illustrated for a 2D crystal in Fig. 10.4. Each partitioned sub-crystal should be equivalent to any other and so the waves that travel from one into the next must be restricted so that each wave arrives identically into each partitioned crystal. Hence,

$$a_j(t) = A\exp[i(Kjd - \omega t)] = a_{j+L}(t) = A\exp[i(K(j+N)d - \omega t)], \quad (10.16)$$

or,

$$\exp(iKNd) = 1. \quad (10.17)$$

For this condition (known as a *periodic boundary condition*) to be satisfied, the allowed modes are restricted to those,

$$K_n = \pm\frac{2\pi}{Nd}, \pm\frac{4\pi}{Nd}, \pm\frac{6\pi}{Nd}, \cdots \pm\frac{\pi}{d} = \pm n\frac{2\pi}{Nd}, \quad n = 1, 2, 3, \ldots N/2,$$

$$(10.18)$$

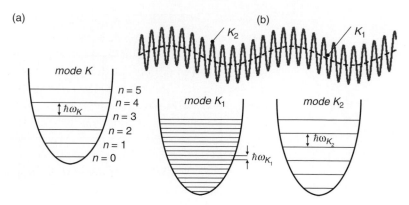

(a)

Figure 10.5 (a) Energy levels of a quantum mechanical harmonic oscillator. The oscillator frequency is determined from the dispersion relation (Eq. (10.9)) for the corresponding wave vector, K. (b) Any elastic wave in a crystal can be decomposed into N^3 particles, vibrating (synchronously) in the manner of a combination of quantized harmonic oscillators, one for each allowed wave vector (i.e. mode). The figure illustrates how an elastic wave, formed by the superpositioning of two modes, is formed by the combined excitation of two corresponding harmonic oscillators.

for which Eq. (10.17) is satisfied. Note that for the 1D crystal, the total number of allowed modes just equals the number of particles, N, in a chain of length $L = Nd$. For a 3D crystal, the allowed modes in each direction are identically restricted so that a crystal of size $L^3 = N^3d^3$ has exactly as many allowed modes as it has particles.

10.1.4 Phonons

As discussed above, the finite extent of the crystal sets restrictions on the allowed wavelengths of elastic waves. Consequently, the frequencies of the waves described by the dispersion relation (Eq. 10.9) are similarly restricted to a limited set of allowed values. What does this mean for the displacements of any given particle? Suppose just one mode of wave number K happens to be active in the crystal. Then, each particle is oscillating at the corresponding frequency, ω_K, given by the dispersion relation for that particular mode, albeit with a differing phase from its neighbor. Individually, each particle acts as a quantum mechanical *harmonic oscillator* and additional energy added must appear in the form of quanta of an amount $\hbar\omega_K$, as illustrated by the energy level diagram in Fig. 10.5a, where the energy levels are given by

$$E_{n_K}^{(K)} = \left(n_K + \frac{1}{2}\right)\hbar\omega_K. \tag{10.19}$$

Collectively, however, the oscillators function as an elastic wave whose amplitude increases when all of these individual harmonic oscillators are simultaneously excited to a higher energy level.

Now let us suppose that waves are present in the crystal at several allowed wave vectors. This is possible because waves superpose – multiple waves can occupy the same medium at the same time. As an illustration, Fig. 10.5b shows what two modes of very different K might look like when both are present. Individually, each particle contains a contribution to its total displacement due to these two modes, oscillates with each of two harmonic frequencies, and its energy is described by the tally of the energy levels of each component (see Fig. 10.5b),

$$E_{\text{Total}} = \sum_K E_{n_K}^{(K)} = \sum_K \left(n_K + \frac{1}{2} \right) \hbar \omega_K. \qquad (10.20)$$

Furthermore, any transition between energy levels (Δn_K) results in a change in the amplitude of the waves corresponding to that given mode.

It is customary, then, to dispense with the notion that a solid is just a collection of individual vibrating atoms and to replace this with the concept of the solid as a box containing very many quantized, propagating elastic waves. Each such elastic wave of a given K is then the result of n_K units of a fundamental quantum, known as a *phonon*, that has both energy,

$$E = \hbar \omega_K, \qquad (10.21)$$

and momentum,

$$\vec{p} = \frac{h}{\lambda} = \hbar \vec{K}. \qquad (10.22)$$

The phonon acts as a *boson* particle (like the photon) and thus several phonons can simultaneously occupy the same quantum state. In this way the limited set of allowed macroscopic waves in the crystal can be viewed as a constructive superposition of very many identical phonons. The implications of this phonon picture will become more apparent when we examine their roles in various scattering processes, in a later section.

Magnons: magnetic spin waves

This is a good spot to discuss a related type of wave-like excitation that appears in ferromagnetic materials, known as a *spin wave*. Recall from Chapter 4, that in a ferromagnetic material the magnetic moments (i.e. spins) are aligned in a common direction: the direction of the net magnetization. This alignment is a consequence of the exchange interaction (see Eq. (4.27),

$$u(r_{ij}) = -2J_{ex}(r_{ij}) \, \vec{S}_i \cdot \vec{S}_j,$$

acting between neighboring spins, which results in a lowering of the energy when spins are aligned.

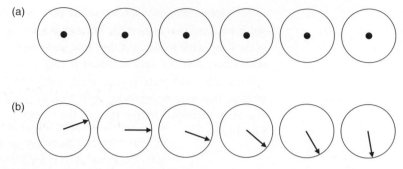

Fig. 10.6 (a) A chain of magnetic spins in the ground state as viewed from above (looking down on the arrowheads). (b) In a spin wave, spins precess about the ground state orientation at a common rate, ω, but with a small difference in phase.

The state in which all the spins are aligned is then the lowest energy state (ground state) for the system of spins. But, small disturbances of the spin orientation can develop in the form of propagating waves. As illustrated in Fig. 10.6a, each spin of the wave precesses about the magnetization direction at a common frequency ω, but with a slightly different phase, ϕ, so as to achieve the appearance of a wave. These spin waves are similar to phonons in that both represent small displacements from equilibrium that are countered by a restoring force. Like phonons, spin waves are also quantized and the quantum of a spin wave is known as a *magnon*.

However, there are important differences. Because the interactions that produce restoring forces are different in each situation, the dispersion relation of a spin wave is quite unlike that for phonons. To obtain this dispersion relation, we approach the problem much as we did before for elastic waves, by considering just a single jth spin. Here we consider how the energy of the spin is altered from its ground state energy when it is oriented by a small angle ϕ relative to its neighbors, as illustrated in Fig. 10.6b. In the ground state configuration, this single spin experiences energy,

$$E_j^o = -2J_{ex}S^2 - 2J_{ex}S^2 = -4J_{ex}S^2,$$

due to interactions with each of two neighboring spins. In the disturbed case, the energy increases (becomes less negative) to

$$E_j = -2J_{ex}S^2 \cos\phi - 2J_{ex}S^2 \cos\phi = -4J_{ex}S^2 \cos\phi.$$

Thus the energy increases by an amount,

$$\Delta E = E_j - E_j^o = \hbar\omega = 4J_{ex}S^2(1 - \cos\phi) = 8J_{ex}S^2\sin^2(\phi/2),$$

and the dispersion relation is

$$\omega = \frac{8J_{ex}S^2}{\hbar}\sin^2(Kd/2). \tag{10.23}$$

Here we see that long wavelength spin waves exhibit a quadratic dependence on K that is much unlike that of acoustic phonons.

10.2 Diatomic basis

We now consider a crystal with a basis containing two dissimilar atoms, one with a small mass m and the other with a larger mass M. Why? Because this two-basis situation is present in many ionic crystals (e.g. NaCl, CsCl) and, as we will see, gives rise to an additional band of phonons that are not present in the monatomic case. Again, we treat the simple cubic lattice and will assume that propagation is along one of the lattice directions for which alternating planes contain either one atom or the other (but not both). Such would be the case of the [100] planes in the BCC structure of CsCl, or the [111] planes of the FCC structure of NaCl. Furthermore, we assume all the spring constants are equivalent between the planes. This is justified for an ionic crystal, as each ion has an equivalent valence and interacts via a Coulomb force that depends only on the charge. A series of planes is illustrated in Fig. 10.7 in which various displacements of the atoms are indicated.

As before, we seek an equation of motion for each plane of atoms. We begin with the smaller atoms of mass m present in the jth plane, which undergo a displacement $a_j(t)$. The net force on this plane due to the plane behind and the plane in front is

$$F_j^m = -C_1(a_j - b_j) + C_1(b_{j+1} - a_j)$$
$$= C_1(b_{j+1} + b_j - 2a_j) = m\frac{d^2 a_j}{dt^2}. \qquad (10.24)$$

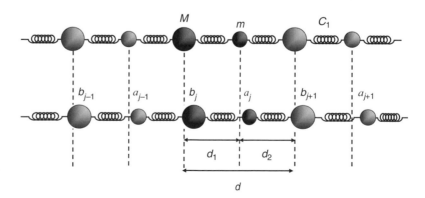

Figure 10.7 Elastic model of a one-dimensional diatomic crystal, consisting of particles of small mass m and large mass M separated by springs of spring constant C_1.

Similarly, the net force on the jth plane containing larger atoms of mass M is

$$F_j^M = -C_1(b_j - a_{j-1}) + C_1(a_j - b_j)$$
$$= C_1(a_j + a_{j-1} - 2b_j) = M\frac{d^2 b_j}{dt^2}, \qquad (10.25)$$

where $b_j(t)$ is the displacement of this plane of larger atoms from equilibrium.

Equations (10.24) and (10.25) represent two simultaneous equations of motion for the two planes. We again assume plane wave solutions,

$$a_j(t) = Ae^{i(Kjd - \omega t)}$$
$$b_j(t) = Be^{i(Kjd - \omega t)}$$
$$a_{j\pm 1}(t) = Ae^{i(K(j\pm 1)d - \omega t)} = a_j(t)e^{\pm iKd} \qquad (10.26)$$
$$b_{j\pm 1}(t) = Be^{i(K(j\pm 1)d - \omega t)} = b_j(t)e^{\pm iKd},$$

and after inserting these solutions, our two equations of motion become,

$$B\{C_1(1 + e^{+iKd})\} = A\{2C_1 - m\omega^2\}$$
$$A\{C_1(1 + e^{-iKd})\} = B\{2C_1 - M\omega^2\}. \qquad (10.27)$$

Substituting one into the other produces our desired dispersion relation,

$$C_1^2(1 + e^{-iKd})(1 + e^{+iKd}) = (2C_1 - m\omega^2)(2C_1 - M\omega^2), \qquad (10.28)$$

which can be arranged as a quadratic equation for ω^2 as,

$$\omega^4 - \frac{2C_1(m + M)}{mM}\omega^2 + \frac{2C_1^2}{mM}(1 - \cos Kd) = 0. \qquad (10.29)$$

The solutions are then,

$$\omega^2 = \frac{C_1(m + M)}{mM} \pm C_1\sqrt{\frac{(m + M)^2}{m^2 M^2} - \frac{2(1 - \cos Kd)}{mM}}. \qquad (10.30)$$

There are two solutions, one for each choice of the sign in the second term in Eq. (10.30), which gives rise to two separate *branches* in the dispersion curve, as illustrated in Fig. 10.8. To investigate the similarities and differences between the waves that are present in each branch, let us examine their behavior in both the low K (long wavelength) and $K \approx \pm\pi/d$ (near the Brillouin zone) limits.

10.2.1 Long wavelength limit

As K approaches zero, we can expand the cosine in Eq. (10.30) to obtain,

$$\lim_{K \to 0} \omega^2 = \frac{C_1(m + M)}{mM} \pm \frac{C_1(m + M)}{mM}\sqrt{1 - \frac{mM}{(m + M)^2}K^2 d^2}. \qquad (10.31)$$

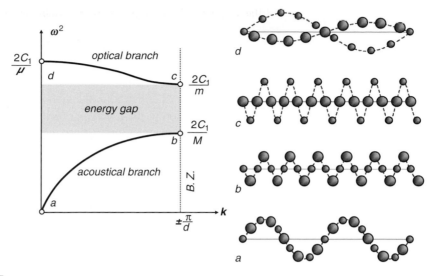

Figure 10.8 Dispersion curve for the one-dimensional diatomic crystal showing the lower frequency acoustical branch, higher frequency optical branch and the energy gap between. The right-hand figures illustrate relative particle displacements for waves at the Brillouin zone (*b* and *c*) and in the long wavelength limit (*a* and *d*).

This can be further approximated by a binomial expansion of the square root, such that

$$\lim_{K \to 0} \omega^2 = \frac{C_1(m+M)}{mM} \left\{ 1 \pm \left(1 - \frac{mM}{(m+M)^2} \frac{K^2 d^2}{2} \right) \right\}. \qquad (10.32)$$

For the solution using the minus sign, we find a modified version of the acoustical modes seen in the monatomic case (see Eq. (10.10)),

$$\lim_{K \to 0} \omega_{\text{lower}} = \sqrt{\frac{C_1 d/2}{(m+M)/d}} \, K = v_o K, \qquad (10.33)$$

wherein the tension and mass per length are modified accordingly for the present diatomic basis set. In addition, we can examine the ratio of the amplitude of two adjacent masses, *A/B*, as given by Eq. (10.27). For these long wavelength acoustical waves, the ratio is approximately unity and indicates that both masses (large and small) move together in forming the wave, as illustrated in Fig. 10.8a.

Now let us examine the other branch associated with the positive sign in Eq. (10.32). In the long wavelength limit, this branch approaches a constant (non-zero) value,

$$\lim_{K \to 0} \omega^2_{\text{upper}} = \frac{2C_1(m+M)}{mM} = \frac{2C_1}{\mu_{\text{effective}}}, \qquad (10.34)$$

and the amplitude ratio between the amplitude of the small mass (A) and the large mass (B) is

$$A/B = -M/m. \tag{10.35}$$

The negative value of the amplitude ratio implies that adjacent masses move exactly out of phase. Furthermore, the amplitudes of the oscillations of each mass are in proportion to the inverse of their mass: the small masses execute larger displacements that are out of phase with the smaller-sized displacement of the larger masses. The wave associated with this long wavelength limit of the upper branch is illustrated in Fig. 10.8d.

10.2.2 Waves near the Brillouin zone

Consider now the form of the dispersion relation (Eq. (10.30)) as K approaches values of either $\pm\pi/d$. In this limit, the cosine term becomes -1, and the expression can be quickly reduced to,

$$\lim_{Kd\to\pm\pi} \omega^2 = \frac{C_1(m+M)}{mM} \pm \frac{C_1}{mM}\sqrt{(m+M)^2 - 4mM}, \tag{10.36}$$

which has two independent solutions:

$$\lim_{Kd\to\pm\pi} \omega^2 = \frac{C_1}{mM}\{(m+M) \pm (M-m)\} = \begin{cases} 2C_1/m, & \text{upper branch} \\ 2C_1/M, & \text{lower branch.} \end{cases} \tag{10.37}$$

Because the two masses are dissimilar, one solution occurs at a higher frequency than the other and hence marks the termination of the upper branch, while the other marks the termination of the lower branch in Fig. 10.8. A consideration of the amplitude ratio in this limit leads to:

$$\frac{A}{B} = \begin{cases} \infty, & \text{upper branch} \\ 0, & \text{lower branch.} \end{cases} \tag{10.38}$$

This means that for the upper branch, the wave consists of small masses undergoing oscillations, while the large masses remain stationary. For the lower branch, the wave appears just the reverse with the large masses oscillating and the small masses at rest (see Fig. 10.8b). Again, both waveforms correspond to standing waves, and one can show (Ex. 2) that the group velocity in each instance vanishes on approach to the Brillouin zone.

10.2.3 Acoustical waves, optical waves and energy gaps

Because the lower branch incorporates at long wavelengths the same sort of acoustical modes found in the monatomic case, this lower branch is commonly referred to as the *acoustical branch*. The upper branch is known as the *optical branch*. This latter description arises from the nature of the oscillations depicted in Fig. 10.8d, in which the two masses move in opposite directions. How would such a wave be excited in the first place? Well, if the crystal is indeed an ionic crystal, as we suggested at the start, then an electromagnetic wave (i.e. a photon) present in the crystal would produce the oppositely directed forces on each anion and cation needed to generate these sorts of optical phonon waves.

Note also the existence of a range of frequencies between the two frequencies in Eq. (10.36) at which no waves can exist. This frequency gap implies a corresponding *energy gap* for phonon energies in the crystal. Just at the Brillouin zone we find two standing waves (constructed from phonons) of equivalent K that are degenerate in their energies. This is foreshadowing. We will see other examples of such degeneracies and energy gaps occurring in regard to the Brillouin zone when we examine the nature of electron waves in later chapters.

10.3 Scattering from phonons

In the crystal, particles reside mainly in fixed locations associated with an ordered lattice, but undergo small displacements from these fixed locations due to their vibrational motion. Because the displacements are small, the crystal planes remain well defined and X-rays continue to produce diffraction peaks consistent with the Bragg law. However, the small displacements do serve to reduce the intensity of these peaks somewhat, as we will show below. Moreover, the scattering of neutrons from the crystal reveals mechanisms at work in which incoming waves (recall from quantum mechanics that neutrons, like other small particles, can be viewed as waves) are scattered as if they exchanged energy and momentum by the creation or annihilation of phonons.

To see this, we start with the dynamic structure factor obtained by the time-dependent generalization of the static structure factor (Eq. (5.20)):

$$S(q,t) = \frac{1}{N}\left\langle \sum_i \sum_j e^{-i\vec{q}\cdot(\vec{r}_i(t)-\vec{r}_j(0))} \right\rangle, \tag{10.39}$$

where the positions of the particles are described in reference to their displacements ($\vec{a}(t)$) from the equilibrium lattice sites,

$$\begin{aligned} \vec{r}_i(t) &= \vec{T}_i + \vec{a}_i(t) \\ \vec{r}_j(0) &= \vec{T}_j + \vec{a}_j(0). \end{aligned} \tag{10.40}$$

Introducing Eq. (10.40) into Eq. (10.39), we obtain,

$$S(q,t) = \frac{1}{N}\sum_i \sum_j e^{-i\vec{q}\cdot(\vec{T}_i-\vec{T}_j)}\left\langle e^{-i\vec{q}\cdot\vec{a}_i(t)}e^{+i\vec{q}\cdot\vec{a}_j(0)} \right\rangle. \tag{10.41}$$

Because the displacements from equilibrium are small, it is appropriate to expand the two exponentials and carry out their multiplication to second order in the displacement $\vec{a}(t)$:

$$\left\langle e^{-i\vec{q}\cdot\vec{a}_i(t)} e^{+i\vec{q}\cdot\vec{a}_j(0)} \right\rangle \approx 1 + \left\langle -i\vec{q}\cdot\vec{a}_i(t) \right\rangle + \left\langle +i\vec{q}\cdot\vec{a}_j(0) \right\rangle + \left\langle [\vec{q}\cdot\vec{a}_i(t)][\vec{q}\cdot\vec{a}_j(0)] \right\rangle$$
$$- \frac{1}{2}\left\langle [\vec{q}\cdot\vec{a}_i(t)]^2 \right\rangle - \frac{1}{2}\left\langle [\vec{q}\cdot\vec{a}_j(0)]^2 \right\rangle.$$

$$(10.42)$$

The second and third terms on the right-hand side will vanish when the ensemble averaging (denoted by the brackets $\langle\cdots\rangle$) is performed, because the displacements $\vec{a}(t)$ involve repeated oscillations (positive and negative) about an equilibrium lattice site. Similarly, the last two terms on the right-hand side are equivalent, non-vanishing, quantities that each represent an average of the *square* of the scattering phase angle associated with the displacements. It is customary to define these as

$$\frac{1}{2}\left\langle [\vec{q}\cdot\vec{a}_i(t)]^2 \right\rangle = \frac{1}{2}\left\langle [\vec{q}\cdot\vec{a}_j(0)]^2 \right\rangle \equiv W, \qquad (10.43)$$

and then Eq. (10.43), when contracted back into the form of an exponential, becomes

$$\left\langle e^{-i\vec{q}\cdot\vec{a}_i(t)} e^{+i\vec{q}\cdot\vec{a}_j(0)} \right\rangle \approx 1 + \left\langle [\vec{q}\cdot\vec{a}_i(t)][\vec{q}\cdot\vec{a}_j(0)] \right\rangle - 2W$$
$$\approx \exp\left\{ \left\langle [\vec{q}\cdot\vec{a}_i(t)][\vec{q}\cdot\vec{a}_j(0)] \right\rangle - 2W \right\}.$$

$$(10.44)$$

Inserting this result back into Eq. (10.40), we can express the dynamic structure factor as

$$S(q,t) = \frac{e^{-2W}}{N} \sum_i \sum_j e^{-i\vec{q}\cdot(\vec{T}_i - \vec{T}_j)} \exp\left\{ \left\langle [\vec{q}\cdot\vec{a}_i(t)][\vec{q}\cdot\vec{a}_j(0)] \right\rangle \right\}$$
$$= \frac{e^{-2W}}{N} \sum_i \sum_j e^{-i\vec{q}\cdot(\vec{T}_i - \vec{T}_j)} \sum_{m=0}^{\infty} \frac{1}{m!} \left\langle [\vec{q}\cdot\vec{a}_i(t)][\vec{q}\cdot\vec{a}_j(0)] \right\rangle^m.$$

$$(10.45)$$

10.3.1 Elastic (Bragg) scattering: the Debye–Waller factor

The dynamic structure factor contains contributions from terms of increasing index m in Eq. (10.45). The leading term, $m = 0$, is

$$S_o(q,t) = \frac{e^{-2W}}{N} \sum_i \sum_j e^{-i\vec{q}\cdot(\vec{T}_i - \vec{T}_j)} = e^{-2W} S(\vec{q} = \vec{G}), \qquad (10.46)$$

and produces a contribution that is time independent and corresponds to the elastic scattering from the crystal planes discussed in Chapter 6 (see Eq. (6.2)). The only difference here is the reduction in the intensity of the Bragg peaks by an exponential factor known as the *Debye–Waller factor*. Again, this factor (defined by Eq. (10.43)) arises from the absolute magnitude of the vibrational displacements, which serve to blur the actual location of the scattering planes.

10.3.2 Inelastic scattering by single phonons

As was emphasized earlier, the displacement of any given particle is the result of a distribution of phonons, of various allowed K, that are simultaneously sharing the crystal medium:

$$\vec{a}_i(t) = \sum_K \vec{A}_K \exp\left[i(\vec{K} \cdot \vec{T}_i \pm \omega_K t)\right] = \sum_K \left[\sqrt{n_K}\vec{a}_K\right] \exp\left[i(\vec{K} \cdot \vec{T}_i \pm \omega_K t)\right]$$
$$\vec{a}_j(0) = \sum_{K'} \vec{A}_{K'} \exp\left[i(\vec{K}' \cdot \vec{T}_j)\right] = \sum_{K'} \left[\sqrt{n_{K'}}\vec{a}_{K'}\right] \exp\left[i(\vec{K}' \cdot \vec{T}_j)\right]$$

$$(10.47)$$

Here, the amplitude ($\vec{A}_K$) of the macroscopic wave associated with a given mode K is the result of n_K phonons, each described by a quantum mechanical wave function of amplitude $\vec{a}_K$. The square root term appears in Eq. (10.47) because it is the square of this quantum mechanical wave function that is proportional to the number of phonons n_K. The direction of the vector $\vec{a}_K$ denotes the polarization (longitudinal or transverse) and is parallel with $\vec{K}$ only for longitudinal waves.

The next leading term in Eq. (10.45) describes the time-dependent modulation of the scattering that results from the creation or annihilation of a phonon. When the expressions for the displacements (Eq. (10.47)) are introduced, this term appears as

$$\left\langle \sum_K \left[\sqrt{n_K}\phi_{qK}\right] \exp\left[i(\vec{K} \cdot \vec{T}_i \pm \omega_K t)\right] \sum_{K'} \left[\sqrt{n_{K'}}\phi_{qK'}\right] \exp\left[i(\vec{K}' \cdot \vec{T}_j)\right] \right\rangle,$$

$$(10.48)$$

where a polarization angle, $\phi_{qK} = \vec{q} \cdot \vec{a}_K$, has been introduced. If the orders of the summations are redistributed, Eq. (10.45) for $m = 1$ then reads as

$$S_1(q, t) = \frac{e^{-2W}}{N} \left\langle \sum_K \sqrt{n_K}\phi_{qK}\, e^{\pm i\omega_K t} \sum_{K'} \sqrt{n_{K'}}\phi_{qK'} \sum_{i,j} e^{i\left[(\vec{K}-\vec{q}) \cdot \vec{T}_i\right]} e^{i\left[(\vec{K}'+\vec{q}) \cdot \vec{T}_j\right]} \right\rangle.$$

$$(10.49)$$

The last double summation is significant. A similar summation appears in Eq. (10.46). There, sharp maxima (i.e. the Bragg reflections) result only for $\vec{q} = \vec{G}_{hkl}$. Here, these same maxima will only result when

$$K = K'$$
$$\vec{q} = \vec{G}_i + \vec{K}$$
$$\vec{q} = \vec{G}_j - \vec{K},$$

$$(10.50)$$

and the double summation will then return the value,

$$\sum_{i,j} e^{i\left[(\vec{K}-\vec{q})\cdot\vec{T}_i\right]} e^{i\left[(\vec{K'}+\vec{q})\cdot\vec{T}_j\right]} = N^2. \tag{10.51}$$

This means that the dominant scattering contribution occurs when Bragg scattered waves gain or lose momentum equivalent to that of one of the allowed phonons. Thus in the scattering, a phonon is either *created* or *annihilated* with its energy, and momentum traded with the scattering wave. In this case, the spectrum of the scattered waves is shifted to produce a doublet for each allowed K for which this phonon scattering can occur,

$$S_1^G(\vec{K},\omega) = Ne^{-2W} n_K \phi_{qK}^2 \left\{ \int_{-\infty}^{+\infty} e^{\pm i\omega_K t} e^{i\omega t} dt \right\} = Ne^{-2W} n_K \phi_{qK}^2 \delta(\omega \pm \omega_K). \tag{10.52}$$

Neutrons are most commonly used to study this inelastic phonon scattering and provide the most direct means of experimentally determining the dispersion curves. However, visible light, which probes near $K = 0$ also observes this interaction in the acoustic branch, where $\omega_K \approx \upsilon K$. In this case, the scattering process is identical with the *Brillouin scattering* discussed in the previous chapter for liquids (with $G_{hkl} = 0$, and $q = \pm K$). In the optical branch, the phonon interaction with visible light is referred to as *Raman scattering*. There the q-dependence vanishes because the group velocity of the optical branch vanishes as $K \to 0$ (see Fig. 10.8).

Summary

- Atoms of a crystal lattice vibrate in a concerted manner to produce a discrete set of allowed vibrational modes of wave vector $\vec{K}$.

- Vibrational energy is quantized into *phonons* with an energy $\hbar\omega$ and momentum $\hbar\vec{K}$. Any macroscopic elastic wave can be viewed as the constructive superpositioning of multiple, identical phonons. The creation and annihilation of phonons results in *inelastic scattering*.

- The first *Brillouin zone* is defined by the Wigner–Seitz primitive cell in reciprocal space. Only phonons described by $\vec{K}$ within the first Brillouin zone are physically meaningful. Waves with $\vec{k}_i$ that terminate on the Brillouin zone boundary undergo multiple scattering (Bragg diffraction).

- Magnons are the magnetic corollary of the elastic phonon and consist of quantized spin waves traveling in a ferromagnetic crystal.

- In scattering events, incoming waves can undergo inelastic scattering, in which both energy and momentum are conserved by the creation or annihilation of a phonon.

Exercises

10.1. Estimate the size of the phonon energy gap in NaCl relative to the energy of an acoustic phonon at the Brillouin zone.

10.2. From Eq. (10.30), demonstrate that the group velocity vanishes on approach to the Brillouin zone for both the acoustical and optical branches.

10.3. Determine the group velocity of a spin wave (i.e. magnon) both in the long wavelength limit and near the Brillouin zone.

10.4. In our development of the dispersion relation in Eq. (10.9) we assumed that interactions between other than nearest neighbors could be ignored. Show that when the higher-order springs (C_j) acting between other than nearest neighbors cannot be ignored, the dispersion relation in Eq. (10.9) is replaced by,

$$\omega^2 = \frac{2}{m} \sum_{j=1}^{\infty} C_j \left(1 - \cos(jKd)\right).$$

10.5. 532 nm laser light is incident in a solid of refractive index 2.1, in which the speed of sound is 5000 m/s. A small fraction of the incident photons undergo Brillouin scattering such that they emerge at a scattering angle $2\theta = 90°$ with slightly lower energy. (a) In what direction are the phonons created in this process initially traveling? (b) What energy do these phonons have? (c) By what fraction is the wavelength of the scattered light increased relative to that of the incident light?

10.6. Real crystals are not constructed of ideal elastic springs for which the potential varies as $U(x) = cx^2$. Rather, the potential includes also anharmonic contributions, $U(x) = cx^2 - gx^3 - fx^4$. We can estimate the thermal expansion of a crystal by computing the thermal average equilibrium separation using Boltzmann statistics as, $\langle x \rangle = \dfrac{\int_{-\infty}^{\infty} x e^{-U(x)/k_B T} dx}{\int_{-\infty}^{\infty} e^{-U(x)/k_B T} dx}$. (a) Show that for a crystal bound with ideal springs, there is no thermal expansion. (b) Show that for an anharmonic interaction, the thermal expansion is given approximately as $\langle x \rangle = \frac{3g}{4c^2} k_B T$. (Hint: expand the integrands for small x as $e^{-U(x)/k_B T} \approx e^{-cx^2/k_B T}(1 + gx^3/k_B T + fx^4/k_B T)$.)

Suggested reading

A discussion of phonons in crystals can be found in most any standard Solid State textbook. These are just a few personal favorites.

C. Kittel, *Introduction to Solid State Physics*, 8th Ed. (John Wiley and Sons, 2005).

J. S. Blakemore, *Solid State Physics*, 2nd Ed. (W. B. Saunders Co., Philadelphia, 1974).

N. W. Ashcroft and N. D. Mermin, *Solid State Physics* (Holt, Rinehart and Winston, New York, 1976).

M. A. Omar, *Elementary Solid State Physics* (Addison Wesley, Reading, MA, 1975).

11

Thermal properties

Introduction

In the previous chapter, we examined the elastic nature of a crystal and replaced the notion of atoms as independent harmonic oscillators with the concept of phonons as quantized pieces of elastic waves propagating within a crystal. In this chapter we bolster our confidence in the reality of these phonons by examining two thermal properties of a crystal: its specific heat and its thermal conductivity. At low temperatures, the specific heat of a crystal decreases as the cube of the temperature. A model (attributed to Einstein) based only on independent harmonic oscillators is unable to account for this particular low temperature dependence, while the Debye model, involving a population of phonons, properly accounts for the temperature dependence. Likewise, the thermal conductivity of a crystal can only be understood using the phonon picture. The thermal conductivity exhibits a sharp division in its temperature dependence between a T^3 variation at low temperatures and a $1/T$ dependence at high temperatures. This division stems from the nature of phonon–phonon collisions, which are only truly successful in retarding heat flow at high temperatures where so-called *Umklapp* processes dominate. In these collisions, the resultant phonon emerging from the collision extends beyond the boundaries of the Brillouin zone and suffers strong Bragg scattering by the lattice.

11.1 Specific heat of solids

Consider a crystal maintained at some finite temperature. Clearly the crystal contains energy in the form of lattice vibrations for which we have now developed two, self-consistent pictures. In one picture, we view this energy as stored in atoms that act as local harmonic oscillators. In the other picture, the energy is stored in a large population of phonons. The phonons appear in a variety of energies consistent with both the dispersion relation and the restricted set of allowed wave vectors imposed by the finite size of the crystal and the boundary of the first Brillouin zone.

Regardless of the picture, thermal equilibrium of the crystal is maintained by a steady inflow and outflow of energy with the surrounding thermal bath and

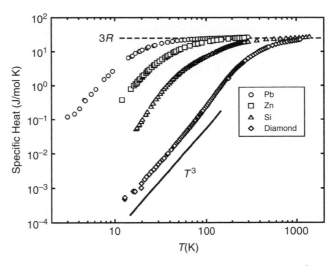

Figure 11.1 Specific heat (per mole) for four crystals as a function of temperature. Each shows T^3 behavior at low temperatures and an approach at high temperatures to the Dulong–Petit law predicted by the classical equipartition theorem. (Data from *Thermophysical Properties of Matter* (1970).)

the crystal assumes some well-defined internal energy, U, as a result. To increase the temperature of the crystal requires the net inflow of additional heat by an amount $dQ = CdT$, where C is the specific heat. From the first law of thermodynamics, $dU = dQ - PdV$, we find

$$C = \frac{dQ}{dT} = \frac{\partial U}{\partial T} + P\frac{\partial V}{\partial T} = C_V + P\frac{\partial V}{\partial T},$$

where,

$$C_V \equiv \frac{\partial U}{\partial T}\bigg|_V \qquad (11.1)$$

is the specific heat at constant volume. Solids exhibit only very minor expansion with temperature and so the specific heat at constant pressure, C_P, is essentially equivalent to C_V. Nevertheless, in the following, we will focus specifically on C_V as defined by Eq. (11.1).

Examples of the temperature dependence of C_V are shown in Fig. 11.1, for typical monatomic crystals. At high temperatures, C_V is seen to approach a temperature-independent limit of $3R$ per mole, where $R = 8.314$ J/K is the gas constant ($R = N_A k_B$). This high temperature limit is in agreement with classical *Dulong–Petit* theory, based upon the *equipartition theorem*, which posits that there should be $\frac{1}{2} k_B T$ of internal energy for each degree of freedom. Since the oscillating atoms in the crystal are constantly shifting energy between potential and kinetic forms, there are in fact, a total of six degrees of freedom per atom for a 3D crystal and hence an internal energy $U = 3RT$ per mole.

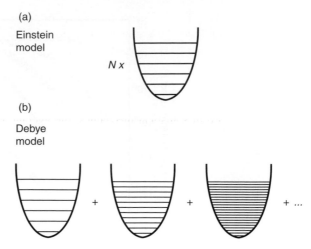

(a)

Einstein
model

$N \times$

(b)

Debye
model

Figure 11.2 The fundamental difference between the Einstein and Debye models for the specific heat is illustrated. (a) In the Einstein approach, the vibrational energy of the crystal is uniformly distributed into N^3 equivalent harmonic oscillators whose vibrational frequency is treated as a fitting parameter. (b) In the Debye approach, the vibrational energy is distributed among all the allowed vibrational modes, each with a frequency approximated by $\omega_K \approx \upsilon_0 K$.

However, this classical result is severely violated at low temperatures, as seen in Fig. 11.1, where the $C_V (T)$ curve vanishes like T^3 as absolute zero is approached. Such breakdowns of classical thermodynamics are not uncommon and are often a signature of the growing importance of quantization effects. This is true for the present situation, and we will see that a correct description of the C_V of the crystal will require a quantum mechanical approach.

In the following, we examine two practical models for the C_V of crystals that take into account the quantization of elastic energy. The first, which is not completely successful in accounting for the behavior in Fig. 11.1, is the *Einstein model*. In this approach, the elastic energy is simply partitioned into N^3 *identical* quantum mechanical harmonic oscillators (one for each particle of the crystal) that oscillate with a common frequency, but independently of one another. In this model, illustrated in Fig. 11.2a, the crystal structure plays no role, as the oscillators are not connected in any manner to each other. The second model we will examine is the *Debye model*. In the Debye model (Fig. 11.2b), the oscillations are not treated as independent of each other, but rather the quantization is introduced in the form of limits on the allowed wavelengths of the propagating phonons that travel within the lattice. As it turns out, this second approach is far more successful in accounting for the C_V curves shown in Fig. 11.1.

11.1.1 Einstein model

To obtain C_V, we first need to develop a statement of the total internal energy of N^3 oscillators. Since the oscillators are treated independently in the Einstein model, it suffices to determine the average thermal energy of a single oscillator and then multiply by N^3. The energy levels of a 1D quantum mechanical harmonic oscillator are given as,

$$E_n = \left(n + \frac{1}{2}\right)\hbar\omega. \tag{11.2}$$

From statistical mechanics, the average energy of such an oscillator is given by,

$$\langle E \rangle = \frac{\sum\limits_0^\infty E_n \exp(-E_n/k_B T)}{\sum\limits_0^\infty \exp(-E_n/k_B T)} = \frac{\hbar\omega}{2} + \frac{\sum\limits_0^\infty [n\hbar\omega]\exp(-[n\hbar\omega]/k_B T)}{\sum\limits_0^\infty \exp(-[n\hbar\omega]/k_B T)}$$

$$= \frac{\hbar\omega}{2} + \hbar\omega \frac{\sum\limits_0^\infty nx^n}{\sum\limits_0^\infty x^n}, \tag{11.3}$$

where $x = \exp(-\hbar\omega/k_B T)$. The two summations are known, so,

$$\langle E \rangle = \frac{\hbar\omega}{2} + \hbar\omega \frac{x/(1-x)^2}{1/(1-x)} = \frac{\hbar\omega}{2} + \hbar\omega \left\{\frac{1}{x} - 1\right\}^{-1} = \frac{\hbar\omega}{2} + \hbar\omega\langle n \rangle, \tag{11.4}$$

where,

$$\langle n \rangle = \left\{\frac{1}{\exp(\hbar\omega/k_B T) - 1}\right\} \tag{11.5}$$

is known as the *Planck distribution function* and equals the average energy level of a harmonic oscillator at temperature T.

For a crystal with a mole of three-dimensional oscillators, the total energy would then be

$$U = 3N_A\langle E \rangle = 3N_A \left\{\frac{\hbar\omega}{2} + \hbar\omega\left[\frac{1}{\exp(\hbar\omega/k_B T) - 1}\right]\right\}, \tag{11.6}$$

and the specific heat (per mole) is given by,

$$C_V = \frac{\partial U}{\partial T} = 3N_A k \left(\frac{\hbar\omega}{k_B T}\right)^2 \frac{e^{\hbar\omega/k_B T}}{\left(e^{\hbar\omega/k_B T} - 1\right)^2} = 3R\frac{y^2 e^y}{(e^y - 1)^2}, \tag{11.7}$$

where $y = \hbar\omega/k_B T$. This equation is a bit awkward, so let us just consider the two limits of high and low temperatures. At high temperatures, y approaches zero and we can expand the exponential terms to obtain,

$$C_V = 3R\frac{y^2(1 + y + \cdots)}{((1 + y + \cdots) - 1)^2} \approx 3R. \tag{11.8}$$

So far, so good. This result agrees with the classical Dulong–Petit result.

At low temperatures, y approaches infinity, and we can rearrange Eq. (11.7) such that in this limit,

$$C_V = 3R \frac{y^2 e^{-y}}{(1 - e^{-y})^2} \approx 3R y^2 e^{-y} = 3R \left(\frac{\hbar \omega}{k_B T} \right)^2 e^{-\hbar \omega / k_B T}. \qquad (11.9)$$

We see that the Einstein model does produce a C_V that vanishes as absolute zero is approached, but which does not quite accommodate the actual T^3 behavior seen in experiments.

11.1.2 Debye model

The basic problem with the Einstein model is that it treats the particles as though they all oscillate identically and independent of one another. In reality, these particles are interconnected by bonds, and as we saw earlier, their oscillations are not independent but give rise to elastic waves (phonons) that propagate through the crystal. Like the independent oscillators, these phonons have energies that are also quantized (see Eq. 10.19), but the quantization arises from constraints on the allowed wavelengths that the phonons can have, as given by Eq. (10.18). Together with the dispersion relation, these constraints lead to a collection of oscillators that now have some distribution of frequencies as opposed to a common frequency of oscillation.

Since the energy is now distributed in the phonons and not in the individual oscillations of the particles, our internal energy is a sum of all the energies of all the phonons that are present at thermal equilibrium,

$$U = \sum_{\text{phonons}} (\text{phonon energy}) \times (\# \text{ phonons of that energy}). \qquad (11.10)$$

Moreover, these phonons are distributed into discrete modes, each represented by a "dot" in K-space, as illustrated in Fig. 11.3. These modes fill the K-space uniformly out to the edge of the first Brillouin zone at $K = \pm \pi/d$, where the index in Eq. (10.18) is $n = N/2$. Hence there are N modes in each spatial direction and N^3 total modes equal to the total number of particles in an $N \times N \times N$ crystal. Given this large number of closely spaced modes, it is appropriate to convert the summation in Eq. (11.10) to an integration over a continuous space of K,

$$U = \int_{\text{modes}} (\hbar \omega_K) \times \left[\frac{1}{\exp(\hbar \omega_K / k_B T) - 1} \right] \times \left(\frac{dN}{dK} \right) dK. \qquad (11.11)$$

Here, the first term is the energy of a given phonon, determined from the dispersion relation based on its value of K. The second term is the Planck distribution which provides the number of identical phonons present at equilibrium for a given temperature. The last term, $(dN/dK) = g(K)$, is known as

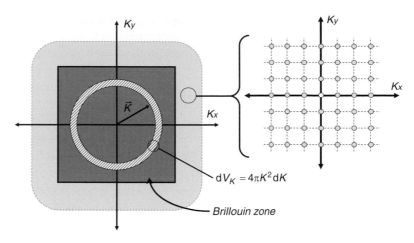

Figure 11.3 The allowed vibrational modes of a three-dimensional crystal (given by Eq. (10.18)) are represented as evenly spaced dots in a three-dimensional K-space. Because there are as many dots as there are particles and their spacing is small, we can treat the allowed modes as a continuum described by a density of states, $g(K)$, such that the number of modes in the range from K to $K + dK$ is given as $g(K)dK$.

the *density of states*. It represents the number of modes (dots in Fig. 11.3) that are included between K and $K + dK$. Since the dots are evenly spaced, and we know that there are N^3 total dots within the Brillouin zone, the density of modes (dots) in this K-space is

$$\rho_K = \frac{N^3}{(2\pi/d)^3} = \frac{(Nd)^3}{(2\pi)^3} = \frac{V}{8\pi^3},$$ (11.12)

and the density of states is then

$$g(K) = \frac{V}{8\pi^3} 4\pi K^2 = \frac{VK^2}{2\pi^2}.$$ (11.13)

The Debye approximation

To carry out the integration, we would now need to express ω_K in terms of the integration variable K using the dispersion relation (Eq. (10.9)). This would, however, produce an integrand containing an exponential of a sine function which would be rather messy to deal with. Instead, we approach the problem using the *Debye approximation*, in which the dispersion is approximated as

$$\omega_K \approx v_o K.$$ (11.14)

Obviously, the Debye approximation in Eq. (11.14) severely overestimates the energy of a large fraction of the modes in the Brillouin zone, especially those that have high frequencies near to the zone boundary. Why then might we think that this approximation is suitable to the integration we are attempting to complete? Although these high frequency modes have substantial energy (contributing to the first term in the integrand of Eq. (11.11)), the actual number of such phonons is sharply diminished by the Planck distribution (the second factor in the integrand of Eq. (11.11)), particularly at low temperatures. Thus, although the integral extends over all modes in the Brillouin zone, many of these modes will not be populated at low temperatures and so the approximation should remain valid in that limit.

Returning to our integral of Eq. (11.11), we now face a predicament. Our integration volume consists of spherical shells in K-space that are inconsistent with the cubic boundary formed by the Brillouin zone. Thus the upper limit of integration is not well determined. However, in the spirit of approximations, we will keep the radial integration but make sure that we limit the radius to a value K_{max} such that only N^3 dots of K-space have been counted. The appropriate K_{max} is then given by,

$$N^3 = \int_{K=0}^{K_{max}} g(K)dK = \frac{VK_{max}^3}{6\pi^2}, \tag{11.15}$$

or,

$$K_{max} = \left(6\pi^2 N^3/V\right)^{1/3} = \left(6\pi^2 n\right)^{1/3}. \tag{11.16}$$

With the Debye approximation, our complete integral is now

$$U = 3 \int_{K=0}^{K_{max}} (\hbar v_o K) \left[\frac{1}{\exp(\hbar v_o K/k_B T)-1}\right] g(K)dK, \tag{11.17}$$

where the factor of three is introduced here to account for the three independent polarizations (two transverse and one longitudinal) that phonons can also possess. The specific heat is then

$$C_V = \frac{\partial U}{\partial T} = \frac{3V}{2\pi^2} (\hbar v_o) \int_{K=0}^{K_{max}} K^3 dK \frac{d}{dy} \left[\frac{1}{e^y-1}\right] \frac{dy}{dT}, \tag{11.18}$$

where $y = \hbar v_o K/k_B T$. After carrying out the differentials, one finds,

$$C_V = \frac{3V}{2\pi^2} (\hbar v_o) \left(\frac{k_B T}{\hbar v_o}\right)^4 \frac{1}{T} \int_{y=0}^{y_{max}} \frac{y^4 e^y}{(e^y-1)^2} dy, \tag{11.19}$$

where $y_{max} = \hbar v_o K_{max}/k_B T$. Substituting Eq. (11.16), this can be rearranged to read,

$$C_V = 9N^3 k \left(\frac{k_B T}{\hbar v_o K_{\max}}\right)^3 \int\limits_{y=0}^{y_{\max}} \frac{y^4 e^y}{(e^y-1)^2}\,dy = 9N^3 k_B \left(\frac{T}{\theta_D}\right)^3 \int\limits_{y=0}^{y_{\max}} \frac{y^4 e^y}{(e^y-1)^2}\,dy,$$

$$(11.20)$$

where

$$\theta_D = \hbar v_o K_{\max}/k_B \qquad (11.21)$$

is known as the *Debye temperature*.

This looks very promising since the factor in front of the integral now has the appropriate T^3 behavior we had hoped to find. To check its performance, we again examine the two temperature limits. At high temperatures, y approaches zero and we can again expand the exponentials,

$$C_V \approx 9N^3 k_B \left(\frac{T}{\theta_D}\right)^3 \int\limits_{y=0}^{y_{\max}} \frac{y^4(1+y+\cdots)}{((1+y+\cdots)-1)^2}\,dy \approx 9N^3 k_B \left(\frac{T}{\theta_D}\right)^3 \int\limits_{y=0}^{y_{\max}} y^2\,dy$$

$$= 9N^3 k_B \left(\frac{T}{\theta_D}\right)^3 \frac{1}{3}\left(\frac{\theta_D}{T}\right)^3 = 3R,$$

$$(11.22)$$

to again obtain the Dulong–Petit law, observed at high temperatures. Now in the low temperature limit, $y_{\max}$ approaches infinity, so we will set the upper limit of integration in Eq. (11.20) to produce a known definite integral,

$$C_V \approx 9N^3 k_B \left(\frac{T}{\theta_D}\right)^3 \int\limits_{y=0}^{\infty} \frac{y^4 e^y}{(e^y-1)^2}\,dy = 9N^3 k_R \left(\frac{T}{\theta_D}\right)^3 \frac{12\pi^4}{45}, \qquad (11.23)$$

which produces the correct T^3 dependence seen experimentally.

11.2 Thermal conductivity

The thermal conductivity, κ_{th}, is defined in reference to the steady state rate at which heat travels through a solid from a region of high temperature to low temperature. As illustrated in Fig. 11.4, two ends of a long bar are maintained at different temperatures. Macroscopically, the bar has a cross-sectional area A, and the heat flux, j_Q, is given as,

$$j_Q = \frac{d(Q/A)}{dt} = -\kappa_{th}\frac{dT}{dx}. \qquad (11.24)$$

Microscopically, heat flow is a diffusive process. Heat in the solid is carried by small bundles of propagating energy (i.e. phonons) that undergo numerous

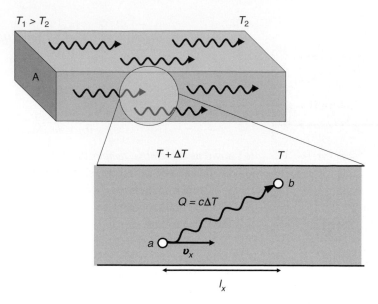

Figure 11.4 A temperature differential is applied to a long block of material with cross-sectional area A causing heat to flow from the hot end (T_1) to the cold end (T_2). Microscopically, the energy is transported by phonons that undergo frequent collisions with imperfections of the crystal lattice and with other phonons. In one instance, shown in the lower figure, a "hot" phonon created at point a is thermalized by numerous collisions as it travels and arrives at point b as a "cooler" phonon. During the journey, a small quantity of heat, Q, is released.

collisions with one another and with imperfections in the crystal lattice. These microscopic collisions serve to retard the flow of heat and appear as a macroscopic thermal resistance, κ_{th}^{-1}. A key measure of the retardation is the mean collision time, τ_{coll}, or the *mean free path*, $l_{mfp} = v\tau_{coll}$, where v is the average speed of the carrier. The mean free path is a measure of the average distance an energy carrier travels before suffering a collision.

With this picture of heat transport, simple arguments from kinetic theory can be used to evaluate the thermal conductivity. Consider, as shown in Fig. 11.4, the microscopic transport of a single heat carrier that travels a distance l_x from a region of high temperature to lower temperature before encountering a collision. During the trip, an amount of heat, $Q = c\Delta T$, is released by the carrier, where c is the specific heat of a single carrier. The rate of thermal flux is then,

$$j_Q = \frac{-c\Delta T}{A\tau} = \frac{-c\Delta T v_x}{Al_x} = Cv_x\Delta T, \qquad (11.25)$$

where C is the (macroscopic) specific heat per volume. Since ΔT is given by the temperature gradient, we find,

$$j_Q = Cv_x l_x \frac{dT}{dx} = -Cv_x^2 \tau_{coll}\frac{dT}{dx} = -C\frac{v^2}{3}\tau_{coll}\frac{dT}{dx} = -\frac{C}{3}vl_{mfp}\frac{dT}{dx}, \qquad (11.26)$$

so that the thermal conductivity is given as,

$$\kappa_{th} = \frac{1}{3} C \upsilon l_{\mathrm{mfp}}. \qquad (11.27)$$

11.2.1 Phonon collisions

The thermal conductivity is mainly controlled by the size of the mean free path, which is in turn sensitive to several sources of collisions. An ultimate limit to the mean free path is the physical boundary of the crystal. However, imperfections in the crystal lattice are typically present at much shorter distances. Beyond the nature of the lattice itself, collisions between phonons provide an important source for thermal resistance and among such collisions there are two significant processes: *normal* (or *N*-processes) and *Umklapp* (or *U*-processes). Each of these is illustrated in Fig. 11.5. In the normal process, two phonons collide to produce a third whose wave vector resides within the Brillouin zone. In the process, momentum is conserved and so

$$\vec{K}_1 + \vec{K}_2 = \vec{K}_3 \ (N\text{-}process). \qquad (11.28)$$

The *N*-process, however, does not lead to a retardation of heat flow, since whatever forward momentum was present before the collision still remains. To see how retardation arises, we must consider the so-called Umklapp process in which two phonons collide to produce a third that lies outside the Brillouin zone. As we have discussed earlier, phonons described by wave vectors outside the Brillouin zone are not physically realistic elastic waves, because they imply a wavelength shorter than twice the particle spacing. Such phonons are instead represented by an equivalent wave vector, residing inside the Brillouin zone, which is obtained by adding an appropriate reciprocal lattice vector:

$$\vec{K}_1 + \vec{K}_2 + \vec{G} = \vec{K}_3 \ (U\text{-}process). \qquad (11.29)$$

To see how this occurs, you may need to treat the collision product that extends past the Brillouin zone as a sum of two vectors: one that ends at the zone boundary and a snippet continuing outside the zone. As we know, the first part is Bragg reflected to the opposite side of the Brillouin zone and, when the snippet is added, the resulting wave vector appears inside the zone, as given by Eq. (11.29). Momentum conservation is maintained during the process as some momentum is transferred to the lattice itself via the Bragg reflection. The significance of the *U*-process is quite obvious as it leads to a reversal of the heat transport and considerable contribution to thermal resistance.

(a)

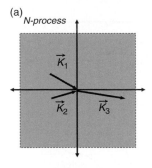

(b)

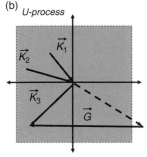

Phonon–phonon scattering processes. (a) In the normal process, two low energy phonons undergo a collision. Each phonon is annihilated with the total energy and momentum appearing in the form of a new phonon ($\vec{K}_3$) that still resides within the Brillouin zone. This process does not lead to retardation of heat transfer since the forward momentum is unchanged. (b) In the Umklapp process, two (high energy) phonons collide to produce what would be a phonon outside the Brillouin zone (dashed arrow). In reality, the new phonon resides inside the Brillouin zone and the missing momentum is distributed to the crystal lattice as a whole. This process leads to a reduction of the phonon momentum and a retardation of heat transfer.

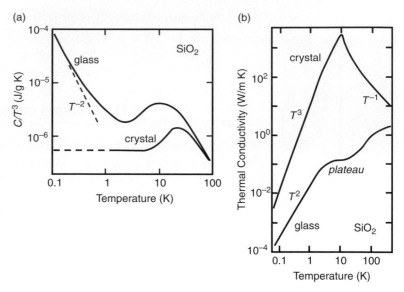

Figure 11.6 (a) Plot of the specific heat (divided by T^3) of crystalline and amorphous SiO_2. Although the crystal approaches the Debye T^3 limit at low temperatures, the glass exhibits an anomalous behavior consistent with the presence of an additional contribution which increases linearly with temperature. (b) Plot of the thermal conductivity of crystalline and amorphous SiO_2. For the crystal, κ_{th} increases as T^3 at low temperatures, but decreases as T^{-1} at high temperatures. For the glass, κ_{th} increases only as T^2 at low temperatures and exhibits an anomalous plateau in the vicinity of 10 K. (Adapted from Phillips (1987).)

With this background, we can now understand the temperature variations of κ_{th} commonly seen in crystals and illustrated in Fig. 11.6. At high temperatures, κ_{th} varies inversely with temperature. In this high temperature regime, average phonon energies are large and most collisions result in U-processes. The mean free path is then limited by the high density of phonons and is proportional to the Planck distribution,

$$l_{mfp} \propto \langle n \rangle^{-1} = \lim_{T \to \infty} \left(\exp(\hbar\omega/k_B T) - 1 \right) \approx \frac{\hbar\omega}{k_B T} \propto T^{-1}. \qquad (11.30)$$

Since at high temperatures the specific heat approaches the constant Dulong–Petit value, the thermal conductivity, given by Eq. (11.27), thus falls off as T^{-1}.

At low temperatures, average phonon energies are small and the U-processes become ineffective as very few phonons possess sufficient momentum to produce a collision product residing outside the Brillouin zone. In this limit, the mean free path is a temperature-independent constant set only by the distance between imperfections of the crystal, and so the temperature dependence of κ_{th} in Eq. (11.27) arises from that of the specific heat, which varies as T^3 at low temperatures.

11.3 Amorphous materials

Disordered materials lack the regular particle positions found in crystals and so irregularities appear in particle positions, as illustrated in Fig. 11.7, that are comprised of intermittent crowding and voids. This has two important consequences for phonons. Firstly, because of the density variations in the amorphous solid, atoms or groups of atoms in a glass can undergo small, local rearrangements following a small energy exchange. These transitions between two (or more) local arrangements can be modeled by an equivalent *two-level system* (TLS) in statistical mechanics and this opens up a novel mechanism for energy exchange with phonons of a similar energy.

Secondly, disorder introduces a relevant length scale, ξ, which characterizes the average distance between adjacent defects in the otherwise ordered structure. Long wavelength phonons, with $\lambda \gg \xi$, span over numerous defects within a single wavelength and are largely unaffected by the disorder. These phonons propagate through the solid like ordinary, extended waves. However, as the phonon wavelength shrinks towards ξ, the phonons become increasingly sensitive to the disorder and undergo significant scattering. In this case, the phonon can be thought of as being multiply reflected between adjacent defects, so as to become trapped in the form of a non-propagating, *localized phonon*.

Because many of the thermal properties in a solid are dominated by the large supply of long wavelength phonons, the specific heat and thermal conductivity of crystals and glasses are much alike at high temperatures. It is only at low

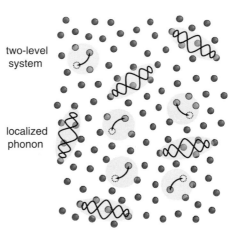

two-level
system

localized
phonon

Figure 11.7 Unlike crystals, disordered materials contain irregularities in particle positions that provide for local relaxations between two (or more) configurations that can be modeled by a two-level system (TLS). Although long wavelength phonons propagate in the form of extended phonons traveling as in a crystal, short wavelength modes can become localized (non-propagating) when their wavelength recedes below some characteristic scale, ξ, where continuum behavior vanishes.

(a)

(b)
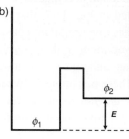

The two-level system. (a) In a disordered material there exist a great many local regions that can undergo some minor rearrangement between two configurational states that are nearby in energy. (b) The kinetics of these transitions can be modeled using an asymmetric double well potential in which the two configurations differ by an energy E and are separated by a barrier.

temperatures, below about 10 K, where disordered materials display anomalies. These anomalies are illustrated in Fig. 11.6 for SiO_2. They include a linear contribution to the specific heat in addition to the Debye law, and a thermal conductivity that increases as T^2 for temperatures below 1 K and exhibits a *plateau* for temperatures from 1 K to roughly 10 K.

11.3.1 Two-level systems

In a glass, we imagine that there exist a great many local regions whose particles are able to undergo a slight rearrangement from one state (ϕ_1) to another (ϕ_2) of different energy. We can model the underlying thermodynamics by a TLS, like that shown in Fig. 11.8, in which the two states are separated by a barrier. In principle, there are a variety of these TLSs in the glass with some distribution of energy differences, but for the moment we consider just a single one. For convenience we take the lower energy level to be zero, so that the average energy of the TLS at equilibrium is given by Boltzmann statistics as,

$$\langle E \rangle = \frac{\sum\limits_{i=1,2} E_i e^{-E_i/k_B T}}{\sum\limits_{i=1,2} e^{-E_i/k_B T}} = \frac{E}{e^{E/k_B T} + 1}. \tag{11.31}$$

Specific heat

This single TLS then contributes to the specific heat by an amount

$$C(E) = \frac{\partial \langle E \rangle}{\partial T} = k_B \left(\frac{E}{k_B T}\right)^2 \frac{e^{E/k_B T}}{\left(e^{E/k_B T} + 1\right)^2}, \tag{11.32}$$

and the cumulative contribution of several such TLSs could then be described as,

$$C = \int C(E) g_{TLS}(E) dE, \tag{11.33}$$

where $g_{TLS}(E)dE$ is the density of states (i.e. the number of TLSs with energy between E and $E + dE$). Let us for the moment assume that there are roughly equal numbers of TLSs of any given energy difference such that $g_{TLS}(E) =$ constant $= N_{TLS}$. In this simple case, the contribution to the specific heat by the TLSs is given by,

$$C = N_{TLS} k_B \int\limits_0^\infty \left(\frac{E}{k_B T}\right)^2 \frac{e^{E/k_B T} dE}{\left(e^{E/k_B T} + 1\right)^2}$$

$$= N_{TLS} k_B^2 T \int\limits_0^\infty \frac{x^2 e^x dx}{(e^x + 1)^2} = \left(\frac{\pi^2}{6}\right) N_{TLS} k_B^2 T, \tag{11.34}$$

and reproduces precisely the linear contribution found experimentally.

Thermal conductivity

As we saw earlier, the main consideration for thermal conductivity is the mean free path of the heat carrier. In crystals at low temperature, the mean free path is limited only by the imperfections of the crystal lattice and so the thermal conductivity increases as T^3. In a glass, the mean free path is limited instead by the density of the TLSs present. These TLSs act as phonon traps that can capture an incoming phonon and later release it in a random direction, thus retarding the flow of heat.

A given TLS of energy E can only trap phonons of that same energy. Consequently, to determine the thermal conductivity we will need to integrate over all the energies these TLS traps might have,

$$\kappa_{th} = \frac{v}{3} \int_0^{\omega_D} C(\omega) l_{\mathrm{mfp}}(\omega) d\omega, \qquad (11.35)$$

where $C(\omega)$ is the contribution to the phonon specific heat (per volume) made by just a narrow band of phonons in the range between ω and $\omega + d\omega$. One can show (Ex. 5) that this contribution is given by

$$C(\omega, T) = 3k_B \left(\frac{g(\omega)}{V} \right) \frac{(\hbar\omega/k_B T)^2 e^{\hbar\omega/k_B T}}{\left(e^{\hbar\omega/k_B T} - 1 \right)^2}, \qquad (11.36)$$

where $g(\omega)$ is the density of phonon states (in the Debye approximation) of Eq. (11.13), only now expressed in terms of ω.

So now, to obtain the thermal conductivity due to TLSs, we need to find the corresponding mean free path or, optionally, the average collision time (or phonon lifetime), τ_{coll}, such that $l_{\mathrm{mfp}} = v\tau_{\mathrm{coll}}$. Consider then, as illustrated in Fig. 11.9, the three processes by which a TLS might interact with a field of phonons of energy E. These include stimulated absorption, stimulated emission, and spontaneous emission, and are synonymous with the exchange processes found in a laser (Silfvast, 2004) wherein a field of photons exchanges energy with an active lasing medium. Spontaneous emission occurs when the TLS transitions from the high energy state ϕ_2 to the lower energy state ϕ_1 to produce a new phonon. This process occurs without the perturbing influence of other phonons present and takes place at a rate, Γ_{21}^o, that is typically much slower than the other two processes. Stimulated emission, like its spontaneous cousin, leads to the generation of a new phonon, but occurs at a faster rate, Γ_{21}, due to the presence of other phonons that serve to "stimulate" the transition. In stimulated absorption, a TLS moves up in energy from state ϕ_1 to ϕ_2 by the annihilation of an existing phonon at a rate, Γ_{12}. As implied by their name, the rates for both stimulated processes are proportional to the density of phonons present that serve to provide the stimulation,

$$\begin{aligned} \Gamma_{12} &= B_{12} n_{\mathrm{ph}}(E) \\ \Gamma_{21} &= B_{21} n_{\mathrm{ph}}(E), \end{aligned} \qquad (11.37)$$

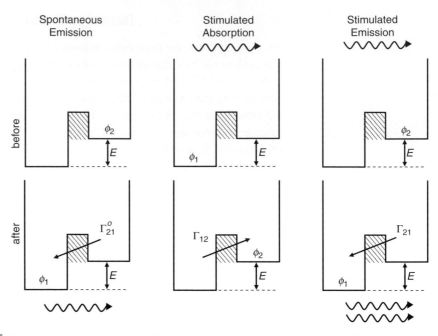

Figure 11.9 Illustration of the three processes by which a TLS can interact with a phonon field. Because the two stimulated processes occur at higher rates than does spontaneous emission, they contribute most in limiting the mean free path of phonons. At equilibrium, detailed balance requires that the upward and downward transition rates (Γ_{12} and Γ_{21}, respectively) be equal.

and hence the change in the density of phonons with energy E is given by the following rate equation,

$$\frac{dn_{ph}(E)}{dt} = N_{TLS}\,p_2\Gamma_{21}^o + N_{TLS}p_2\Gamma_{21} - N_{TLS}\,p_1\Gamma_{12}, \qquad (11.38)$$

where p_1, p_2 are the equilibrium probabilities that a TLS is in state ϕ_1 or ϕ_2, respectively. In equilibrium, the upward and downward stimulated transition rates are equal (a condition known as *detailed balance*), and so $B_{12} = B_{21} = B(E)$. Furthermore, since both stimulated processes occur far more rapidly than the spontaneous emission process, we ignore the latter and write,

$$\frac{dn_{ph}(E)}{dt} \approx -N_{TLS}B(E)(p_1 - p_2)n_{ph}(E). \qquad (11.39)$$

The solution to this differential equation is an exponential decay with a phonon lifetime (i.e. average collision time) given by,

$$\tau_{ph}^{-1} \approx N_{TLS}B(E)(p_1 - p_2). \qquad (11.40)$$

The probability of a TLS being in its excited state is given by the corresponding Boltzmann factor, $p_2 = p_1 e^{-E/k_B T}$, and, because $p_1 + p_2 = 1$, it follows that

$$(p_1 - p_2) = \frac{(1 - e^{-E/k_B T})}{(1 + e^{-E/k_B T})} = \tanh(E/2k_B T), \qquad (11.41)$$

so that

$$\tau_{ph}^{-1} \approx N_{\text{TLS}}B(E)\tanh(E/2k_BT). \qquad (11.42)$$

What is $B(E)$? Defined in Eq. (11.37), we see that $B(E)$ is a measure of the likelihood that a phonon of energy E will trigger the transition of a two-level system. Physically, the phonon does this by "rocking" the TLS as it passes by. The phonon is an elastic wave and produces a local stress at the site of the TLS. For low frequency acoustic phonons, this stress is nearly imperceptible. But, as the wavelength of the phonon decreases, the local stress becomes more severe. It reasons then that $B(E) \approx A\hbar\omega$, where A is some undetermined constant, and increases with decreasing wavelength so that

$$l_{ph}^{-1} = \frac{1}{\upsilon}\tau_{ph}^{-1} = \left(\frac{N_{TLS}A\hbar\omega}{\upsilon}\right)\tanh(\hbar\omega/2k_BT). \qquad (11.43)$$

The thermal conductivity is then given from Eq. (11.35) as

$$\kappa_{th} = \frac{k_B^3 T^2}{2\pi^2\hbar^3 N_{\text{TLS}}A\upsilon} \int\limits_{0}^{y_{\max}\approx\infty} \frac{y^3 e^y}{(e^y-1)^2}\coth(y/2)dy \approx \left(\frac{k_B^3}{2\hbar^3 N_{\text{TLS}}A\upsilon}\right)T^2,$$

$$(11.44)$$

and varies at low temperatures, as seen experimentally.

11.3.2 Phonon localization

As mentioned previously, disordering gives rise to a characteristic length scale, ξ, that separates phonon behavior into two extremes. For phonon modes with $\lambda \gg \xi$, the phonon appears as an "extended" elastic wave that propagates in the glass much the same as it would in a crystalline solid. At large wavelengths, both the crystal and disordered media appear as a continuum solid. However, phonon modes with $\lambda < \xi$ sample a non-continuous medium, punctuated by irregularities in particle positions that severely limit the establishment of a propagating elastic wave. The characteristic length scale thus operates much like a mean free path, $l_{\text{mfp}} \approx \xi$, wherein the incipient phonon is multiply scattered to produce a non-propagating, localized phonon. The crossover condition at which this occurs,

$$\lambda \approx \xi \quad \text{or} \quad Kl_{\text{mfp}} \approx 1, \qquad (11.45)$$

is often referred to as the *Ioffe–Regel criterion*.

Phonon localization is particularly relevant in certain self-similar materials such as aerogels and epoxy resins which possess a self-similar or fractal structure at length scales smaller than the correlation length, ξ. In these materials, the correlation length marks an interface between normal extended phonon modes, as described by the Debye density of states $g(\omega) \approx \omega^2$, and

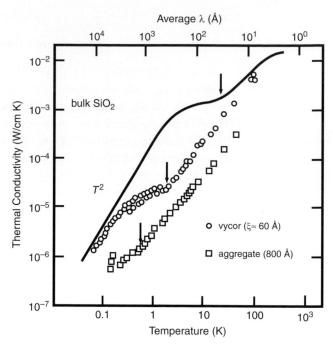

Measurements of the thermal conductivity of amorphous silica for a bulk sample and two samples formed by sintering of silica aggregates of differing cluster size, ξ. In each instance a kink is located (arrows) where the conductivity plateau first emerges with decreasing temperature. (Adapted from Graebner and Golding, 1986.)

localized phonons, here referred to as *fractons* (Aharony *et al.* (1987)), that exhibit a fractal density of states,

$$g(\omega) \approx \omega^{d_s - 1}, \tag{11.46}$$

where d_s is the fracton or spectral dimension closely related to the fractal dimension discussed in Chapter 8.

Although the effects of phonon localization in glasses possessing only short-range order are not fully understood, there is compelling evidence that localization does impact the thermal properties of glasses at low temperatures. For example, studies suggest a strong connection between the high temperature onset of the plateau in thermal conductivity discussed earlier and satisfaction of the Ioffe–Regel condition by the average phonon wavelength. Studies of the thermal conductivity of certain amorphous SiO_2 aggregates, shown in Fig. 11.10, indicate that the high temperature edge of the plateau shifts to lower temperatures with increasing particle size, ξ. One can show (Ex. 6) that the temperature at which this occurs corresponds to an average phonon wavelength,

$$\langle \lambda \rangle = h\upsilon / 2.70 k_B T, \tag{11.47}$$

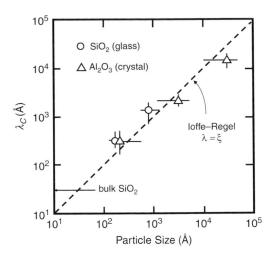

Figure 11.11 The average phonon wavelength, corresponding to the temperatures in Figure 11.10 where the kink in the thermal conductivity occurs, are graphed against the aggregate cluster size. Data for crystalline aggregates of alumina (which exhibits a discontinuity in $\kappa_{\text{th}}(T)$) are also included. Equivalence of these two quantities (dashed line) indicates that the Ioffe–Regel criterion (Eq. (11.45)) is satisfied and the onset of phonon localization has occurred. (Adapted from Graebner and Golding, 1986.)

which is seen in Fig. 11.11, to exactly conform to the Ioffe–Regel criterion. The inference drawn for bulk SiO_2 suggests localization on a scale of $\xi \approx 30\text{Å}$.

Summary

- The specific heat of a solid results from a collection of phonons with a distribution of allowed energies. The thermal population of any given type of phonon is given by the Planck distribution, $\langle n \rangle = \{\exp(\hbar\omega/k_B T) - 1\}^{-1}$.

- The *Debye model* for the specific heat approximates the phonon dispersion as $\omega_K = v_o K$ and is successful at accounting for the T^3-dependence of C_V at low temperatures.

- Thermal resistance in solids arises from *phonon collisions*, either with lattice imperfections, or with other phonons through the Umklapp scattering process.

- Anomalies in the thermal properties of amorphous solids arise from the presence of voids in the structure. These voids promote *phonon localization* (i.e. non-propagating phonon modes) and are the source for *two-level system* processes that couple with the phonon field.

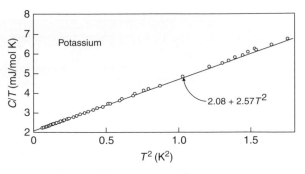

Figure 12.1 Specific heat (divided by T) of potassium plotted against T^2 at low temperatures. The fit to a straight line is evidence for an additional contribution (linear in temperature) due to mobile electrons. (Adapted from Lien and Phillips, 1964.)

arrangement. In the case of a metal, there are then two contributions to the specific heat. The first is the Debye contribution which is associated with the phonons that surf the elastic waves superimposed on the lattice of ion cores. The second arises from the motion of the electrons, which when freed from the atoms, possess an additional set of translational degrees of freedom. Moreover, because the electrons possess charge, their motion can be influenced by an externally applied electric field to produce a measurable conductivity.

These features are evident in measurements conducted on monovalent potassium, shown in Fig. 12.1 and Fig. 12.2. In Fig. 12.1, an excess contribution to the specific heat is seen which is proportional to the temperature. Together with the Debye contribution the total varies as $C = AT + BT^3$. In Fig. 12.2, the electrical resistivity of potassium increases linearly with temperature over a large range, with the exception of very low temperatures, where it approaches a fixed value dependent upon the purity of the sample. Altogether, the electrical resistivity of a sample follows an empirical formula,

$$\rho(T) \approx \rho_{\text{imp}} + \rho_{\text{phonon}}(T), \tag{12.1}$$

known as *Matthiessen's rule*.

12.1.1 The classical (Drude) model

Because the electrons in a metal form something akin to an ideal gas (i.e. point-like particles whizzing about randomly and occasionally colliding with other electrons or the ion cores), it is quite reasonable to apply classical *kinetic theory* of gases to this problem. In this view, the electrons are treated as point particles and move around randomly with an average thermal speed given by the equipartition theorem,

$$U/N = \frac{1}{2} m v_{\text{therm}}^2 = \frac{3}{2} k_B T, \tag{12.2}$$

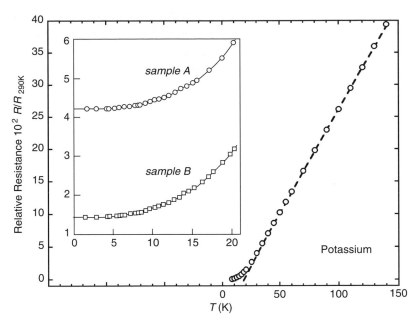

Figure 12.2 Electrical resistivity (relative to that at 290 K) of potassium below 150 K (data from Dugdale and Gugan (1962)) illustrating overall linear temperature dependence. Inset illustrates how samples exhibit a limiting constant resistivity at low temperatures, depending on the purity of the sample. (Adapted from MacDonald and Mendelssohn, 1950.) Solid lines are fits to the Matthiessen rule of Eq. (12.1).

and contribute to the specific heat by a temperature-independent amount $C_V^e = (3/2)R$ per mole. Already we spy a problem. Measurements like those shown in Fig. 12.1 suggest that the true contribution is only about 1% of this magnitude near room temperature and, moreover, decreases linearly with decreasing temperature.

What does the classical picture predict regarding the conductivity of the electrons? Although the electrons whiz about with an average thermal speed, they are unable to travel very far before encountering an obstacle and undergoing some form of elastic collision, as shown in Fig. 12.3. Much like a particle undergoing a random walk, the electron dashes about in random directions following each collision. The mean free path of the electron, l_{mfp}, is given by the product of the average thermal speed and a mean collision time,

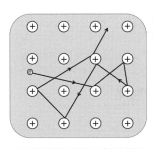

$$l_{\mathrm{mfp}} = v_{\mathrm{therm}} \tau_{\mathrm{coll}}. \qquad (12.3)$$

Since the electrons are treated as point particles, the most likely candidates for collisions are the more sizable ion cores for which the mean free path would be comparable to the atomic spacing. Consider now what happens when an electric field is introduced. We know that Ohm's law predicts that the electric field will produce a steady-state current density, given by

Figure 12.3

Classical model of electrical conductivity. Electrons undergo numerous collisions characterized by a mean free path, l_{mfp}.

$$J = \sigma E, \qquad (12.4)$$

where σ is the conductivity and E the magnitude of the applied electric field. But under the influence of an electric field, each electron is individually *accelerated* and so the resulting current should continue to increase with time, in violation of Ohm's law. In order to stem this acceleration, we must assume the collisions that the electrons experience are such that they provide an effective *drag force* that competes with the accelerating electric field. Much like the drag experienced by a parachute, this would then lead to a steady-state 'drift' velocity of the electrons and a constant current. This is the idea behind the *Drude model* of conduction wherein the average impulse (change in momentum) experienced in collisions is comparable to the average incident momentum,

$$\Delta p \approx F_{\mathrm{drag}}\tau_{\mathrm{coll}} \approx -m v_{\mathrm{drift}}. \tag{12.5}$$

Under steady-state conditions, the accelerating force is just cancelled by the drag force,

$$F_{\mathrm{net}} = F_{\mathrm{drag}} - eE = 0, \tag{12.6}$$

and the drift speed is given as,

$$v_{\mathrm{drift}} = \frac{-eE\tau_{\mathrm{coll}}}{m}. \tag{12.7}$$

The current density is then given by,

$$J = nq v_{\mathrm{drift}} = \left(\frac{n_e e^2 \tau_{\mathrm{coll}}}{m}\right) E, \tag{12.8}$$

where n_e is the density of the mobile valence electrons. We see in Eq. (12.8) the form of Ohm's law with the resistivity given by,

$$\rho = \frac{1}{\sigma} = \frac{m}{n_e e^2 \tau_{\mathrm{coll}}} = \frac{m v_{\mathrm{therm}}}{n_e e^2 l_{\mathrm{mfp}}}. \tag{12.9}$$

Measurements of the temperature dependence of the resistivity of metals display a linear temperature dependence over a wide range, which is simply not supported by the Drude model. In the Drude model, the temperature dependence enters through the average thermal speed, which in Eq. (12.2) is seen to vary only as the square root of the temperature.

12.2 Free electron model

As you might have already surmised, the classical model fails mainly because it is "classical" and so ignores the wave nature of the electron. When we treat the electron as a wave, we encounter some familiar features that are identical with those of the elastic waves studied in previous

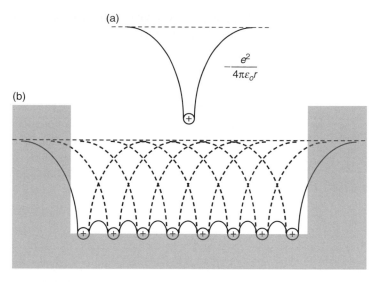

(a)

$$-\frac{e^2}{4\pi\varepsilon_o r}$$

(b)

(a) The Coulomb potential of an isolated ion core. (b) The collective potential of an ordered one-dimensional crystal of ion cores, illustrating reduction of the potential in the interior due to overlap. The gray boundary suggests the free electron model approximation, in which the true potential is replaced by that of an infinite square well potential.

chapters. Namely, because of the finite extent of the crystal, the electron waves are bounded and so limited to a discrete set of allowed wavelengths and energies. But there are also important differences between phonons and electrons. Firstly, the electron is a *fermion*. Unlike the phonon (which is a boson), the electron must obey *Pauli exclusion* and only one electron can occupy a given quantum mechanical state at any given time. This restriction alone leads to significant changes in the distribution or occupancy of the electron energy levels. Secondly, the electron wave function is a measure of the electron density in space and (unlike phonons) is not restricted by the limits of the Brillouin zone boundary. While it was meaningless to discuss an elastic wave with a wavelength shorter than $2d$, it is perfectly legitimate to discuss an electron density existing within this distance (i.e. outside the first Brillouin zone). And finally, the electron possesses the property of electrical charge which gives rise to measurable electrical properties of the material.

Let us begin by considering a single, metallic atom that has Z valence electrons. Examples would include Na with $Z = 1$ or Ca with $Z = 2$. In either instance, the valence electrons are loosely bound to the remaining ion core and experience a potential energy something like that sketched in Fig. 12.4a, arising from the Coulomb attraction:

$$V(r) = -Ze^2/4\pi\varepsilon_o r. \qquad (12.10)$$

The bound energy states of the electron would be given by the time-independent Schrödinger equation:

$$H\psi_n = \left\{ -\frac{\hbar^2}{2m}\nabla^2 + V(r) \right\}\psi_n = E_n\psi_n, \qquad (12.11)$$

and would mimic those of a hydrogen-like atom.

But now, consider the Coulomb potential of a one-dimensional crystal, as sketched in Fig. 12.4b. The ion cores are uniformly spaced and each produces an individual Coulomb potential, given as in Eq. (12.10). However, note that when these individual potentials are superimposed and combined to form the net potential, the overlap between adjacent potentials results in significant reduction of the potential inside the crystal. The valence electrons are no longer bound to any single ion core, but are able to move throughout the crystal and are bounded only by the edges of the crystal, where the potential well rises sharply. Inside the crystal, the net potential appears as a periodic series of small potential energy "bumps."

As a first stab at a quantum mechanical description of the electron's motion, we make a brash approximation. We assume that the bumps in Fig. 12.4b are much smaller than the walls and treat the overall potential as that of an infinite well. In this approximation, known as the *free electron model*, the behavior of the electrons reduces to that of the familiar "particle-in-a-box" problem in which the allowed electron wavelengths are given by the standing wave patterns:

$$\lambda_n = 2L/n, \qquad (12.12)$$

and the allowed energy levels are given by,

$$E_n = p^2/2m = n^2\frac{h^2}{8mL^2} = n^2 E_1. \qquad (12.13)$$

Although Eq. (12.13) is a familiar result, beware, the energy levels are much more closely spaced than those in the hydrogen atom because the size of the box (L) now corresponds to the macroscopic size of the crystal (Ex. 1).

12.2.1 Fermi level

A crystal with N atoms has a total of $N_e = ZN$ free electrons that must be distributed into the energy levels in accordance with the Pauli exclusion principle. Since an electron has two independent spins (up and down), each energy level, described by Eq. (12.13) for a 1D crystal, can support no more than two electrons. At absolute zero, these electrons will settle into the lowest energy configuration possible (i.e. a ground state configuration), which corresponds to a *stacking* of electron pairs into the energy levels from the lowest, then upwards until all the electrons have been accounted

Fermions

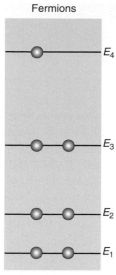

Bosons

A comparison of the occupancy of energy levels by fermions and bosons. Fermions obey Pauli exclusion and thus only two electrons (spin up and spin down) are allowed in each energy level. Bosons (e.g. phonons and photons) are not restricted by Pauli exclusion and any number can occupy a given energy level.

for, as illustrated in Fig. 12.5. The uppermost filled energy level is then given by,

$$n_F = N_e/2. \tag{12.14}$$

This uppermost filled level is known as the *Fermi level* and has a corresponding energy, given for the 1D case from Eq. (12.13) as,

$$E_F = \left(\frac{N_e}{2}\right)^2 E_1, \tag{12.15}$$

known as the *Fermi energy*. For typical metals, shown in Table 12.1, the Fermi energy is in the range 2 to 10 eV (or Fermi temperatures, $T_F = E_F/k_B$ in the range 20,000 to 100,000 K). The electrons in this Fermi level have a corresponding average speed, v_F, given by,

$$E_F = \frac{1}{2}mv_F^2, \tag{12.16}$$

which is in the range of 2×10^6 m/s or about 1% the speed of light. This is rather fast, but still amenable to a non-relativistic treatment.

12.2.2 Specific heat

We will return to the issues of electron conduction shortly, but first let us see what our free electron model predicts for the specific heat contribution associated with the electron gas. Because the electrons are fermions, their thermodynamic properties are governed by the *Fermi–Dirac energy distribution*:

$$P_{FD}(E) = \{\exp[(E - \mu)/k_B T] + 1\}^{-1}, \tag{12.17}$$

where μ is known as the *chemical potential*. This distribution function expresses the probability that a level of energy E will be occupied at a given temperature. As seen from consideration of Eq. (12.17), μ marks the energy of the level that is 50% filled and, for most modest temperatures, this chemical potential remains very nearly equal to the Fermi energy ($\mu \approx E_F$).

At absolute zero, this distribution assumes the shape of a step function in which all the levels below E_F are filled (each with two electrons) and all those above E_F are empty. As the temperature is increased, the distribution function spreads somewhat, as shown in Fig. 12.6, and at extremely high temperatures, the distribution becomes quite asymmetric near E_F.

Now, before we go any further, we should take a small reality check. As is evident from the melting points included in Table 12.1, most metals remain crystalline only at temperatures well below about 2000 K and so the only distributions shown in Fig. 12.6 that will have any relevance for crystalline *solids* are those ($y = T_F/T > 30$) for which the distribution remains

Table 12.1 Electrical properties of several metallic elements. (Values obtained from *Handbook of Chemistry and Physics* (1983) and from C. Kittel (2005).)

	Z	n_e $(10^{28}$ m$^{-3})$ at 293 K	E_F (eV)	T_F $(10^4$ K)	v_F $(10^6$ m/s)	σ $(10^7$ mho/m) at 293 K	R_H $(10^{-10}$m^3/C)	T_m (K)	ϕ (eV)	l_{mfp} (Å)
Li	1	4.63	4.74	5.51	1.29	1.1	−1.7	453	2.9	105
Na	1	2.54	3.24	3.77	1.07	2.1	−2.07	371	2.75	305
K	1	1.34	2.12	2.46	0.86	1.4	−4.23	336	2.3	310
Rb	1	1.08	1.85	2.15	0.81	0.83	−5.04	312	2.16	201
Cs	1	0.85	1.59	1.84	0.75	0.5	−	301	2.14	148
Cu	1	8.45	7.00	8.16	1.57	5.9	−0.54	1336	4.65	390
Ag	1	5.86	5.49	6.38	1.39	6.2	−0.9	1234	4.26	526
Au	1	5.77	5.53	6.42	1.40	4.5	−0.72	1337	5.1	376
Be	2	24.7	14.3	16.6	2.25	2.5	+2.43	1551	4.98	80
Mg	2	8.62	7.08	8.23	1.58	2.3	−0.83	922	3.66	145
Ca	2	4.63	4.69	5.44	1.28	2.9	−	1112	2.87	245
Zn	2	13.2	9.47	11.0	1.83	1.7	−	693	4.33	82
Cd	2	9.26	7.47	8.68	1.62	1.4	−	594	4.22	89
Al	3	18.1	11.7	13.6	2.03	3.8	+1.02	933	4.28	148
Ga	3	15.3	10.4	12.1	1.92	0.71	−	303	4.2	25
In	3	11.5	8.63	10.0	1.74	1.2	+1.6	430	4.12	64

symmetric about E_F and which display only a minimal deviation from the step function observed at absolute zero. Near E_F, one finds (Ex. 5) the slope of $P_{FD}(E)$ is $1/4k_BT$ and hence, for a wide range of "terrestrial" temperatures, the only electrons that can be excited beyond the Fermi energy are those within about $2k_BT$ of the Fermi level. How many electrons meet this condition? Consider the energy levels just below n_F, as illustrated by the shaded triangle in Fig. 12.6. The energy of these levels can be expressed in terms of the Fermi energy as

$$E_n = n^2 E_1 = (n_F - \Delta n)^2 E_1 = (1 - \Delta n/n_F)^2 E_F. \qquad (12.18)$$

Since Δn is much smaller than n_F, we can use the binomial expansion to obtain the energy near E_F as

$$E_n \approx (1 - 2\Delta n/n_F)E_F, \qquad (12.19)$$

and for an energy difference of $2k_BT$, the range of excitable levels is given by,

$$2k_BT = E_F - E_n \approx 2\Delta n E_F/n_F. \qquad (12.20)$$

Since there are roughly $N_{ex} \approx \Delta n/2$ electrons, represented by the shaded triangle in Fig. 12.6, the total number of excitable electrons is given by,

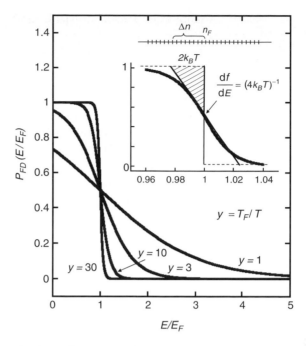

The Fermi–Dirac distribution function plotted against energy (scaled to the Fermi energy) is shown for various temperature ratios $y = T_F/T$. For $y > 10$, the distribution is well approximated by a step function. Inset shows an enlarged view of the distribution for $y = 100$ and the linear approximation to the shape near E_F. The number of excited electrons is roughly proportional to the area of the shaded triangle.

$$N_{ex} \approx \frac{\Delta n}{2} = \frac{n_F}{2} \frac{T}{T_F} = \frac{N_e}{4} \frac{T}{T_F}. \qquad (12.21)$$

Only these N_{ex} electrons are able to access thermal energy and participate in the exchange of heat with an external environment needed to produce a contribution to the specific heat. All other electrons are "locked" into their energy states and cannot participate. The amount of additional energy gained by these N_{ex} electrons is roughly $3k_BT/2$ for each, and the corresponding contribution to the internal energy is

$$U \approx 3N_{ex}\,k_BT/2 \approx \frac{3N_e k_B}{8}\frac{T^2}{T_F}. \qquad (12.22)$$

Hence, the contribution to the specific heat is of order,

$$C_V = \mathrm{d}U/\mathrm{d}T \approx \frac{3}{4}R\left(\frac{T}{T_F}\right) \qquad (12.23)$$

per mole of electrons. This result matches well with experiment. It contains both the linear temperature dependence and, because $T_F \approx 30{,}000$ K, accounts for roughly 1% of the classical value near room temperatures, as observed.

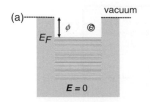

(a)

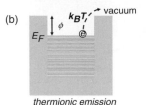

(b)

thermionic emission

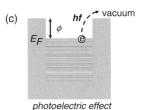

(c)

photoelectric effect

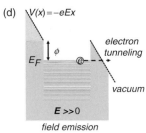

(d)

field emission

Figure 12.7

Electron emission phenomena are illustrated for (b) thermionic emission, (c) the photoelectric effect and (d) the field effect. Note that the field effect involves a quantum mechanical tunneling of the electron through a barrier that is controlled by the strength of the applied electric field.

A more exact calculation yields nearly the same result to within a factor of order unity.

12.2.3 Emission effects

The simple free electron model also helps to explain a number of electron emission phenomena, including *thermionic emission, field emission*, and the *photoelectric effect*. These three processes are sketched in Fig. 12.7, where the electron levels are indicated. The uppermost level is the Fermi level that resides at some energy difference, ϕ, below the zero potential of the outside vacuum. This energy difference is known as the *work function* and is a measure of the minimum energy input required to emit an uppermost electron to the exterior.

In *thermionic emission*, shown in Fig. 12.7b, thermal energy (k_BT) supplied to the crystal is sufficient to overcome the work function. In the *photoelectric effect*, shown in Fig. 12.7c, incident photons provide the needed energy. Because photons are energy quanta, photoelectric emission will occur only for photons whose energy meets or exceeds the work function. Excess photon energy, beyond that of the work function, appears as kinetic energy of the emitted electron.

The *field emission* process involves *tunneling* of the electron from the Fermi level into the vacuum. In order for this to happen, a narrow potential energy barrier must be formed through which the electron has a significant probability for tunneling. By applying an external electric field, the potential outside the crystal is distorted, as shown in Fig. 12.7d. On one side the potential is raised, while on the other side it is lowered and forms a barrier whose width is controlled by the magnitude of the applied field.

12.2.4 Free electron model in three dimensions

To deal with real crystals, we must extend our quantum mechanical model to a three-dimensional space, where the Schrödinger equation is now expressed as,

$$-\frac{\hbar^2}{2m}\nabla^2\psi_K = E_K\psi_K. \tag{12.24}$$

We then assume corresponding plane wave solutions for the electron wave function of the form,

$$\psi_K(\vec{r}) = Ae^{i\vec{K}\cdot\vec{r}} = Ae^{iK_xx}e^{iK_yy}e^{iK_zz}, \tag{12.25}$$

and find that the allowed energies are given by,

$$E_K = \frac{\hbar^2}{2m}\left\{K_x^2 + K_y^2 + K_z^2\right\} = \frac{\hbar^2|\vec{K}|^2}{2m}. \tag{12.26}$$

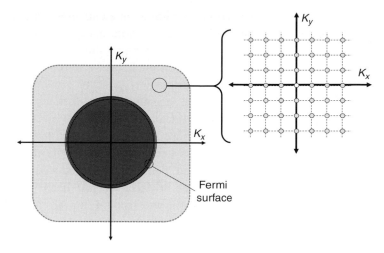

Figure 12.8 A representation of the K-space for allowed electron wave functions in a three-dimensional crystal. Each allowed mode is represented by a dot in the upper right figure. At low temperatures, electrons occupy modes nearest the origin, filling K-space in an approximately spherical fashion. The highest energy electrons reside on the surface of this sphere known as the Fermi surface.

However, the electrons are still contained in a box formed by the boundaries of a macroscopically large, three-dimensional crystal. To enforce this limitation, we again apply the periodic boundary conditions, introduced previously in Chapter 10 for the restriction of phonon modes in a three-dimensional crystal. Specifically, we demand that the wave functions repeat themselves after some large distance, $L=Na$, such that

$$\psi_K(\vec{r}) = \psi_K(\vec{r} + L\hat{x}) = \psi_K(\vec{r} + L\hat{y}) = \psi_K(\vec{r} + L\hat{z}). \qquad (12.27)$$

Application of the periodic boundary condition to the wave function in Eq. (12.25) leads to the constraint

$$e^{iK_x Na} = e^{iK_y Na} = e^{iK_z Na} = 1, \qquad (12.28)$$

and the allowed electron modes are then given as

$$K_x, K_y, K_z = \pm n\frac{2\pi}{Na}, \quad n = 1, 2, \cdots \infty. \qquad (12.29)$$

Note that these modes are virtually *identical* with those of the phonons discussed in Chapter 10 and can again be represented as a uniformly distributed set of discrete points in a three-dimensional K-space, as shown in Fig. 12.8. Again, the density of states is

$$g(K)dK = (VK^2/2\pi^2)dK. \qquad (12.30)$$

The only differences are that, (1) while phonons were only meaningful for K *inside* the Brillouin zone (i.e. for $\lambda > 2d$), the electron waves outside the Brillouin zone remain meaningful, and (2) while an unlimited number of phonons could occupy a given mode, only *two* electrons (one spin up and the other spin down) are allowed to occupy a mode. Consequently, in the ground state, electrons will fill in the lowest energy levels nearest the origin in Fig. 12.8, expanding radially outwards until all N_e electrons are accounted for, forming a spherical pattern in K-space centered at the origin. The outermost electrons on the surface of this sphere are those with the largest kinetic energy and so the outer surface forms what is known as the *Fermi surface*. This surface is simply the 3D analog of the Fermi level introduced earlier. To determine the Fermi energy at this surface, we just need to fill the modes from the center outwards until we account for all $N_e = ZN$ electrons. Thus, for two electrons per mode,

$$N_e = \int_{K=0}^{K_F} 2\left(\frac{V}{8\pi^3}\right)(4\pi K^2 dK), \tag{12.31}$$

from which

$$K_F = \left(\frac{3\pi^2 N_e}{V}\right)^{1/3} = \left(3\pi^2 n_e\right)^{1/3} \tag{12.32}$$

and

$$E_F = \frac{\hbar^2 K_F^2}{2m} = \frac{\hbar^2}{2m}\left(3\pi^2 n_e\right)^{2/3}. \tag{12.33}$$

12.2.5 Conduction in the free electron model

How does electron conduction appear in this three-dimensional free electron model? Since we are now treating the electron as a wave, we need to consider how an external electric field might affect the electron wave function. The momentum of the electron is given by the product of its mass and group velocity,

$$p = mv_g = m\frac{d\omega}{dK} = \frac{m}{\hbar}\frac{dE}{dK} = \hbar K, \tag{12.34}$$

and, in this instance of a free electron, matches that given by the deBroglie relation. Under the application of a constant uniform electric field, Newton's second law reads as,

$$\vec{F} = \frac{d\vec{p}}{dt} = \hbar\frac{d\vec{K}}{dt} = -e\vec{E}, \tag{12.35}$$

from which we see that the mode (K) of the electron increases at a rate proportional to the strength of the field:

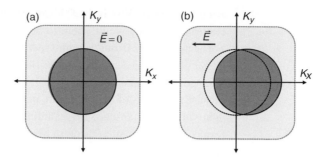

Figure 12.9 (a) In the absence of an external electric field, the occupied electron modes are symmetrical about the origin of K-space. As a result, there is no net momentum to produce an electrical current. (b) When an external field is applied, electrons near the Fermi surface are promoted to modes of higher energy and a net momentum results in the direction opposite the field.

$$\frac{\mathrm{d}\vec{K}}{\mathrm{d}t} = -e\vec{E}/\hbar. \qquad (12.36)$$

Because the electron states are discrete, this means that the electric field forces the electron to "jump" from one dot in K-space to the next adjacent dot by either shortening or lengthening its wavelength. Since this applies to all the electrons in the crystal, the entire sphere of occupied modes is *shifted* at a constant rate in the direction opposite to the field, as illustrated in Fig. 12.9. Without the field present, the sphere is initially unshifted and the momenta of all the electrons are balanced such as to produce no net momentum and hence no net flow of charge. Conversely, in the presence of the electric field, this balance is removed and a net momentum, produced by a small fraction of electrons near the Fermi surface traveling in a direction opposite to the field, produces a net current in the crystal.

Of course, Eq. (12.36) implies that this current would increase continuously over time, in an apparent violation of Ohm's law, and so to compensate, we must assume (as in the classical picture) that the electrons still undergo collisions of some sort. What are some of the objects that might interfere with electron propagation? They certainly include the ion cores of the lattice, as well as the other electrons and impurities that might be present. Returning to the Drude expression for the conductivity, we now replace the thermal average speed by the corresponding speed for electrons near the Fermi surface and find,

$$\sigma = \frac{n_e e^2 \tau_{\mathrm{coll}}}{m} = \frac{n_e e^2}{m v_F} l_{\mathrm{mfp}}. \qquad (12.37)$$

When the mean free path is evaluated from measurements of the conductivity of metals (see Table 12.1), one may be surprised to find that the mean free path is quite large – indeed far larger than the typical spacing between the ion cores. It seems then that these ion cores do not serve as any impediment to electron motion. Why is that? There are at least as many ion cores as there are electrons and they would seem to be an easy target!

The answer lies in our quantum mechanical treatment, where the electron is no longer viewed as a classical particle, but rather as a wave. Waves, as we have seen repeatedly in Chapter 6 and Chapter 7, travel largely *unscattered* through a crystal because the atoms of the crystal are perfectly arranged in a periodic manner. For light waves, this absence of scattering is interpreted as a consequence of the Ewald–Oseen extinction theorem of classical optics. For X-rays, scattering only occurs for a discrete set of waves that happen to satisfy the Bragg scattering condition ($\vec{q} = \vec{G}_{hkl}$). All other waves that do not satisfy this Bragg condition (including electron waves) are simply unaffected by a perfectly periodic lattice.

Since the ion cores are ineffective in scattering the electrons, what then determines the mean free path? Since the scattering of the electrons is not caused by the periodic arrangement of ion cores, it must be a result of *imperfections* in that periodicity. These imperfections could be dislocations at which the periodic pattern is abruptly shifted, or impurities in the crystal that likewise alter the periodicity. This would then account for the rather large mean free path seen experimentally.

But what of the temperature dependence of the resistivity which increases linearly with temperature? This feature arises from the wealth of phonons present in the crystal, which not only disturb the perfect periodicity of the lattice, but also participate in scattering with the electrons. As we saw in the previous chapter, the mean free path for phonon scattering decreases with increasing phonon density. At high temperatures, this phonon density increases linearly with temperature (see Eq. (11.30)), and from Eq. (12.36), one sees that the scattering of electrons by the phonon gas produces a linearly temperature-dependent contribution to the resistivity. Together with the impurities discussed above whose mean free path is temperature independent, these two mechanisms account well for the observed Matthiessen's rule of Eq. (12.1).

12.2.6 Hall effect

The Hall effect provides a reliable method for determining the density of conduction electrons in a metal and, as we will see in a moment, raises some important problems for the free electron model. Consider a rectangular bar of metal, as illustrated in Fig. 12.10, that is placed in crossed electric and magnetic fields as shown. The electric field directed along the length of the bar propels electrons in the opposite direction, while their motion in the magnetic field forces the electrons to veer off towards the side of the bar. After some time, a steady-state situation will develop in which the electron density is higher on one side of the bar than on the other, producing a detectable voltage difference.

The equation of motion of an electron under these conditions is given by,

$$\vec{F} = \frac{d\vec{p}}{dt} = -e(\vec{E} + \vec{v} \times \vec{B}) - \frac{\vec{p}}{\tau_{coll}}, \tag{12.38}$$

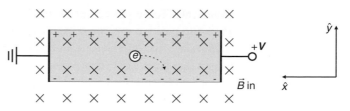

View of a conducting metallic bar situated in crossed magnetic and electric fields. Electrons entering from the left travel to the right under the influence of the applied electric field, but are forced downwards by the magnetic field. Under steady-state conditions, a higher density of electrons is found flowing along the lower side of the bar than along the upper side, creating a vertical potential known as the Hall field, E_y.

where $-\dfrac{\vec{p}}{\tau_{\text{coll}}}$ is the effective drag force (see Eq. (12.5)) that retards the electron's acceleration. Because there is no applied force in the vertical direction, Eq. (12.38) results in two simultaneous equations for motion of the electron in the horizontal plane of the bar:

$$m\left[\frac{\mathrm{d}}{\mathrm{d}t} + \frac{1}{\tau_{\text{coll}}}\right]v_x = -e(E_x + v_y B)$$
$$m\left[\frac{\mathrm{d}}{\mathrm{d}t} + \frac{1}{\tau_{\text{coll}}}\right]v_y = -e(E_y - v_x B). \tag{12.39}$$

When steady-state conditions are reached, the accelerations vanish and $v_y \to 0$ such that the above two equations each reduce to

$$v_x = -eE_x\tau_{\text{coll}}/m$$
$$v_x = E_y/B. \tag{12.40}$$

The *Hall field*, E_y, associated with the vertical voltage difference, is then given as,

$$E_y = -eE_x\tau_{\text{coll}}B/m. \tag{12.41}$$

By knowing the magnitudes of the two applied fields and measuring the Hall field and current density along the length of the bar, one can determine the *Hall coefficient*,

$$R_H \equiv E_y/J_x B, \tag{12.42}$$

which, upon substitution of Eq. (12.41) and Eq. (12.8), can be expressed as,

$$R_H \equiv E_y/J_x B = \frac{-eE_x\tau_{\text{coll}}B/m}{\left(\frac{n_e e^2 \tau_{\text{coll}}}{m}E_x\right)B} = -\frac{1}{n_e e}. \tag{12.43}$$

Measurements of the Hall coefficient for a variety of metals are presented in Table 12.1. Although several values are negative, as they should be for conduction by electrons, a handful of materials exhibit anomalous *positive* coefficients. These present a real puzzle, as our picture of the metal is one in which the positive ion cores are affixed to an immobile lattice and only the electrons are free to propagate around. The positive Hall coefficients appear to suggest that there exist some form of positively charged particles that provide the conduction in some metals. To resolve this puzzle will require a revision of our free electron model, in which the periodic bumps in the potential (see Fig. 12.4b) formed by the ion cores are *not* ignored.

Summary

- Electrons are fermions and, unlike phonons, are constrained by Fermi–Dirac statistics. In the ground state, the electrons of a crystal fill in lowest energy states up to the Fermi level, E_F.

- Conduction electrons in a metal produce a small contribution to the specific heat, in addition to the Debye specific heat caused by phonons. Only those electrons near the Fermi level participate in this contribution.

- Like phonons, the quantum mechanical wave function of the electron is restricted by boundary conditions to assume one of a discrete set of allowed modes, K. Each mode can be occupied by no more than two electrons (with spins opposed).

- The finite resistivity of a metal results from electron collisions, either with lattice imperfections or scattering by phonons. Together, these two processes account for Matthiessen's rule, $\rho(T) \approx \rho_{\text{imp}} + \rho_{\text{phonon}}(T)$.

- The Hall effect allows both the density and sign of charge carriers to be determined. Instances of a positive Hall coefficient indicate the dominance of anomalous positive charge carriers in the overall conduction of some materials.

Exercises

12.1. Compute the ground state energy of a 1D crystal of size $L = 1$ cm and compare this with the ground state energy of a hydrogen atom (13.6 eV) and to the value of $k_B T$ at room temperature.

12.2. The kinetic energy of an electron gas is given by $U = \int_0^\infty E P_{FD}(E) g(E) dE$. Show that the kinetic energy of a 3D gas of N_e free electrons at 0 K is $U_o = (3/5) N_e E_F$.

12.3. Show (from your finding in Exercise 12.2) that the pressure of an electron gas is two-thirds of the kinetic energy per unit volume.

12.4. A ^{3}He nucleus has spin ½ and, like electrons, must also obey the Pauli exclusion such that it fills a box with two atoms per quantum state (one with spin up and the other spin down). Calculate the corresponding Fermi energy and Fermi temperature for liquid ^{3}He if it has a density of 0.059 g/cc.

12.5. Consider the Fermi–Dirac distribution function given in Eq. (12.17) with μ taken to equal E_F. (a) Show that the slope of $P_{FD}(E)$ at E_F is $-(4k_BT)^{-1}$. (b) Show that $-\frac{\partial P_{FD}}{\partial E}$ is symmetrical and that its integral is unity. How does this function compare with that of the Dirac delta function?

12.6. Pure gold is commonly labeled as "24 karat". A lower purity form of gold containing 4.2% of impurities is labeled as "23 karat". (a) Suppose that the impurities in a 23 karat gold specimen are evenly distributed within the gold crystal. What mean free path is associated with these impurities? (b) Assume that the conductivity of pure gold, given at room temperature in Table 12.1, is that due to phonon scattering alone. Use Matthiessen's rule to determine the conductivity you would expect to find for the 23 karat specimen.

12.7. Assuming that Eq. (12.23) properly expresses the specific heat contribution of the conduction electrons, show that the contribution of these electrons to the thermal conductivity is $\kappa_{\text{th}} = nk_B^2 T \tau_{\text{coll}}/2m$.

12.8. In Chapter 4, we found that paramagnetic materials exhibit the Curie law, in which the magnetic susceptibility increases with decreasing temperature, as given by Eq. (4.20): $\chi_m = \frac{\mathrm{d}M}{\mathrm{d}H} = n\mu_o \frac{g^2\mu_B^2 J(J+1)}{3k_BT} = \frac{C}{T}$. For metals containing N_v mobile conduction electrons, we would expect to find a corresponding paramagnetic contribution, only, with n replaced by $n_e = N_e/V$. However, the fermion character of these mobile electrons limits their response to an applied magnetic field. Show that in this case of so-called Pauli spin paramagnetism, the susceptibility associated with the conduction electrons is temperature independent and given as $\chi_m = n_e\mu_o \frac{\mu_B^2}{4E_F}$.

Suggested reading

A comparable introduction to the free electron model can be found in almost any standard Solid State textbook. Below are just a few personal favorites.

C. Kittel, *Introduction to Solid State Physics*, 8th Ed. (John Wiley and Sons, 2005).

N. W. Ashcroft and N. D. Mermin, *Solid State Physics* (Holt, Rinehart and Winston, New York, 1976).

M. A. Omar, *Elementary Solid State Physics: Principles and Applications* (Addison-Wesley, Reading, MA, 1975).

J. S. Blakemore, *Solid State Physics*, 2nd Ed. (W. B. Saunders Co., Philadelphia, 1974).

13 Electrons: band theory

Introduction

In the last chapter, we took a brash and somewhat unrealistic approach to treating the motion of electrons in a crystal. Although we know that the electron travels through a periodic potential caused by the regular arrangement of ion cores, we disregarded this "bumpy terrain" and considered instead only the barest consequences of the electron being trapped in the crystal "box" as a whole. In spite of its simplicity, this free electron model provided insightful explanations, not only for the origin of the small electronic contribution to specific heat and the temperature dependence of the electrical resistivity, but also for a host of emission phenomena, including the photoelectric effect.

However, the free electron model fails to provide any insight into additional questions regarding electrical conduction, such as (1) the anomaly of positive Hall coefficients that would imply positive charge carriers, and (2) the peculiar pattern of conductors, insulators and semiconductors that is found in the periodic table. In this chapter, we examine the *nearly free electron model* as a natural extension in which a weak, periodic potential is introduced. As a direct consequence of this addition, the continuum of electron energies in our free electron model now becomes separated into bands of allowed electron energy, separated by disallowed energy gaps. This separation of the electron energy into bands and gaps is key to understanding the division of materials into conductors, insulators and semiconductors, as well as providing a natural interpretation for the positive Hall coefficients.

13.1 Nearly free electron model

13.1.1 Bloch functions

In the free electron model of the previous chapter, the potential was uniformly zero inside the crystal, because we ignored the small bumps created by the overlap of the Coulomb potential of neighboring ion cores. As a consequence of ignoring these bumps, the electron wave function assumed a simple plane

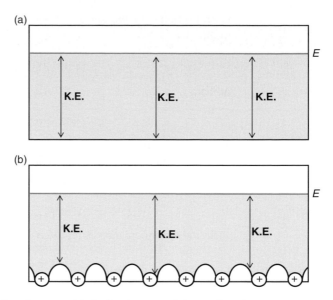

Figure 13.1 (a) In the free electron model, the kinetic energy of the traveling electron remains constant. (b) In the nearly free electron model, the traveling electron must contend with a spatially varying potential energy causing the electron's kinetic energy to vary accordingly.

wave constrained only by the physical boundary of the crystal itself, and its energy was completely kinetic. In reality, these bumps exist and so the potential is not quite exactly zero inside the crystal. As illustrated in Fig. 13.1, this non-zero potential forces the kinetic energy of the electron to now vary as it travels through the crystal and will require some corresponding modification of the electron wave function to achieve this spatial dependence.

One feature in our favor is that the potential inside is *periodic* and so repeats itself at regular multiples of the lattice spacing:

$$V(x) = V(x + d). \tag{13.1}$$

Without going into a full-scale assault on the Schrödinger equation for this situation, we can infer a number of properties of the electron wave function that would result from such a periodic potential. There are two conditions we can place on the allowed electron wave functions. The first is just the restriction set by the cyclic boundary conditions that reflect the finite extent of the crystal,

$$\psi(x) = \psi(x + Nd). \tag{13.2}$$

The other condition arises from the crystal symmetry. Although the electron is propagating through the crystal, its time-averaged electron density should appear the same in every unit cell of the crystal. In other words, there should not occur any cell that appears different from another by virtue of having some additional accumulation of negative charge. Since the electron density is proportional to the square of the electron wave function, this condition means that,

$$|\psi(x)|^2 = |\psi(x+d)|^2, \tag{13.3}$$

or that,

$$\psi(x+d) = e^{i\phi}\psi(x). \tag{13.4}$$

The type of wave functions that are consistent with these two conditions are known as *Bloch wave functions*, and have the form of a modulated traveling wave:

$$\psi_K(x) = u_K(x)e^{iKx}, \tag{13.5}$$

such that the modulation is repeated for every unit cell of the lattice,

$$u_K(x+d) = u_K(x). \tag{13.6}$$

13.1.2 Bragg scattering and energy gaps

As we migrate over from the free to the nearly free electron model, we in a sense are "turning on" the lattice potential. Clearly this will distort the electron wave function to assume a periodic modulation in the form of a Bloch wave, but it also has another important consequence. As the periodic lattice appears, so too do the familiar Brillouin zones and their boundaries, at which waves satisfy the Bragg scattering condition.

Imagine an electron traveling along the (100) direction of a simple cubic lattice with $\vec{K}_x = +\pi/d$. Because this $\vec{K}_x$ coincides with the edge of the first Brillouin zone, the electron is strongly scattered to $\vec{K}' = \vec{K}_x - \vec{G}_{100} = -\pi/d$, and appears (almost magically it seems) on the opposite side of the Brillouin zone. That is, the incident wave spawns a scattered wave of the same wavelength but traveling in the opposite direction. But wait, this new wave also lands on the edge of the Brillouin zone and so is strongly scattered back into the original wave! Multiple scattering reflections quickly evolve into equal portions of electron probability traveling in both directions, and so the electron assumes the properties of a standing wave.

In all respects, this standing wave feature is not new. You may recall we observed a similar standing wave feature appear for phonons at the first Brillouin zone boundary. Moreover, you might recall that in the case of a diatomic lattice, the phonon energy at the boundary was split into two possible values depending on which atom of the diatomic lattice (the light or the heavy one) was in motion. As illustrated again in Fig. 13.2a, the high energy state corresponded to motion of the lighter atom and formed the terminus of the optical branch, while the low energy state,

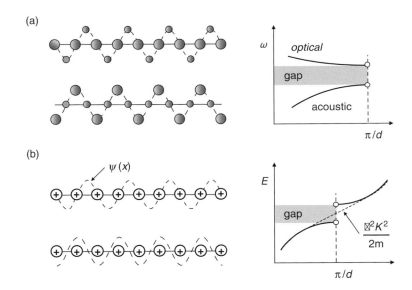

Figure 13.2 (a) Phonons in a diatomic crystal suffer an energy gap at the Brillouin zone edge due to two possible standing wave patterns. (b) A similar energy gap develops for electrons near the Brillouin zone edge.

corresponding to motion of the heavier atom, formed the terminus of the acoustic branch. Each wave shares a common wavelength of $\lambda = 2d$, but differs in phase angle by a quarter cycle.

   In the case of the standing electron wave, we can again anticipate a splitting of the energy into two values, corresponding to the two equivalent standing wave patterns that differ only by a quarter cycle in phase. As illustrated in Fig. 13.2b, the energy of the electron is lower when the electron density is centered on the ion core and higher when the electron density is centered between the ion cores.

13.2 Kronig–Penney model

For those not swayed by the simple arguments above, let us also illustrate the effects of a periodic potential by tackling Schrödinger's equation for a simple, 1D model that was first presented by Kronig and Penney (1931). In this simplistic model, the bumps of the potential are simulated by a series of small, narrow, periodically spaced barrier potentials, as illustrated in Fig. 13.3. Each barrier is described by

$$V(x) = \begin{cases} V_o, & -b < x < 0 \\ 0, & 0 < x < a, \end{cases} \quad (13.7)$$

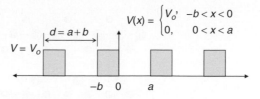

Figure 13.3 The periodic potential in the Kronig–Penney model.

and the electron wave function must satisfy the one-dimensional Schrödinger equation,

$$-\frac{\hbar^2}{2m}\frac{d^2\psi_K}{dx^2} + V\psi_K = E\psi_K, \tag{13.8}$$

which for each of the two regions of the barrier potential becomes

$$(I) \quad \frac{d^2\psi_K^I}{dx^2} + \frac{2m}{\hbar^2}[E - V_o]\psi_K^I = 0, \quad -b < x < 0$$

$$(II) \quad \frac{d^2\psi_K^{II}}{dx^2} + \frac{2m}{\hbar^2}E\psi_K^{II} = 0, \qquad\qquad 0 < x < a. \tag{13.9}$$

On introducing the two quantities:

$$\alpha^2 = \frac{2m}{\hbar^2}E \tag{13.10}$$

$$\beta^2 = \frac{2m}{\hbar^2}[V_o - E], \tag{13.11}$$

the Schrödinger equation, as it appears in each region, can be conveniently expressed as,

$$(I) \quad \frac{d^2\psi_K^I}{dx^2} - \beta^2\psi_K^I = 0, \quad -b < x < 0$$

$$(II) \quad \frac{d^2\psi_K^{II}}{dx^2} + \alpha^2\psi_K^{II} = 0, \qquad 0 < x < a. \tag{13.12}$$

Our electron wave function is a Bloch function of the form given in Eq. (13.5), and so these can quickly be reduced to statements regarding the form of the modulation, $u_K(x)$ as,

$$(I) \quad \frac{d^2 u_K^I}{dx^2} + 2iK\frac{du_K^I}{dx} - (\beta^2 + K^2)u_K^I = 0, \quad -b < x < 0$$

$$(II) \quad \frac{d^2 u_K^{II}}{dx^2} + 2iK\frac{du_K^{II}}{dx} + (\alpha^2 - K^2)u_K^{II} = 0, \qquad 0 < x < a. \tag{13.13}$$

The solutions to these two differential equations have the form,

$$u_K^I(x) = Ae^{(\beta - iK)x} + Be^{-(\beta + iK)x}$$
$$u_K^{II}(x) = Ce^{i(\alpha - K)x} + De^{-i(\alpha + K)x}, \tag{13.14}$$

where the various coefficients are determined by matching conditions at the two boundaries:

$$(i) \quad u_K^I(0) = u_K^{II}(0)$$

$$(ii) \quad u_K^I(-b) = u_K^{II}(a)$$

$$(iii) \quad \left.\frac{du_K^I}{dx}\right|_{x=0} = \left.\frac{du_K^{II}}{dx}\right|_{x=0} \tag{13.15}$$

$$(iv) \quad \left.\frac{du_K^I}{dx}\right|_{x=-b} = \left.\frac{du_K^{II}}{dx}\right|_{x=a}.$$

After applying the matching conditions, we arrive at four simultaneous equations that determine the coefficients,

$$(i) \quad A + B = C + D$$

$$(ii) \quad Ae^{-(\beta-iK)b} + Be^{(\beta+iK)b} = Ce^{i(\alpha-K)a} + De^{-i(\alpha+K)a}$$

$$(iii) \quad (\beta - iK)A - (\beta + iK)B = i(\alpha - K)C - i(\alpha + K)D \tag{13.16}$$

$$(iv) \quad (\beta - iK)Ae^{-(\beta-iK)b} + -(\beta + iK)Be^{(\beta+iK)b}$$
$$= i(\alpha - K)Ce^{i(\alpha-K)a} + -i(\alpha + K)De^{-i(\alpha+K)a},$$

which can be solved (Ex. 1) to obtain the following *dispersion relation* (a relation between the energy of the wave, given by α^2, and the wave vector K):

$$\frac{\beta^2 - \alpha^2}{2\alpha\beta} \sinh(\beta b) \sin(\alpha a) + \cosh(\beta b) \cos(\alpha a) = \cos[K(a + b)]. \tag{13.17}$$

This dispersion relation is a little complicated, but we can reduce it somewhat by narrowing the barriers (shrinking b) in such a way that the area ($V_0 b$) of the barrier is not altered. In this limit, the hyperbolic functions can be expanded (Ex. 7) as a Taylor series in first order and the relation reduced to,

$$P\left[\frac{\sin(\alpha d)}{\alpha d}\right] + \cos(\alpha d) = \cos[Kd], \tag{13.18}$$

where the lattice spacing $d \approx a$ and

$$P \equiv \frac{mV_0 ba}{\hbar^2}. \tag{13.19}$$

13.2.1 Energy bands and gaps

In Fig. 13.4, we have plotted the left-hand side of Eq. (13.18) as a function of the "energy", actually αd, for an arbitrarily chosen value of $P = 10$. One sees that the function oscillates in a damped-like manner. Also shown in the figure are the limits of the cosine function for which the right-hand side of Eq. (13.18) restricts the allowed solutions of the electron wave function. Apparently there

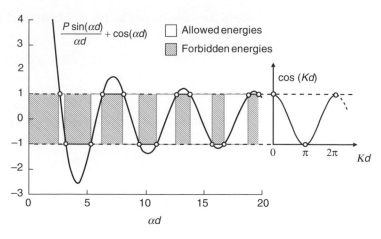

A plot of Eq. (13.18) for $P = 10$ (solid curve) shows undulations as a function of the energy parameter αd. Only those segments (white regions) of the curve between ± 1 can satisfy Eq. (13.18) while other segments (hashed regions) represent forbidden energies.

are some portions of the oscillating function that are clipped off and reside outside the limits of the cosine. These clipped sections of the curve therefore represent sections of the energy axis (αd) that are *forbidden*. Sections of the oscillating curve that lie within the limits of the cosine represent *allowed* energies.

Moving outwards along the energy axis, we see that the first allowed energy band (corresponding to the first Brillouin zone extending from $K = 0$ to $K = \pm \pi/d$) begins well above zero energy and extends over a rather narrow range of energies. This is followed by a forbidden band (or *energy gap*) that terminates at the start of the next range of K, extending from $K = \pm \pi/d$ to $K = \pm 2\pi/d$. This second band corresponds to the second Brillouin zone. Additional allowed energy bands correspond to higher-order Brillouin zones.

Using the set of discrete K enumerated as in Eq. (12.29), we can determine the corresponding energy of the electron wave and plot these energies now as a function of K, as is done in Fig. 13.5a to represent the dispersion relation more as we did for phonons in a previous chapter. In Fig. 13.5a, one sees the overall appearance of an envelope,

$$E_{\text{free}} = \frac{\hbar^2 K^2}{2m}, \tag{13.20}$$

which corresponds to the energy of electrons in the free electron model (where the periodic "bumps" in the potential were ignored), but which is now punctuated by energy gaps with two degenerate states at the boundaries of the Brillouin zones. A common, space-saving representation of this energy diagram is that of the reduced zone representation, shown in Fig. 13.5b, in which each band is folded back into the first Brillouin zone by appropriate addition of a reciprocal lattice vector, $\vec{G}_{hkl}$. In this representation we clearly see the

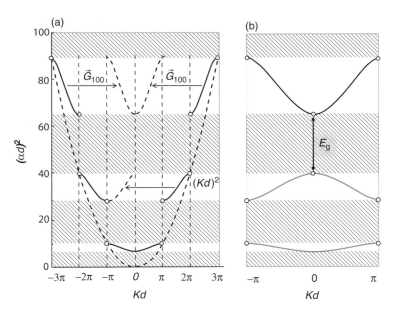

Figure 13.5 (a) The allowed energy levels of Figure 13.4 are plotted against the wave vector extending out to the 3rd Brillouin zone. Dashed gray curve shows the corresponding energy of a free electron (Eq. (12.26)). (b) In the reduced zone representation, the dispersion curves in each allowed band are folded back (by appropriate addition or subtraction of a reciprocal lattice vector, G_{hkl}) into the first Brillouin zone.

development of an energy-level diagram, consisting of bands of allowed energy separated by forbidden energy gaps.

We can now consider two extremes: (1) $P \to 0$, in which the bumps in the potential vanish, and (2) $P \to \infty$, in which the bumps in the potential grow to form a series of neighboring infinite well potentials. When the bumps vanish, Eq. (13.18) reduces to,

$$\cos(\alpha d) = \cos[Kd] \quad \text{or} \quad \alpha = K, \tag{13.21}$$

for which we recover the free electron result of Eq. (13.20). The energy gaps disappear as the bumps disappear. In the opposite limit, $P \to \infty$, the oscillating left-hand side of Eq. (13.18) makes vertical cuts through the allowed range of the cosine on the right-hand side of Eq. (13.18) at $K = \pm n\pi/d$. The allowed energy bands then narrow to discrete energy levels, such that

$$\alpha d = n\pi, \tag{13.22}$$

or,

$$E_{\text{bound}} = n^2 \frac{h^2}{8md^2}. \tag{13.23}$$

In this case, the electrons become bound to the ion cores and assume the energy levels of a particle bound in a small box of size d.

13.2.2 Mott transition

Increasing P in the Kronig–Penney model basically amounts to increasing the height of the bumps until they penetrate the energy band of the conducting electrons, and force a narrowing of the band into a discrete level. An obvious means for increasing the height of the bumps would be to increase the lattice spacing, d, as this would decrease the overlap between neighboring Coulomb wells (see Fig. 12.4). Several years ago, Sir Nevill Mott speculated on what would happen to the electron wave function as a lattice was expanded, and concluded that at some point, the Bloch wave functions that describe the electron as an *extended* object, with electron density evenly distributed throughout the entire crystal, would eventually collapse to *localized* wave functions consistent with the electron bound to a single ion core. This metal-to-insulator transition is often referred to as a *Mott transition* (see Mott, 1990).

As an extended wave function, each of the N_e mobile electrons in the metal can be thought of as contributing $1/N_e$ of their electron charge density in each unit cell of the lattice. Thus each cell of the lattice has, on average, Z units of electron charge and together with the ion core, retains charge neutrality. However, as the lattice constant increases to some unrealistically large value (say 1 meter), it is clear that the "lattice" is no longer a metal, but rather a collection of isolated atoms. At some point, each extended electron wave must have collapsed into one of the atoms.

Moreover, this transition should be rather abrupt as charge neutrality requires that either all or none of the electrons become localized. To see this, consider if only one electron became localized to a cell. This cell would still contain $(N_e - 1)/N_e$ contributions of electron density from the remaining extended state electrons, and would essentially contain $Z + 1$ electrons. The strong Coulombic repulsion generated by the close proximity of these electrons would be energetically unfavorable and so the localized electron would be prompted back into an extended state. Only when all the extended states collapse into localized states (one in each cell) would a favorable lowering of energy appear.

13.3 Band structure

We have learned much about the nature of energy bands and gaps in crystals from our study of the Kronig–Penney model. However, real crystals are far more complicated and the lattice potential, while still periodic, is shaped much by the specific crystal structure and contents of the unit cell. Moreover, real crystals are three-dimensional and the electrons within must be referred to a corresponding 3D K-space replete with a 3D Brillouin zone. The first Brillouin zone is formed by the Wigner–Seitz unit cell of reciprocal space, and typically

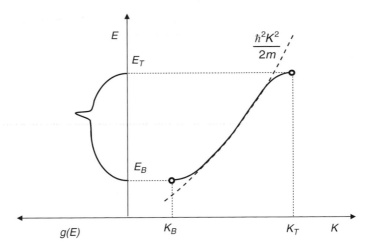

Figure 13.6 Characteristic features of a typical energy band and its density of states.

assumes the shape of a polyhedron for both the FCC and BCC lattices. For the simple cubic lattice, the first Brillouin zone is simply a cube.

In spite of all this inherent complexity, we can still draw some general conclusions regarding the band structure (i.e. the form of *E(K)*) of real crystals:

(1) ***E(K)* resembles that of the free electron model**. Since the lattice potential is assumed to be sufficiently weak as to allow the electron to exist in extended (traveling) states, the overall shape of *E(K)* should mimic that of a free electron, namely $E(K) \approx \frac{\hbar^2 K^2}{2m}$. This is illustrated in Fig. 13.6.

(2) **Energy gaps appear near the Brillouin zone boundaries**. The primary exception to the free electron energy spectrum of item (1) above, occurs near the boundary of any of the Brillouin zones. Here, the electron wave undergoes multiple Bragg scattering to form a standing wave pattern and an energy gap that is roughly centered about the average energy (i.e. that associated with a free electron whose *K* matches a BZ edge).

(3) **d*E*/d*K*→0 near a Brillouin zone boundary**. Because the Brillouin zone boundary corresponds to a standing wave pattern of the electron state, it follows that the group velocity of the electron must approach zero at the boundary. Since $v_g = \frac{d\omega}{dK} = \frac{1}{\hbar}\frac{dE}{dK}$, this implies that *E(K)* must curve flat as it either approaches or recedes from the Brillouin zone boundary.

(4) **Curvature of *E(K)* defines the electron's *effective mass***. Although the effect of an applied force on an electron as represented in *K*-space still results in a steady advance of the electron from one dot to the next in accordance with Eq. (12.36), the *real space* acceleration of the electron is given as,

$$a = \frac{dv_g}{dt} = \frac{dv_g}{dK}\frac{dK}{dt} = \frac{1}{\hbar}\frac{d^2E}{dK^2}\frac{F_{\text{ext}}}{\hbar} = \frac{F_{\text{ext}}}{m^*}, \tag{13.24}$$

where in comparison with Newton's second law, we introduce the notion of an effective mass,

$$m^* = \left[\frac{1}{\hbar^2}\frac{d^2E}{dK^2}\right]^{-1}, \tag{13.25}$$

whose magnitude depends on the curvature of the energy band.

(5) **E(K) is quadratic near the Brillouin zone boundary**. Near a Brillouin zone boundary, the electron energy deviates from that of a free electron because of the increasing propensity for scattering to occur. On the bottom of an energy band, for example, we might express this departure in a series expansion,

$$E(K) \approx E_B + A(K - K_B) + B(K - K_B)^2 + \cdots, \tag{13.26}$$

where the linear term must vanish if item (3) above is to be satisfied. Thus, to leading order, the energy band near a gap is quadratic. At the bottom of the band, the energy has positive curvature and thus a positive effective mass. Here the energy band is described as,

$$E(K) \approx E_B + \frac{\hbar^2}{2m_B^*}(K - K_B)^2 + \cdots. \tag{13.27}$$

At the top of the band, the curvature is negative and the effective mass is negative. In a similar fashion the band energy near the top of the band can be described as,

$$E(K) \approx E_T - \frac{\hbar^2}{2|m_T^*|}(K_T - K)^2 + \cdots. \tag{13.28}$$

(6) **The density of states, g(E), varies as $\sqrt{E}$ near the Brillouin zone boundary**. In the free electron model, electron energy was not afflicted by the Brillouin zones. In that situation, the Fermi surface assumed the shape of a sphere centered about $K = 0$, and the density of electron states (as a function of the electron energy) would be given as,

$$dN = g(K)dK = 2 \times \left(\frac{V}{8\pi^3}\right)4\pi K^2 dK = \left(\frac{V}{2\pi^2}\right)\left(\frac{2m}{\hbar^2}\right)^{3/2}E^{1/2}dE. \tag{13.29}$$

In real crystals, however, the Brillouin zones do affect the electron by creating gaps of energy where the density of states must vanish.

Imagine then a simple cubic lattice whose valence electrons were taken out and are now being gradually returned. As the electrons are returned, they stack into the lowest energy levels (two per level) and our Fermi surface initially proceeds outwards from $K = 0$ in the shape of a sphere. In this range of K-space the density of states is given as that of Eq. (13.29) (except that m is

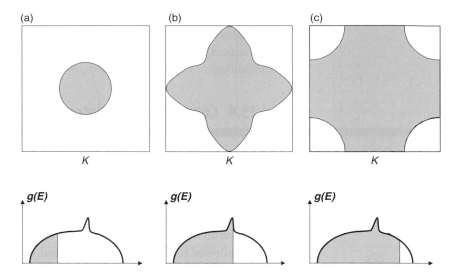

Figure 13.7 Filling the electron states of a real crystal. Top panel shows how electrons fill K-space within the confines of the Brillouin zone. (a) Initial addition of electrons reside near the bottom of a band and produce a spherical Fermi surface. (b) As electrons continue to fill the band, electron states near the Brillouin zone boundary are filled and the Fermi surface is distorted by the specific shape of the Brillouin zone. (c) Near the top of the band, vacant states vanish again in the form of a spherical surface. Lower panels display the corresponding density of states.

replaced by m^*) and increases as the square root of the energy difference from the bottom of the band.

However, as we add more electrons the Fermi sphere will first begin to contact the Brillouin zone boundary near the six walls of the cube which form the first Brillouin zone (see Fig. 13.7b). Since we know that the slope of $E(K)$ flattens near these walls, the number of dots in K-space that have similar electron energy (to within dE) will be enhanced here and the density of states, $g(E)$, will exhibit some corresponding discontinuity as a result.

For the simple cubic example, the highest energy states are those near the eight corners of the Brillouin zone cube (where $|K_x| = |K_y| = |K_z| = \pi/d$). Consequently, as we continue to add electrons these corners are the last dots of K-space to become occupied. Here the Fermi surface again assumes the form of a sphere, albeit inverted, 1/8th of which is located at each of the eight corners of the Brillouin zone. Here again, one can show (Ex. 2) that on approaching the top of the energy band at E_T, the density of states vanishes as the square root of the energy difference,

$$dN = g(E)dE = \left(\frac{V}{2\pi^2}\right)\left(\frac{2m}{\hbar^2}\right)^{3/2}(E_T - E)^{1/2}dE. \qquad (13.30)$$

In between these extremes, the energy contour is very sensitive to the specific shape of the (3D) Brillouin zone. The contours deviate considerably

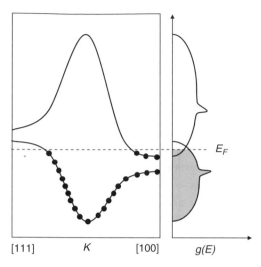

Figure 13.10 In some divalent crystals, conduction results from an overlap of neighboring energy bands.

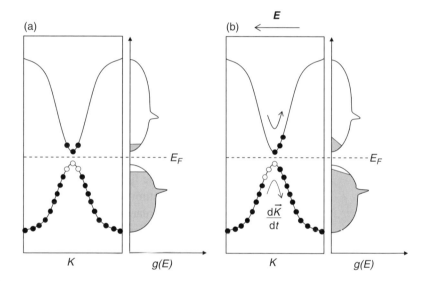

Figure 13.11 A semiconductor. (a) In a semiconductor the narrowness of the energy gap allows some levels of the conduction band to be occupied by electrons at finite temperatures. (b) This frees up modes in the valence band allowing conduction to occur under application of an electric field.

This then allows for the required imbalance of electron momentum (see Fig. 13.11) needed to create a net electrical current.

13.4.1 Holes

Quite often in a semiconductor, only a small fraction of the valence band electrons are thermally populated in the conduction band and thus only an equivalently small fraction of vacant modes exist in the valence band. In these

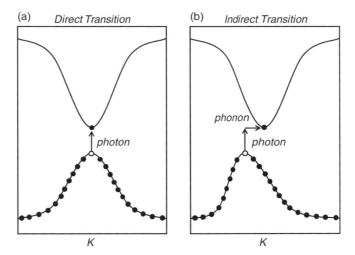

Figure 13.12 (a) In a direct bandgap transition, an incident photon excites an electron across the energy gap without altering its momentum. (b) In an indirect transition, excitation of the electron across a staggered energy gap is assisted by the production or annihilation of a phonon, whose main purpose is to appropriately alter the momentum of the electron.

instances, it is often more practical to discuss the motion of these vacancies as the motion of quasi-particles in the valence band, rather than the collective motion of all of the remaining electrons. These so-called *holes* act in the presence of an electric field as though they had a positive charge $+e$ and, depending on the detailed shape of the band structure, can dominate the overall conductivity to produce positive Hall coefficients.

Holes also play a significant role in the photonic properties of semiconductors. Consider the simple band structure shown in Fig. 13.12a in which a single electron has been excited into the conduction band by absorption of a photon. In being excited, the electron leaves behind a vacancy in the valance band at K_e which corresponds to a hole with wave vector $K_h = -K_e$. Thus the photon is responsible for the creation of two conducting charges, an *electron–hole pair*, both of which contribute to the conductivity under the application of an electric field. This direct photon transition is contrasted with the indirect transition that appears in more complex band structures, illustrated in Fig. 13.12b. Here, the creation of an electron–hole pair can occur by the absorption of a photon, together with the creation of a phonon. Phonon energies are characteristically small compared with the gap energy, and mostly provide the required momentum conservation since the momentum of a photon is characteristically small compared with that of a phonon.

13.4.2 Intrinsic semiconductors

Given the widespread popularity and variety of semiconductor-based devices found today in all sorts of solid state electronics, it is fitting to examine the properties of semiconductors in greater detail. We begin by looking at so-called *intrinsic* semiconductors (e.g. Si or Ge) in which very little of any impurity is present. In these intrinsic semiconductors, the small energy gap ($E_g \approx 1$ eV) permits a small concentration, n, of electrons to be thermally excited into the conduction band, as illustrated in Fig. 13.11a. Since these electrons are excited from a filled valence band, their absence results in the formation of an equal number of holes in the valence band and a hole concentration, p.

In contrast to a normal metal, whose resistivity increases with increasing temperature due to increased occurrence of scattering of the electrons by phonons, the resistivity of an intrinsic semiconductor actually decreases quite strongly with increasing temperature. In the case of the semiconductor, the reduction of conductivity due to phonon scattering is completely offset by the increase due to increased numbers of mobile charge carriers, achieved by raising the temperature. We can thus determine the temperature dependence of the conductivity of an intrinsic semiconductor by determining how the concentrations of both conduction electrons and holes increase with increasing temperature.

The number of electrons in the conduction band is given as,

$$N = \int_{E_C}^{E_T} P_{FD}(E)g(E)\mathrm{d}E = \int_{E_C}^{E_T} \left\{ \frac{1}{\exp[(E - E_F)/k_B T] + 1} \right\} g(E)\mathrm{d}E, \quad (13.31)$$

where E_C is the lower edge of the conduction band. Since $P_{FD}(E)$ decreases very rapidly above E_F at any meaningful temperatures, we can (1) replace the density of states by its quadratic form valid near the bottom of a band, and (2) extend the upper limit of integration to infinity. Furthermore, for a typical intrinsic semiconductor whose energy gap is around 1 eV, the exponential in the denominator of $P_{FD}(E)$ is large compared to unity, and we can rewrite Eq. (13.31) as,

$$N = \left(\frac{V}{2\pi^2}\right)\left(\frac{2m_C^*}{\hbar^2}\right)^{3/2} \int_{E_C}^{\infty} (E - E_C)^{1/2} \exp[-(E - E_F)/k_B T]\mathrm{d}E,$$

or,

$$n = N/V = 2\left(\frac{m_C^* k_B T}{2\pi\hbar^2}\right)^{3/2} \exp[(E_F - E_C)/k_B T]. \quad (13.32)$$

A similar derivation can be performed for the number density of holes, p, in the valence band whose occupancy is given as $1 - P_{FD}(E)$. This yields,

$$p = 2\left(\frac{m_V^* k_B T}{2\pi\hbar^2}\right)^{3/2} \exp[(E_V - E_F)/k_B T], \quad (13.33)$$

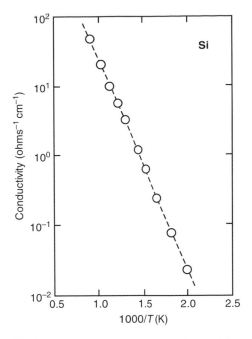

Figure 13.13 Intrinsic conductivity of Si plotted against inverse temperature (adapted from Moran and Maita (1954)).

where E_V is the upper edge of the valence band. The product of both concentrations can thus be expressed in terms of only the energy gap as,

$$np = 4\left(\frac{k_B T}{2\pi\hbar^2}\right)^3 (m_V^* m_C^*)^{3/2} \exp\left[-E_g/k_B T\right]. \qquad (13.34)$$

The temperature dependence arising from the exponential dominates over that from the prefactor and we observe that for an intrinsic semiconductor, the conductivity exhibits an Arrhenius dependence. An example of this for Si is shown in Fig. 13.13, where the slope equals half the band gap energy.

13.4.3 Extrinsic semiconductors

Because of their simple design, the intrinsic semiconductors discussed above always maintain equal populations of both conduction electrons and holes. However, in many semiconductor applications it is beneficial to alter this distribution, so as to make a semiconductor material with a majority of one type of carrier or the other. In practice, this is achieved by doping a small amount of impurity into the semiconductor and this produces what is known as an *extrinsic semiconductor*. Extrinsic semiconductors come in two varieties depending on whether electrons or holes are the majority carrier. The *n-type*

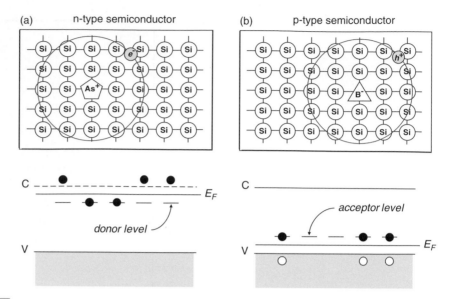

Figure 13.14 (a) An n-type semiconductor formed by doping arsenic (As) into silicon. The excess electron is loosely bound to the As impurity and occupies energy levels just below the conduction band. (b) A p-type semiconductor formed by doping boron (B) into silicon. The missing electron equates to an excess hole that is loosely bound to the B impurity and occupies energy levels just above the valence band.

extrinsic semiconductors have an excess of electrons in the conduction band, while *p-type* semiconductors have an excess of holes in the valence band.

A common example of an n-type material is that resulting from the addition of arsenic to silicon, as illustrated in Fig. 13.14a. Since Si resides in the fourth column of the periodic table and As resides in the fifth column, As substitutes for Si in the crystal lattice with a leftover electron not participating in the covalent bond formation. This leftover electron remains tethered loosely to the As site, whose presence in the lattice appears as a net charge of $+1e$. The state is often referred to as a donor state because the electron is loosely bound and can be easily donated to the conduction band.

A common example of a p-type material is that resulting when boron is doped into Si, as illustrated in Fig. 13.14b. Residing in the third column of the periodic table, B is just shy of the four electrons it needs to covalently bond with four adjacent Si atoms. Although the B enters the network, its net charge is $-1e$ and the missing electron behaves identically as a hole that is again, loosely tethered to the B site. The state is often referred to as an acceptor state, because the hole can readily be moved into the valence band when the site accepts a valence band electron.

Clearly the excess electron in the n-type material is not quite in the conduction band because it remains tethered. Likewise, the hole is not quite in the valence band. How loosely are these charge carriers tethered to their sites and where should their states be located in an energy level diagram? To answer this, we picture both situations as mimicking the features of a hydrogen atom, in that the charge carrier is orbiting the site as a result of the Coulombic attraction. In the Bohr model of the hydrogen atom, the ground state energy is

$$E_1 = \frac{m_o e^4}{2(4\pi\varepsilon_o)^2 \hbar^2} = 13.6\,\text{eV}, \tag{13.35}$$

and the orbital radius is the Bohr radius,

$$r_1 = a_o = \frac{4\pi\varepsilon_o \hbar^2}{m_o e^2} = 0.53\,\text{Å}. \tag{13.36}$$

In applying this to our orbiting charge carriers, we need to make two corrections. Firstly, the mass of the carrier is to be replaced by the corresponding effective mass, m^*. This effective mass is typically about one-tenth the electron mass, m_o. Secondly, the orbiting motion occurs not in free space, but within a material with a dielectric permittivity, $\varepsilon = \varepsilon_r \varepsilon_o \approx 10\varepsilon_o$, that is, about 10 times larger. Making these corrections we find that the charge carriers are tethered to their sites with a binding energy,

$$E_1 = \frac{1}{\varepsilon_r^2} \frac{m^*}{m_o} \times 13.6\,\text{eV} \approx 0.01\,\text{eV}, \tag{13.37}$$

and orbit about the site at a distance of roughly

$$r_1 = \varepsilon_r \frac{m_o}{m^*} \times 0.53\,\text{Å} \approx 50\,\text{Å}. \tag{13.38}$$

Since this binding energy is much smaller than the gap energy (recall, $E_g \approx 1$ eV), the donor states are located just below the conduction band, while the acceptor states are located just above the valence band. Even at modest temperatures, almost all of the donor electrons are excited into the conduction band and almost all of the acceptor sites are occupied by electrons, leaving an equal number of holes in the valence band.

The pn-junction

The most important application of extrinsic semiconductors undoubtedly stems from the combination of an n-type semiconductor with a p-type in the formation of a *pn-junction*. As illustrated in Fig. 13.15, the junction is formed by joining these two materials in a manner such that the crystalline lattice is preserved across the junction. The practical significance of the pn-junction is its ability to operate as a rectifier or diode: passing current only in one direction and not the other.

To see how this rectifying property arises, let us consider what would happen if we "glued" a section of n-type material to a section of p-type material. At the contact point, electrons would rapidly flood into the p-type

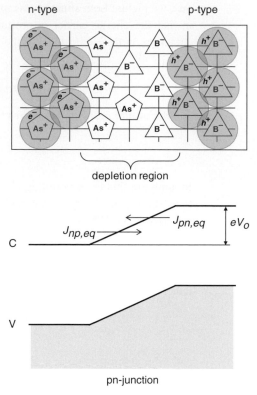

Figure 13.15 A pn-junction. Joining n-type and p-type materials results in a depletion region emptied of extrinsic charge carriers. An internal electrostatic potential develops in the depletion region, that restricts the flow of electrons from n-type to p-type. A steady-state potential V_o is maintained by balanced flows of electrons across the depletion region.

material, filling and thus annihilating the holes of the acceptor states just inside the junction. As a result, the region just near the contact point ceases to be electrically neutral. On the n-type side, the impurities are relieved of their excess electron and leave this region with a net positive charge. On the p-type side, the impurities have gained their missing electron needed for completing the covalent bond requirements and exhibit a net negative charge. After the two materials are joined, a potential difference quickly develops across the contact region which stems the flow of further electrons from the n-type to p-type material. The junction reaches a steady-state configuration, with potential difference V_o.

After this steady-state situation has been achieved, our pn-junction exists with a potential drop present across the junction region, as illustrated in Fig. 13.15. For the energy bands, which indicate the energy of an electron, this potential serves to raise the energy band diagram on the p-type side relative to that on the n-type side. An electron attempting to cross the junction (in the conduction band) from the n-type side to the

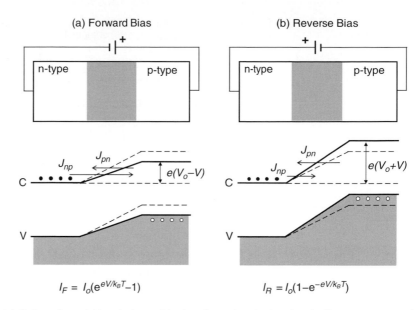

(a) Forward Bias   (b) Reverse Bias

$$I_F = I_o(e^{eV/k_BT} - 1)$$

$$I_R = I_o(1 - e^{-eV/k_BT})$$

Figure 13.16   (a) Under a forward bias, the internal barrier of a pn-junction is reduced, allowing an increased flow of electrons from the n-type to the p-type material. (b) Under a reverse bias, the internal barrier of a pn-junction is increased causing a decrease of electrons flowing from the n-type to p-type material. In this instance, a small flow of electrons from p-type to n-type dominates.

p-type side must surmount an energy barrier of eV_o. At equilibrium, a flow of electrons from n-type to p-type (J_{np}) is balanced by an opposing flow from p-type to n-type (J_{pn}). Since the flow from n-type to p-type is limited by the energy barrier, its equilibrium contribution to the net current is of the order,

$$J_{np,eq} \propto e^{-eV_o/k_BT} \approx J_{pn,eq}. \tag{13.39}$$

The flow of electrons in the other direction is not at all restricted by the barrier. A conduction electron that appears on the p-type side near the junction will readily roll down the energy hill. Here the flow is only limited by the repopulation of the conduction band in the p-type material, through thermal production of electron–hole pairs.

To now see how the pn-junction functions as a rectifier, we consider the effect of applying either a forward or reverse bias to the device. By forward bias, we mean that a battery is applied such that the p-type material is biased more positive than the n-type (see Fig. 13.16a). In the case of forward biasing, we see that the bias potential of the battery, V, serves to reduce the energy barrier height across the depletion region. As a result, the current of electrons from the n-type side is increased to

$$J_{np} \propto e^{-e(V_o - V)/k_BT}, \tag{13.40}$$

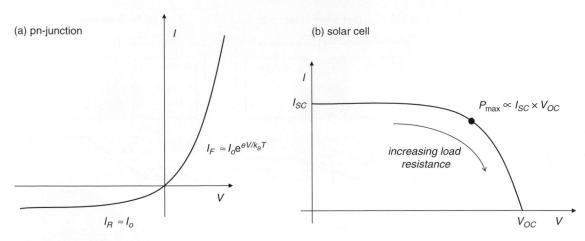

(a) pn-junction

(b) solar cell

Figure 13.17 (a) The IV characteristic curve for an ideal pn-junction illustrating its rectification property. (b) The IV characteristic curve for a photovoltaic cell under fixed illumination with changing load impedance. Maximum power is realized when operated near the knee of this curve.

while the counter flow of electrons from the p-type side remains unchanged. A net flow of electrons develops from the n-type to the p-type producing a conventional current,

$$I_F = I_o \left(e^{eV/k_B T} - 1 \right),$$ (13.41)

directed oppositely (i.e. from p-type to n-type). So as the forward bias increases, the forward current increases exponentially.

In the instance of a reverse bias, the electron energy barrier is increased by the applied voltage, and the flow of electrons from n-type to p-type is reduced below that of its equilibrium value. In this instance a net current,

$$I_R = I_o \left(1 - e^{-eV/k_B T} \right),$$ (13.42)

flows in the direction from n-type to p-type.

Although our discussion above has focused only on the electron currents, an identical, but inverted set of statements can be derived regarding the flow of holes. Taken together, the current–voltage curve for the pn-junction, shown in Fig. 13.17a, demonstrates its rectifying property.

Photovoltaic cells

As one last example of the practical applications of the pn-junction in semiconductor devices, consider what happens when the depletion region of a pn-junction is exposed to a large flux of photons. If the photon energy is sufficient to generate large numbers of electron–hole pairs, electrons appearing in the conduction band of the p-type materials will be rapidly swept (downhill) across the depletion region into the n-type side, while holes, appearing in the valence band of the n-type, will be rapidly swept to the p-type side.

When this pn-junction is open circuited (not connect to any external circuit), this pumping of charges by virtue of electron–hole pair production creates a forward bias of the junction which in turn lowers the junction barrier by an amount V_{OC}. In a steady-state condition, the forward bias current (Eq. (13.41)) is balanced by the current arising from pair production, such that,

$$I_o(e^{eV_{oc}/k_B T} - 1) = I_{SC},$$
(13.43)

where I_{SC} represents the current (flowing from n-type to p-type because of electrons traveling oppositely) produced by the photoexcitation of carriers and is proportional to the photon flux. At steady state, the bias potential across the junction appearing in an open circuit configuration is then

$$V_{oc} = \frac{k_B T}{e} \ln\left(\frac{I_{SC} + I_o}{I_o}\right).$$
(13.44)

Using a high load impedance, changes in this voltage can be measured in response to changes in the photon flux, and the junction then performs as a photodetector.

In the short circuit configuration, the current I_{SC} is merely pumped around a circuit without any potential difference. The complete IV curve of a photovoltaic cell generated by a varying load impedance is shown in Fig. 13.17b, and suggests that the maximum electric power that could be realized by this cell is roughly the product of the short circuited current (I_{SC}) and the open circuited voltage (V_{OC}).

13.5 Amorphous metals: the Anderson transition

In disordered materials, irregularities in the electron potential spawn wave localization effects, much like those seen for phonons discussed in Chapter 11. However, unlike phonon localization, which was controlled by a characteristic length scale, electron localization stems from a degree of variability in the *height* of the periodic "bumps" in the potential, as illustrated in Fig. 13.18, that result from disordering. For the crystal, the electron potential is periodic and each local potential well is equivalent. Together, this periodic potential gives rise to a band of allowed energies encompassing an energy interval $\Delta E = E_T - E_B$. By comparison, the electron potential of a severely disordered material has a wide variety of local potential wells, described by a distribution with a characteristic energy range W.

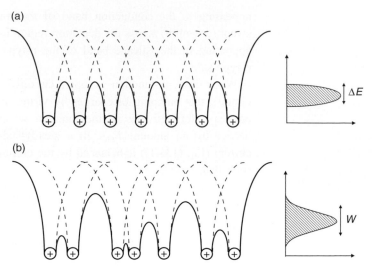

Figure 13.18 The periodic potential of a crystal gives rise to a band of allowed electron energy states of width ΔE. In a glass, disorder introduces variations in the potential characterized by an energy spread W. In cases of extreme disorder, the spread in energy exceeds the corresponding crystalline bandwidth to produce localization of the electron states.

On theoretical grounds, Anderson (1958) demonstrated that when the disorder parameter, roughly $W/\Delta E$, exceeds unity, *all* the electron states in the valence band become localized. These localized electron states are no longer described by propagating (Bloch) wave functions, but rather by wave functions of the form,

$$\psi(\vec{r}) \propto e^{-|\vec{r}-\vec{r}_o|}, \qquad (13.45)$$

in which the electron density vanishes except in the region near $\vec{r} \approx \vec{r}_o$. Thus, a sufficiently disordered conducting metal can be transformed into an insulator via the *Anderson transition*.

However, even modest amounts of disorder, for which $W/\Delta E < 1$, can have pronounced effects on the energy level structure. For real crystals, the density of states often exhibits irregular maxima, so-called van Hove discontinuities, which result from peculiarities of the crystal structure. In a glass, these irregularities are not only smoothed over, but the density of states can even extend into the, previously forbidden, energy gap region, as illustrated in Fig. 13.19. Although states form within the gap, these states are localized and are sharply divided from the remaining, extended states, by a *mobility edge*. Electrons excited above the mobility edge of the conduction band (and holes formed below the edge in the valence band) are mobile and able to contribute to conduction.

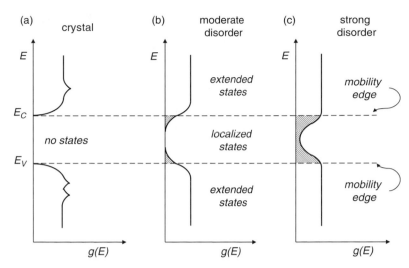

(a) The density of states for a crystal exhibits van Hove discontinuities and vanishes within the energy gap. (b) The introduction of moderate amounts of disorder leads to a reduction of the van Hove discontinuities and the formation of some localized electron states just inside the energy gap. (c) In the case of increased disorder, localized states can fill the entire gap.

Summary

- When effects of a periodic potential are included, the free electron wave function is replaced by a Bloch wave function that is modulated with the periodicity of the lattice.

- Solutions of Schrödinger's equation in the case of a periodic potential give rise to energy bands and energy gaps. Each allowed band can accommodate only $2N^3$ total electrons. As a consequence, monovalent metals tend to exhibit high conductivity, while divalent crystals often behave as poor conductors or insulators.

- Instances of a positive Hall coefficient indicate the dominance of *holes* in the overall conduction.

- Extrinsic semiconductors are those doped with impurities to provide a supply of excess charge carriers.

- The pn-junction is a fundamental element in modern solid state electronics that functions both as a rectifier and as a photocell.

Exercises

13.1. Obtain the dispersion relation of the Kronig–Penney model given in Eq. (13.17) by solving the four simultaneous equations in Eq. (13.16). (Caution: this exercise is straightforward but contains a tedious amount of algebra.)

13.2. Show that near the top of an energy band the density of states, $g(E)dE$ is given as in Eq. (13.30).

13.3. Estimate the energy $(\hbar\omega)$ and momentum $(\hbar K)$ for (a) a typical acoustic phonon near the Brillouin edge, and (b) a typical photon capable of exciting across the bandgap of Si. In so doing, verify that in an indirect bandgap transition, the momentum is chiefly supplied by the phonon, and the energy by the photon.

13.4. Determine the energy gap of Si using the data provided in Fig. 13.13. Express your answer in electron volts and compare with literature sources.

13.5. Discuss the operation of a light-emitting diode or LED in light of what we discussed regarding the operation of a photovoltaic cell.

13.6. Show that, for a simple cubic lattice, the kinetic energy of a free electron at the corner of the first Brillouin zone (111) is higher than that of an electron at the midpoint of a side face of the zone (100), by a factor of 3. Discuss what significance this result might have for the conductivity of divalent metals.

13.7. Show how Eq. (13.18) arises from Eq. (13.17) as the width of the barriers in the Kronig–Penney model is decreased.

13.8. (a) For the Kronig–Penney model with $P \ll 1$, show that, at $K = 0$ the energy of the lowest energy band is approximately $E_o \approx P\frac{\hbar^2}{md^2}$. (b) Likewise, find the first band gap at $K = \pi/d$, and show that it is approximately $2E_o$. (Hint: in this situation, the first term on the left-hand side of Eq. (13.18) is a minor contribution to the second term, and so Taylor expansions about the K in question are warranted.)

13.9. Imagine a 1D crystal with an energy band described by $E(K) = E_B + (E_T - E_B)\sin^2(Kd/2)$ that contains only a single electron. (a) Determine how the effective mass and group velocity depend on K and sketch a graph of this dependence. (b) Based on the results of the previous part, describe how the location of the electron (in real space) would evolve in time under the influence of a steady electric field. Assume there are no sources for scattering other than Bragg reflections occurring whenever the electron encounters the Brillouin zone edge. (c) If the lattice spacing is 2 Å and the field strength is 200 N/C, what is the period of oscillation?

Suggested reading

A nice introduction to both the Mott transition and the Anderson transition can be found in the book by Zallen.

C. Kittel, *Introduction to Solid State Physics*, 8th Ed. (John Wiley and Sons, 2005).

N. W. Ashcroft and N. D. Mermin, *Solid State Physics* (Holt, Rinehart and Winston, New York, 1976).

J. S. Blakemore, *Solid State Physics*, 2nd Ed. (W. B. Saunders Co., Philadelphia, 1974).

R. Zallen, *The Physics of Amorphous Materials* (John Wiley and Sons, New York, 1983).

14 Bulk dynamics and response

Introduction

Up to now, we have considered only those inherent microscopic dynamics in a material that are present at equilibrium and are driven by the thermal energy content of the material itself. Here, in our last chapter dealing with dynamics, we consider instead the macroscopic, bulk dynamics of materials in non-equilibrium situations where an external force is applied or removed. Examples include the stretching or bending of a solid that results from application of a mechanical force, or the polarization of a dielectric material resulting when an external electric field is applied.

Several common features emerge in the response of a material to an external force or field. In all cases, there is some aspect of *elasticity* by which application of the force results in the *storage* of potential energy, that is returned when the force is removed. In all cases, this storage of energy is accompanied by some element of viscous drag or *damping* by which a portion of the work done during the deformation is *lost* in the form of heat. Like friction, this damping is a microscopic feature inherent in the thermodynamic fluctuations, and the energy lost during the deformation is returned to the same thermal bath from which it was derived. In fact, we will show that an important theorem exists, known as the *fluctuation–dissipation theorem*, which relates the macroscopic dissipation of energy in these bulk, non-equilibrium, processes directly to the inherent microscopic fluctuations present at equilibrium.

14.1 Fields and deformations

What happens when you bend a pencil? Obviously, if you bend it too forcefully it will break. But if you bend it only slightly, it will return to its original state after you release it. This behavior is much like that of an ideal spring. For small displacements of the spring, the force is proportional to the degree of stretching. However, if we stretch the spring too far, certain permanent deformations result which will disrupt the proportionality. Throughout this chapter, we will be considering how materials respond to applied forces or fields in

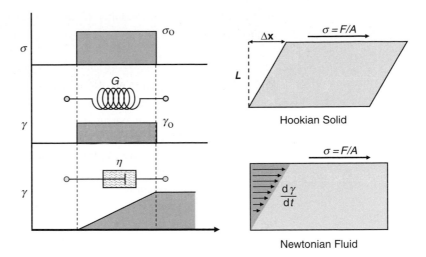

Figure 14.1 Ideal response of a Hookian solid and a Newtonian fluid. Under an applied stress, the solid behaves like an ideal spring and experiences a nearly instantaneous strain that vanishes when the stress is removed. The liquid behaves like an ideal dashpot, whose strain rate is proportional to the applied stress.

what is known as the *linear response regime*. This is merely the range of small deformations of a specimen that remains proportional to the force.

14.1.1 Mechanical deformations

We begin with the simple example of bulk deformations associated with the application of a mechanical force, such as occurs when a block of material is subjected to a shearing force. If a force is applied tangentially to the top of the block whose surface area is A, it produces a *shear stress* $\sigma = F/A$. As a result of this shear stress, the material undergoes a lateral deformation, as illustrated in Fig. 14.1, which is characterized by a *shear strain* $\gamma = \Delta x/L$, where L is the height of the block.

There are two extremes of how the strain will respond to the applied stress, depending on the nature of the material. Firstly, if the material is an elastic solid, we anticipate that it will stretch to some equilibrium strain and, when the stress is removed, it will return to where it started. A material exhibiting such ideal behavior is known as a *Hookian solid*, as it obeys Hooke's law for ideal springs. Alternatively, if the material is a liquid, we anticipate that it will flow under the application of the shear stress and will not return to its original state when the stress is removed. A material exhibiting such ideal behavior is known as a *Newtonian fluid*.

For an ideal Hookian solid, under sufficiently small loads, the limiting strain produced is proportional to the applied stress and the ratio is described alternatively by a shear modulus, G, or its inverse known as a shear compliance, J, as,

$$\sigma = G\gamma \quad \text{or} \quad \gamma = J\sigma. \tag{14.1}$$

In the absence of any damping, the applied stress produces a nearly instantaneous deformation of fixed magnitude (see Fig. 14.1) and the work performed is stored in the form of elastic potential energy. If we were to suddenly release the stress, the material would oscillate at some fundamental frequency indefinitely. As indicated in Fig. 14.1, we could represent this elastic response as similar to that of an ideal spring with a spring constant proportional to G.

For an ideal Newtonian liquid, application of the shear results in a continuous motion of the upper surface of the material relative to the lower surface. As the uppermost layer of fluid is pulled by the shearing force, it experiences a viscous drag due to contact with the layer of fluid immediately below. A steady-state flow is attained such that the applied shear is balanced by a *drag force* proportional to the rate of strain,

$$\sigma = \eta \frac{d\gamma}{dt}, \tag{14.2}$$

where η is the shear viscosity. Because the rate of strain is fixed for a fixed load, the deformation of a Newtonian liquid increases continuously (see Fig. 14.1) and is not recovered when the stress is later removed. No potential energy is stored during the deformation and the work performed is completely lost in the form of energy dissipated into heat. As suggested in Fig. 14.1, we could represent this viscous response as similar to that of an ideal dashpot with a damping force proportional to η.

Viscoelastic behavior

The response of real materials is neither purely elastic nor purely viscous. A solid is mostly elastic by virtue of the strong bonds that hold it together. But if we stretch a solid and release it, we will not find it oscillating indefinitely. The bonds between particles in a solid are not perfect springs and in each cycle of the oscillation damping forces present in the material dissipate away the stored energy. Similarly, although a liquid is primarily viscous, weak bonds between the particles do provide an element of elasticity. Hence, real materials are generally *viscoelastic* and exhibit a mixed response to applied forces that incorporates elements of elasticity (with energy storage) and viscous flow (with energy dissipation).

A prime example of a viscoelastic material would be the polymers discussed in Chapter 9. Recall that a polymer-based material consists of a collection of long polymer chains that are individually coiled up, but also entangled with other chains. Under stress, we might initially observe the elastic response associated with the entropic stretching of the individual polymer coils themselves. However, over time, the stress will act to uncoil the chains and unravel

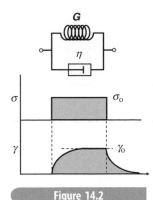

Figure 14.2

A simple model for viscoelastic relaxation consists of a parallel combination of a spring and dashpot. Under an applied stress, the viscous damping provided by the dashpot results in a gradual (time-dependent) approach to and from a stressed state of deformation.

the entanglements and a gradual, fluid-like deformation will result producing a complicated time-dependent *viscoelastic relaxation*.

To illustrate the nature of viscoelastic relaxations, let us examine a simple model consisting of a parallel combination of a Hookian spring and a Newtonian dashpot shown in Fig. 14.2. The equation of motion for this system is given by

$$\sigma = G\gamma + \eta\frac{d\gamma}{dt}. \tag{14.3}$$

Suppose now a fixed stress, σ_o, is instantly applied and maintained for sufficient time that an equilibrium strain, $\sigma_o = G\gamma_o$, is produced. On application of the stress,

$$\sigma_o = G\gamma + \eta\frac{d\gamma}{dt}, \tag{14.4}$$

whose solution is,

$$\gamma_{\text{on}}(t) = \gamma_o\left(1 - e^{-t/\tau}\right), \tag{14.5}$$

where the characteristic relaxation time is $\tau = \eta/G$. In this process, some of the work performed is stored in the spring while some is dissipated to viscous forces present in the dashpot.

If we now remove the stress, the equation of motion becomes

$$G\gamma + \eta\frac{d\gamma}{dt} = 0, \tag{14.6}$$

whose solution is,

$$\gamma_{\text{off}}(t) = \gamma_o e^{-t/\tau}. \tag{14.7}$$

In this case, the elastic potential energy previously stored in the spring is completely dissipated.

This model is far too simple to describe the response of most real materials, but could be extended by adding additional viscous and elastic elements (in parallel and series) in more complex combinations, to arrive at time dependencies that mimic experiment. Instead, we emphasize just two important aspects of viscoelastic relaxation: (1) the deformation is *not instantaneous* but rather described by a time-dependent relaxation function, $\gamma(t)$, and (2) some *energy is always dissipated* in the process due to the presence of microscopic damping forces.

14.1.2 Electric and magnetic deformations

Electric and magnetic fields also produce deformations of a sort. These fields act to align electric dipoles or magnetic moments present in a material so as to

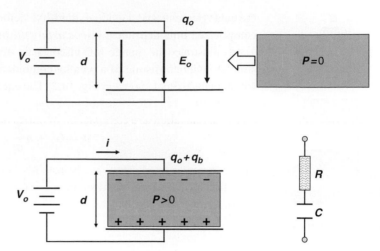

Figure 14.3   In the upper figure, an unpolarized specimen is about to be inserted (instantaneously) into the space between the plates of a parallel-plate capacitor. Upon insertion, a bound charge density, $\pm q_b$, forms at the top and bottom of the specimen as a result of polarization. To maintain the original field, a measurable current, i, is fed from the battery. Analogous to that presented for Figure 14.2, the polarization response can be modeled by a resistor and capacitor in series, whose characteristic relaxation time equals RC.

produce a finite polarization or magnetization, respectively. Although there is no actual (mechanical) deformation occurring in either the dielectric or magnetic material, both the polarization and magnetization are the result of an applied field, and in the linear response regime, are proportional to the magnitude of the field.

Consider an experiment in which a battery is connected to a parallel-plate capacitor, as illustrated in Fig. 14.3. This produces an electric field in the space between the capacitor plates $E_o = V_o/d$, and an accumulation of charge $q_o = (\varepsilon_o A/d)V_o$ on the plates, where V_o is the battery voltage, A the area of the plates and d the spacing. We next insert a dielectric material between the plates at time $t = 0$. What happens? Firstly, the material contains atoms composed of nuclei surrounded by an electron cloud. In the presence of the electric field, the cloud experiences a force and is displaced from equilibrium forming an induced electric dipole and net polarization of the material. This polarization happens rather quickly, but the material may also contain permanent electric dipoles, which, in the electric field, experience a torque and may reorient against viscous forces in such a way as to gradually become polarized.

In any event, the polarization starts at zero and increases over time to a new equilibrium value. As it develops, a layer of bound charge, q_b, forms at the top and bottom of the sample creating an interior electric field opposite to that originally introduced. From electrostatics, the polarization, P, equals this bound charge surface density. Since the battery is a constant voltage source, charges are delivered from it to the plates of the capacitor to offset the bound charge and maintain the original electric field, as shown in Fig. 14.3. When the material is removed from the electric field, the opposing field present in the material returns the polarization to a randomized state over time, and the bound charge vanishes.

In parallel with our discussion of mechanical relaxation, we could model the electrical relaxation as a combination of things that store energy (i.e. capacitors) and dissipate energy (i.e. resistors), as illustrated in Fig. 14.3. On application of the battery, Kirchhoff's sum rule yields the following equation of motion:

$$V_o = \frac{q_b}{C} + R \frac{dq_b}{dt},\qquad(14.8)$$

which is exactly analogous with the mechanical analog of Eq. (14.4). In this instance, the characteristic relaxation time is given by the familiar, $\tau = RC$.

Although it would be natural to associate the "electrical deformation" with either the polarization or the bound surface charge density, it is easiest to measure the *free* charge moving to or from the battery, and more customary to associate the deformation with the displacement field,

$$D(t) = \varepsilon_o E_o + P(t) = \varepsilon_o(1 + \chi_e(t))E_o = \varepsilon(t)E_o,\qquad(14.9)$$

where ε_o is the permittivity of free space and χ_e is known as the dielectric susceptibility.

In the analogous magnetic experiment, a magnetic material is introduced into the core of a long solenoid and develops a magnetization, M. Again, we could associate M (or its corresponding bound current density) with the "magnetic deformation" caused by an applied field H_o. However, it is customary to associate the deformation with the magnetic induction, B, for which,

$$B(t) = \mu_o H_o + \mu_o M(t) = \mu_o(1 + \chi_m(t))H_o = \mu(t)H_o,\qquad(14.10)$$

where μ_o is the permeability of free space and χ_m is known as the magnetic susceptibility.

14.1.3 A generalized response

Hopefully by now you will have gleaned that a pattern is emerging in how a material responds to an applied force or field in the linear response regime. In each of the above situations of mechanical, electrical and magnetic relaxation, the application or removal of a generic field, F_o, results in a time-dependent deformation of the form,

$$\Delta X(t) = R(t)F_o,\qquad(14.11)$$

where *R(t)* is a time-dependent *response function* that describes the relaxation of the system from one state of equilibrium to another. To complete this

Table 14.1 Deformations, fields, and their corresponding response functions.

	Deformation	Field	Response function
Mechanical	γ	σ	$J(t)$
Electrical	D	E	$\varepsilon(t)$ or $\chi_e(t)$
Magnetic	B	H	$\mu(t)$ or $\chi_m(t)$
General	ΔX	F	$R(t)$

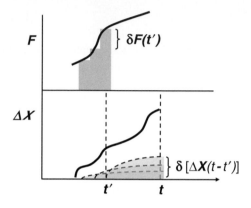

Figure 14.4   For an arbitrary time-dependent field, the resulting time-dependent deformation can be determined by Boltzmann superposition. The applied field is regarded as a combination of sequential stepwise changes of an infinitesimal amount δF, occurring at time t'. Each step produces a corresponding infinitesimal viscoelastic deformation at later times, and the net deformation is obtained by superposition.

generalization, the assignment of the variables in Eq. (14.11) for mechanical, electrical and magnetic relaxations is summarized in Table 14.1.

14.2 Time-dependent fields

Although viscoelastic relaxation can be studied in the time domain using the sort of step loads illustrated in the examples above, it is far more common to examine the response in the frequency domain as it appears when a sinusoidally varying field is applied over a range of frequencies. Consider first, an arbitrary time-dependent field, as illustrated in Fig. 14.4. We can construct this time dependence with a series of small step loads of size dF, occurring at time t', each of which produces a corresponding advance in the deformation at a later time t, given from Eq. (14.11) as,

$$\delta[\Delta X(t)] = \delta F(t')R(t - t'). \tag{14.12}$$

The cumulative effect of the sequential step loads is the time-dependent field and so the net resulting deformation is obtained by an integration process known as *Boltzmann superposition*,

$$\Delta X(t) = \sum_{\text{steps}} \delta F(t')R(t-t') = \int_{-\infty}^{t} \frac{dF(t')}{dt'} R(t-t') dt'. \tag{14.13}$$

Integration by parts yields,

$$\Delta X(t) = \int_{-\infty}^{t} F(t')\phi_R(t-t') dt', \tag{14.14}$$

where

$$\phi_R(t) \equiv \frac{dR(t)}{dt}, \tag{14.15}$$

is known as the *pulse response function*. The pulse response function is most fundamental as it describes the deformation following application of a delta function *pulse*, $F(t') = F_o\delta(t')$, which in Eq. (14.14) yields,

$$\Delta X(t) = \int_{-\infty}^{t} F_o\delta(t')\phi_R(t-t') dt' = F_o\phi_R(t). \tag{14.16}$$

In the case of an exponential relaxation, where either $R_{\text{on}}(t) = R_o\left(1 - e^{-t/\tau}\right)$ or $R_{\text{off}}(t) = R_o e^{-t/\tau}$, it is the function $\phi_R(t) = \pm R_o e^{-t/\tau}/\tau$ that contains the fundamental kernel of the relaxation process present, both when loads are applied and when they are removed.

14.2.1 Alternating fields and response functions

Now let us consider a very common type of experiment in which the time-dependent field is sinusoidal,

$$F(t) = F_o e^{-i\omega t}, \tag{14.17}$$

and for which the deformation is now described by a frequency-dependent response function, $R^*(\omega)$, such that,

$$\Delta X(t) = R^*(\omega)F(t). \tag{14.18}$$

Why is $R^*(\omega)$ a complex quantity? Recall that the response to any deformation is not instantaneous and so we anticipate, as illustrated in Fig. 14.5, that while the deformation will oscillate at the same frequency of the applied field, it will most likely be delayed by some phase angle. That is,

$$R^*(\omega) \equiv \frac{\Delta X(t)}{F(t)} = \{R'(\omega) + iR''(\omega)\} = |R^*(\omega)|e^{i\delta}, \tag{14.19}$$

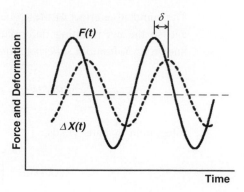

Illustration of an alternating applied field and its corresponding deformation, showing both the differing amplitude and phase.

where,

$$|R^*(\omega)| = \sqrt{[R'(\omega)]^2 + [R''(\omega)]^2} \quad \text{and} \quad \tan\delta = \frac{R''(\omega)}{R'(\omega)}. \qquad (14.20)$$

From Eq. (14.14), we can now express the response function as,

$$R^*(\omega) = \frac{\Delta X(t)}{F(t)} = \frac{\displaystyle\int_{-\infty}^{t} F_o e^{-i\omega t'} \phi_R(t-t')dt'}{F_o e^{-i\omega t}} = \int_{-\infty}^{t} e^{-i\omega(t'-t)}\phi_R(t-t')dt'. \qquad (14.21)$$

Using a change of variables to $t'' = t - t'$, this can be reduced to,

$$R^*(\omega) = \int_{0}^{\infty} \phi_R(t'')e^{+i\omega t''}dt''. \qquad (14.22)$$

Equation (14.22) expresses a remarkable result. The frequency-dependent response function obtained in an ac experiment is simply the one-sided Fourier transform of the time-dependent pulse response function, which, as we observed, contains the fundamental kernel of the viscoelastic relaxation. Because of this Fourier transform relationship, we find that $R^*(\omega)$ obtained through ac experiments contains the *same information* about the inherent dynamics that one would obtain through a static (dc) experiment using fixed loads.

To develop some understanding of the information afforded by $R^*(\omega)$, let us consider again the simple case in which the relaxation is exponential and the pulse response function is

$$\phi_R(t) = R_o e^{-t/\tau}/\tau. \qquad (14.23)$$

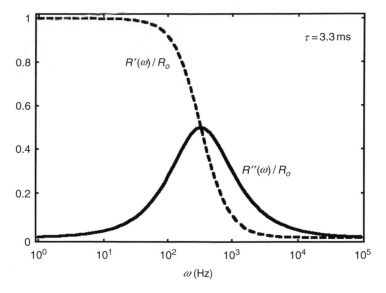

Figure 14.6 Frequency dependence for the exponential response function of Eq. (14.25) for a relaxation time of 3.3 milliseconds. Note that the peak in the imaginary component occurs near $\omega = \tau^{-1} = 300$ Hz.

In this case, the one-sided Fourier transform yields,

$$R^*(\omega) = \frac{R_o}{\tau} \int_0^\infty e^{-t/\tau} e^{+i\omega t} dt = \frac{R_o}{1 - i\omega\tau}, \qquad (14.24)$$

whose real and imaginary parts are given as,

$$
\begin{aligned}
R'(\omega) &= R_o \left(\frac{1}{1 + \omega^2\tau^2} \right) \\
R''(\omega) &= R_o \left(\frac{\omega\tau}{1 + \omega^2\tau^2} \right),
\end{aligned}
\qquad (14.25)
$$

and are each plotted in Fig. 14.6.

What can we discern from these two curves? Firstly, $R'(\omega)$ is a measure of how large a deformation is achieved throughout the cycle. At high frequencies where $\omega \gg \tau^{-1}$, this deformation is small, because the field is flipping directions far too quickly for any deformation to keep pace. By comparison, at low frequencies the field changes slowly enough ($\omega \ll \tau^{-1}$) that an equilibrium deformation is maintained in each cycle.

Secondly, $R''(\omega)$ is seen to exhibit a peaked behavior with the peak coinciding with $\omega = \tau^{-1}$. The location of the peak thus provides a direct measure of the characteristic relaxation time scale, and can be monitored as a function of temperature or composition. As we will demonstrate shortly, $R''(\omega)$ is a measure of the energy dissipated in each cycle. At high frequencies, where the field

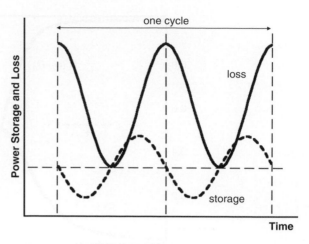

one cycle

loss

storage

Power Storage and Loss

Time

Figure 14.7 Variation of the power stored and dissipated versus time during one cycle of an alternating field.

reverses too rapidly to establish much deformation, the dissipated energy is small and $R''(\omega)$ begins from zero. At low frequencies, the deformation is carried out so slowly that the drag forces (and work performed against them) vanish. Again, $R''(\omega)$ decreases towards zero in this limit. Only when the field is reversing at a rate comparable to the relaxation time, is the energy dissipation maximum.

14.2.2 Energy dissipation

To demonstrate the relationship between $R''(\omega)$ and the energy dissipation per cycle, we start by recognizing that the instantaneous power is given by the product of the force and the velocity, or for our generalized fields and deformations as,

$$\text{Power} = \text{Re}\{F(t)\}\frac{d}{dt}\text{Re}\{\Delta X(t)\} = F_o \cos \omega t \frac{d}{dt}\text{Re}\{\Delta X(t)\}, \qquad (14.26)$$

where $\text{Re}\{\}$ denotes only the real part of a complex quantity. From Eq. (14.18),

$$\begin{aligned}
\text{Re}\{\Delta X(t)\} &= \text{Re}\{R^*(\omega)F(t)\} \\
&= F_o \text{Re}\{(R'(\omega) + iR''(\omega))(\cos \omega t - i \sin \omega t)\} \qquad (14.27) \\
&= F_o(R'(\omega) \cos \omega t + R''(\omega) \sin \omega t),
\end{aligned}$$

and the time derivative is,

$$\frac{d}{dt}\text{Re}\{\Delta X(t)\} = F_o \omega (R''(\omega) \cos \omega t - R'(\omega) \sin \omega t). \qquad (14.28)$$

Thus, from Eq. (14.26), the instantaneous power is obtained as,

$$\text{Power} = F_o^2 \omega \left(R''(\omega) \cos^2 \omega t - R'(\omega) \sin \omega t \cos \omega t\right). \qquad (14.29)$$

This is plotted in Fig. 14.7 and shows that a portion of the power input during one quarter cycle is returned in the next. This part is proportional to $R'(\omega)$ and

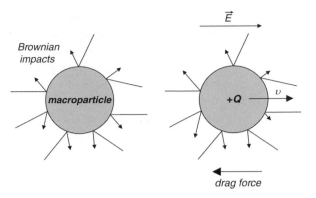

Figure 14.8

In the absence of a motive force, a Brownian particle executes a random walk as a result of random collisions by solvent molecules. However, when the Brownian particle is pulled through the solvent by an external force, it suffers more devastating impacts on its front side than on its rear, which mimic the effects of a drag force proportional to the particle's velocity.

is a measure of the energy storage in the form of elastic potential energy. Another portion of the power input during the cycle is not returned. Averaged over a cycle, this dissipated energy is,

$$\int_0^{T=2\pi/\omega} (\text{Power}) \; \mathrm{d}t = \pi F_o^2 R''(\omega). \tag{14.30}$$

14.3 The fluctuation–dissipation theorem

The energy dissipation in bulk relaxation processes obviously arises from inherent damping, but what is the microscopic source for this damping? Consider once more the Brownian motion of a large particle suspended in a solvent. From our earlier discussion of Brownian motion, we know that the driving force for the erratic motion of the particle arises from the random impacts it experiences from molecules of the solvent. These random impacts are, in turn, generated by the inherent thermodynamic fluctuations present in any system at equilibrium. But suppose we could apply an external force to our particle. Imagine, for example, that our Brownian particle is charged and propelled through the solvent by an external electric field. We know it will suffer a frictional drag force and, as illustrated in Fig. 14.8, the source of this drag force is again seen to be a consequence of the impacts produced by the solvent molecules. Because the Brownian particle moves forward with a finite speed, those impacts occurring on the forward side involve a greater relative speed of approach than those on the rear and impart a correspondingly higher impulse on the Brownian particle.

The upshot of this example is that the random impacts of the solvent molecules serve two purposes: they are the source of the *random* driving force for the Brownian particle, and the source for the *systematic* frictional drag when the Brownian particle is displaced by an external field. It is this inherent connection between the random and systematic aspects of the microscopic fluctuations that lie at the heart of a very general theorem, known as the *fluctuation–dissipation theorem*. There are many various mathematical expressions of this theorem, but the most common is the following relationship:

$$R''(\omega) = \frac{\omega}{2kT} S_X(\omega), \qquad (14.31)$$

between the imaginary part of the (macroscopic) response function and the (microscopic) *power spectrum*, $S_X(\omega) = \int\limits_{-\infty}^{\infty} \langle \Delta X(t) \Delta X(0) \rangle_{eq} e^{i\omega t} dt$. This power spectrum is microscopic because it is merely the Fourier transform of the correlation function associated with the incessant fluctuations in the deformation at equilibrium when no external field is present. The significance of the theorem is that it associates the dynamics occurring for macroscopic deformations caused by an external field with those thermodynamically driven, microscopic fluctuations present in thermal equilibrium. Given the importance of this theorem, we present a short derivation.

Firstly, imagine a thermodynamic system at equilibrium in the absence of the applied field. In general, it may exhibit a non-zero equilibrium value of the quantity X that would be obtained by Boltzmann statistics as,

$$\langle X \rangle_{eq} = \frac{\sum\limits_{i} X_i(t) e^{-\beta E_i^o}}{\sum\limits_{i} e^{-\beta E_i^o}}, \qquad (14.32)$$

where $\beta \equiv 1/k_B T$ and E_i^o is the corresponding energy associated with microstates of $X_i(t)$ in the absence of the field.

What happens when the field is applied? We know that an applied electric field, $\vec{E}_o$, promotes the alignment of electric dipoles, resulting in a polarization P and a lowering of the energy by an amount, $\Delta E = -E_o P \ (= -F_o X_i)$. Likewise, for magnetic materials, the alignment of magnetic moments results in a net magnetization and a lowering of the energy by an amount, $\Delta E = -H_o M \ (= -F_o X_i)$. Thus, it reasons that when a generic field, F_o, is applied for some long time, a net deformation develops and the energy of each microstate is lowered to

$$E_i = E_i^o - F_o X_i(0). \qquad (14.33)$$

Suppose that our field has been applied for a long time such that a fixed deformation has been established. We next imagine turning off the applied

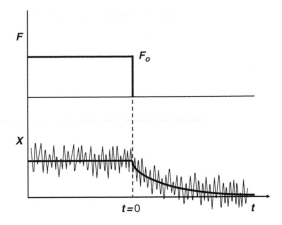

Figure 14.9 A deformation initially produced in a system is accompanied by thermodynamic fluctuations. When the applied field is removed at time $t = 0$, the macroscopic decay back to equilibrium is conditioned by these same microscopic fluctuations.

field at $t = 0$ and monitoring the decay of the quantity X back to its equilibrium value over time, as shown in Fig. 14.9. The decay of the non-equilibrium deformation is then given as,

$$\Delta X(t) = \frac{\sum_i X_i(t) e^{-\beta \left[E_i^o - F_o X_i(0) \right]}}{\sum_i e^{-\beta \left[E_i^o - F_o X_i(0) \right]}} - \langle X \rangle_{eq}. \qquad (14.34)$$

For the moment, let us suppose that our field is small and produces a tiny deformation such that $F_o X_i(0) << k_B T$. In this instance, we can expand the exponential in Eq. (14.34) to obtain,

$$\Delta X(t) \approx \frac{\sum_i X_i(t) e^{-\beta E_i^o} (1 + \beta F_o X_i(0))}{\sum_i e^{-\beta E_i^o} (1 + \beta F_o X_i(0))} - \langle X \rangle_{eq}. \qquad (14.35)$$

One can show (Ex. 1) that the denominator is,

$$\frac{1}{\sum_i e^{-\beta E_i^o} (1 + \beta F_o X_i(0))} = \frac{1 - \beta F_o \langle X \rangle_{eq}}{\sum_i e^{-\beta E_i^o}}, \qquad (14.36)$$

so that,

$$\Delta X(t) \approx \frac{\sum_i X_i(t) e^{-\beta E_i^o} \left[1 + \beta F_o X_i(0) - \beta F_o \langle X \rangle_{eq} + O(F_o^2) \right]}{\sum_i e^{-\beta E_i^o}} - \langle X \rangle_{eq}.$$

$$(14.37)$$

The first term in the summation is given by Eq. (14.32) as $\langle X \rangle_{eq}$, and so,

$$\Delta X(t) \approx \beta F_o \left\{ \langle X(t)X(0) \rangle_{eq} - \langle X \rangle_{eq}^2 \right\}. \tag{14.38}$$

One can show (Ex. 2) that for an equilibrium fluctuation in the quantity X given by,

$$\Delta X_{eq}(t) = X(t) - \langle X \rangle_{eq}, \tag{14.39}$$

the macroscopic deformation is,

$$\Delta X(t) \approx \beta F_o \langle \Delta X(t)\Delta X(0) \rangle_{eq}. \tag{14.40}$$

Compare this result with that obtained in Eq. (14.11). The earlier expression was developed to describe the linear response to rather large, but linear, deformations. But, of course, this expression must also apply equally well for any tiny deformation, since they too are within the linear regime. It then reasons that Eq. (14.40) holds for any macroscopic, but linear, deformation and that

$$R(t) = \frac{1}{k_B T} \langle \Delta X(t)\Delta X(0) \rangle_{eq}. \tag{14.41}$$

This is the essential feature of the fluctuation–dissipation theorem: any externally imposed perturbation in the linear regime decays in the same fashion as any spontaneous microscopic fluctuation occurring at equilibrium (in the absence of the field).

To obtain the version of the fluctuation–dissipation theorem of Eq. (14.31), we can apply a time derivate to both sides and recognize that $dR/dt = -\phi_R(t)$ in this situation that a field or load is being removed. We then have,

$$\phi_R(t) = -\frac{1}{k_B T}\frac{d}{dt} \langle \Delta X(t)\Delta X(0) \rangle_{eq}, \tag{14.42}$$

and next apply a Fourier transform. The term on the left-hand side (see Eq. (14.22)) becomes the frequency-dependent response function, $R^*(\omega)$, obtained in a macroscopic ac experiment, while the right-hand side must be integrated by parts to arrive at,

$$R^*(\omega) = \frac{1}{k_B T} \left\{ \langle \Delta X^2(0) \rangle_{eq} + i\omega \int_0^\infty \langle \Delta X(t)\Delta X(0) \rangle_{eq} e^{i\omega t} dt \right\}. \tag{14.43}$$

The second term on the right-hand side is half the power spectrum, $S_X(\omega)$, associated with fluctuations in $X(t)$ that occur in the absence of the external field, and thus we arrive at the usual statement of the fluctuation–dissipation theorem, given above in Eq. (14.31).

Summary

- Bulk dynamics refers to the non-equilibrium response of a material to an external field.

- The linear response regime implies that the deformation, ΔX, caused by an applied field, F_o, is proportional to the field.

- In general, the viscoelastic response of a material is non-instantaneous and accompanied by dissipation of energy.

- In ac experiments, the response function, $R^*(\omega)$, is a Fourier transform of the underlying relaxation, as described by the pulse response function, $\phi_R(t)$.

- The fluctuation–dissipation theorem relates the energy dissipation in a bulk relaxation to inherent fluctuations in the corresponding deformation present in the absence of an external field.

Exercises

14.1. Show that the denominator of Eq. (14.35) can be expressed as given in Eq. (14.36).

14.2. Show that for any fluctuation defined as in Eq. (14.39), the autocorrelation can be expressed as $\langle \Delta X(t) \Delta X(0) \rangle_{eq} = \langle X(t) X(0) \rangle_{eq} - \langle X \rangle_{eq}^2$.

14.3. The viscoelastic element shown in Fig. 14.2 is known as a Kelvin element and is representative of how stresses add together. A corollary to this is the Maxwell element, formed by a spring and dashpot connected *in series*, which is representative of how strains add together. For this situation, the equation of motion is given by,

$$\frac{d\gamma}{dt} = \frac{1}{G}\frac{d\sigma}{dt} + \frac{\sigma}{\eta}. \qquad (14.44)$$

(a) Suppose a Maxwell element starts with an overall length x_o and is instantly stretched to $x_o + \gamma_o$. Determine the form of $\sigma(t)$ which results. (b) After some very long time, the Maxwell element is now compressed back to its original length (x_o). Determine the form of $\sigma(t)$ which results in this instance.

14.4. Ac dielectric measurements were performed on a polymer liquid at five temperatures. The frequency dependence of the dielectric loss is shown in Fig. 14.10 below.

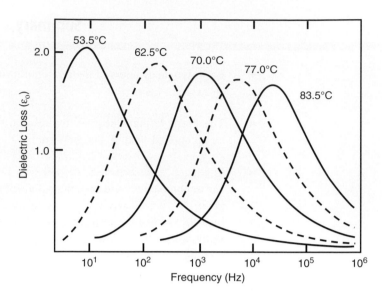

Figure 14.10

(a) For each temperature, provide an estimate of the characteristic relaxation time and complete the table below.

T (°C)	T (K)	1/T (K)	τ (sec)
53.5			
62.5			
70.0			
77.0			
83.5			

(b) Use your results to estimate an activation energy, E, such that $\tau = \tau_o \exp(E/k_B T)$.

14.5. The relaxation of glassforming materials is commonly non-exponential. In this case, the response function is highly asymmetric in comparison with that seen in Fig. 14.6, and is often modeled using a modified form of Eq. (14.24), such as the Cole–Cole function:

$$R^*(\omega) = \frac{R_o}{1 + (i\omega\tau)^\alpha},$$

where $0 < \alpha < 1$ is a parameter whose smallness is a measure of the non-exponentiality. Show that, for this modified form, the real and imaginary components of the response function are:

$$R'(\omega) = R_o \left\{ \frac{1 + (\omega\tau)^\alpha \cos(\pi\alpha/2)}{1 + 2(\omega\tau)^\alpha \cos(\pi\alpha/2) + (\omega\tau)^{2\alpha}} \right\},$$

and,

$$R''(\omega) = R_o \left\{ \frac{(\omega\tau)^\alpha \cos(\pi\alpha/2)}{1 + 2(\omega\tau)^\alpha \cos(\pi\alpha/2) + (\omega\tau)^{2\alpha}} \right\}.$$

14.6. In the falling sphere method for determining viscosity, a metal sphere of known radius b and mass m is sealed in a long tube containing a liquid. Upon inverting, the sphere experiences drag given by the Stokes coefficient, $\zeta = 6\pi\eta b$, and falls at its terminal velocity. Describe how the viscosity can be determined using this method and derive an expression for the viscosity in terms of the measured terminal velocity and the mass and radius of the sphere.

14.7. Consider a Brownian particle that flows at a constant speed $\langle v_x(t) \rangle_{NE}$ under the influence of an external force $F_o \hat{x}$, as described by its *mobility* (the inverse of the drag coefficient),

$$\mu = \frac{1}{\zeta} = \frac{\langle v_x(t) \rangle_{NE}}{F_o}.$$

Use the Einstein relation (Eq. (8.19)) to show that the diffusivity (at long times) can be expressed by the following *Green–Kubo relation*:

$$D = \frac{1}{3} \int_0^\infty \langle v(0) \cdot v(t') \rangle dt'.$$

Suggested reading

The first text on this list provides a good review of bulk relaxation in polymers. The subject of fluctuation and dissipation is covered in some detail by Chaikin and Lubensky as well as by Chandler.

N. G. McCrum, B. E. Read and G. Williams, *Anelastic and Dielectric Effects in Polymeric Solids* (Dover Publications, New York, 1991).

P. M. Chaikin and T. C. Lubensky, *Principles of Condensed Matter Physics* (Cambridge University Press, New York, 2003).

D. Chandler, *Introduction to Modern Statistical Mechanics* (Oxford University Press, New York, 1987).

R. Kubo, Rep. Prog. *Phys.* **29**, 255 (1966).

G. Strobl, *Condensed Matter Physics* (Springer-Verlag, Berlin, 2004).

PART IV

TRANSITIONS

On my bulletin board I have a picture of a recent U.S. president that someone has photo shopped to include a text balloon that says, "The ice caps are not melting. The water is being liberated." Although intended to be humorous, there is an element of truth to this statement. Phase transitions are in many respects the result of a competition or war between two opposing forces. On one side are the attractive interactions between particles that act to bind them together and force them into a more ordered structure – a world governed by potential energy. On the other side is thermal energy that acts to break these bonds and liberate the particles so that they are free to move about – a world dominated by kinetic energy. There is then a point of transition where one world order trumps the other and it is this phenomenon that we consider in this final set of chapters.

There are many sorts of phase transitions, but the two prominent examples we will consider are the gas-to-liquid transition and the transition of paramagnet to ferromagnet. In spite of the obvious differences between these two systems, we will emphasize a remarkable level of similarity in how their phase transitions proceed and how order develops in both situations.

Beginning in Chapter 15, we lay some groundwork regarding the fundamental nature of phase transitions and explore the meanings behind various phase diagrams used to describe the transitions between different phases. Here we examine the competition between inter-particle interactions and thermodynamic forces in determining the conditions for phase transitions to occur and emphasize the special role played by thermodynamic fluctuations near so-called "critical" points, where certain thermodynamic quantities tend to diverge.

Next, in Chapter 16, we investigate percolation theory as a simple, random process that generates self-similar structures with similar diverging behavior near a transition point. The percolation transition serves as an important template for understanding second-order phase transitions and the percolation structures bear a strong resemblance to the fluctuations present near a critical point.

Theoretical attempts to understand phase transitions near a critical point are developed in Chapter 17. These approaches begin from a fluctuation-free assumption, known as mean field theory, but are extended using renormalization techniques that exploit the self-similarity of the fluctuations. Finally, in Chapter 18, we wrap up our survey of phase transitions with a look at superconductors. Here again, critical-like features arise in the transition from a normal conducting material to a superconducting phase.

15 Introduction to phase transitions

Introduction

In this chapter, we develop some fundamental understanding of the nature of phase transitions by examining two well-studied examples: the vapor-to-liquid transition of fluids and the paramagnetic-to-ferromagnetic transition in magnetic materials. Here, our focus is on the experimentally observed features of these two transitions and how to interpret and navigate the many phase diagrams that describe them. The theoretical interpretation will be tackled later in Chapter 17. We will find that, in general, a phase transition is accompanied by some change in the amount of order as when, for example, liquid water freezes into crystalline ice. Moreover, we can describe this amount of ordering quantitatively by introducing an appropriate *order parameter*, whose value changes significantly only during the transition. Based upon the manner in which the order parameter changes, we can distinguish two different types of phase transitions: those of first order for which the order parameter changes discontinuously and those of second order for which it changes continuously. Second-order transitions are possible for both the vapor-to-liquid transition and the paramagnetic-to-ferromagnetic transition and are of interest due to the way in which many properties diverge near the transition in a similar, power law manner.

15.1 Free energy considerations

Phase transitions are governed by thermodynamics and thermodynamics is governed by two laws. The first of these, $dU = TdS - PdV$, is a statement of conservation of energy during a reversible process. The second law is notoriously imprecise, but for our current purposes can be interpreted to mean that a system will strive to minimize its Gibbs free energy, $G = U + TS - PV$. In a conceptual way we can use this minimization principle to understand why and where a phase transition occurs. Recall that changes in the free energy are given by,

$$dG = VdP - SdT, \tag{15.1}$$

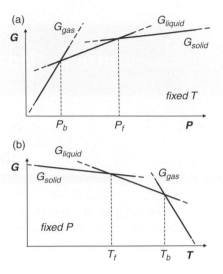

A schematic representation of the free energy of a fluid as a function of (a) pressure and (b) temperature, illustrating how minimization of the free energy necessitates the occurrence of well-defined phase transitions.

from which it follows that

$$V = \left(\frac{\partial G}{\partial P}\right)_T, \tag{15.2}$$

and,

$$S = -\left(\frac{\partial G}{\partial T}\right)_P. \tag{15.3}$$

Each of these last two equations is a statement regarding how the slope of either $G(P)$ or $G(T)$ depends on the system's volume or entropy, respectively. Consider then how pressure affects a system of particles to produce either gas, liquid or solid phases. A schematic representation of the $G(P)$ diagram would appear something like that shown in Fig. 15.1a, where the slope of each line segment is consistent with Eq. (15.2) and the experimental observation that the volume of the gas is much larger than that of the liquid, and the volume of the liquid slightly larger than that of the solid (i.e. $V_{gas} \gg V_{liquid} > V_{solid}$). At low pressures, the minimum free energy corresponds to the gas phase and hence the system assumes the properties of a gas at these pressures. As the pressure increases to the value P_b shown in the figure, the system is equally content to assume either the gas or the liquid phase, since both share a common minimum free energy at this point where the two line segments (labeled G_{gas} and G_{liquid}) intersect. At this transition point, the two phases are said to *coexist*. Above P_b, $G_{gas} > G_{liquid}$ and so the system liquifies. At even higher pressures above P_f, the system freezes into the solid phase, whose free energy is now the smallest of the other phases.

A similar pattern emerges in Fig. 15.1b, where $G(T)$ is developed using Eq. (15.3) for a situation at a fixed pressure. Here the slopes are arranged according to the ranking of entropies: $S_{gas} > S_{liquid} > S_{solid}$. In this instance the gas–liquid transition occurs at T_b and the liquid–solid transition at T_f.

15.2 Phase diagrams for fluids

The state of a fluid is defined entirely by its pressure, temperature and density. Thus a truthful representation of the phase diagram would require three independent axes and would be challenging to display. It is far more common to see the phase diagram depicted only in a series of two-dimensional perspectives obtained by a projection of the corresponding 3D diagram.

15.2.1 PT diagram

One familiar diagram is the *PT* diagram formed by projection onto the *PT*-plane. A typical example is shown in Fig. 15.2 and appears as a set of line segments known as *coexistence curves*. These curves mark the boundaries between various phases as well as indicating the conditions of P and T at which two or more phases can coexist in equilibrium. The first line segment extending out from the origin is known as the *sublimation* curve and marks the gas–solid transition. At the point labeled *TP*, this curve branches into two other curves. The lower branch is known as the *vaporization* or *condensation*

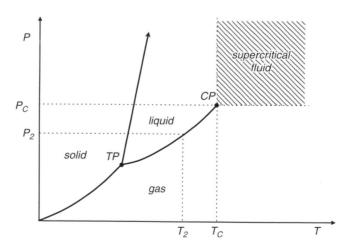

Figure 15.2   The *PT* phase diagram of a typical fluid. The point labeled *TP* is the triple point where all three phases coexist in equilibrium. Point *CP* is the critical point above which there is no gas-to-liquid transition. In this region the system remains in a gas–fluid state known as a supercritical fluid.

curve. It represents the gas–liquid transition and terminates abruptly at the point labeled *CP*. The upper branch is known as the *melting* or *freezing* curve and represents the liquid–solid transition. For many materials it is unclear whether this last curve ever terminates.

In principle, the location of these line segments is quite precise. As an example, imagine a container with a mixture of ice and water maintained at 0 °C and atmospheric pressure. This mixture is stable indefinitely. The ice will neither shrink nor grow over time. However, if the temperature or pressure is disturbed ever so slightly (as much as even, say, a part in a thousand), one or the other phase will dominate. Either the ice will melt completely or the water will freeze completely as a result of heat entering or exiting the container.

Returning to Fig. 15.2, there are two additional features to note. The first of these is the unique point labeled *TP* in the figure at which three lines/segments intersect. This intersection marks the *triple point* where all three phases (gas, liquid and solid) are able to coexist indefinitely. The second point of interest is the point labeled *CP* in Fig. 15.2, where the vaporization curve abruptly terminates at some P_c and T_c. This point is referred to as the *critical point*, above which condensation will not occur. Above T_c, the bonding between particles needed for condensation to take place is thwarted by excessive thermal agitation. As the pressure increases above P_c at these temperatures, the system just continues to densify without the particles sticking together. In the region above P_c and T_c, the phase is known as a *supercritical fluid*. Here the system exhibits density like a liquid, but remains compressible like a gas. Lacking cohesion, supercritical fluids also lack surface tension and are often employed as "dry" cleaning solvents.

15.2.2 PV diagram

In Fig. 15.3, we present a typical example of the projection of the fluid phase diagram onto the *PV*-plane for a series of fixed temperatures, increasing from T_1 to T_3. In actuality, we have normalized the volume by the number of particles to obtain a specific volume (i.e. the volume per particle, $v \equiv V/N = n^{-1}$), the inverse of the number density. This provides an intensive variable, independent of the actual size of the system. Each curve of fixed temperature is known as an *isotherm* and we see that at the highest temperatures the shape is approximating that predicted by the ideal gas law, where $P \propto n \approx v^{-1}$. However, as temperature decreases near the critical temperature, T_c, we see significant distortions develop. These distortions are a reflection of how the attractive forces between the particles begin to dominate over the diminishing thermal agitation. Below T_c, a gas–liquid phase transition is present and indicated by the horizontal stretches of the isotherm where P remains constant.

To understand the information contained in the *PV* diagram, we consider an experiment in which a gas starting at T_3 is cooled at a fixed pressure, P_2. To achieve this fixed pressure, the gas is contained in a cylinder with a movable

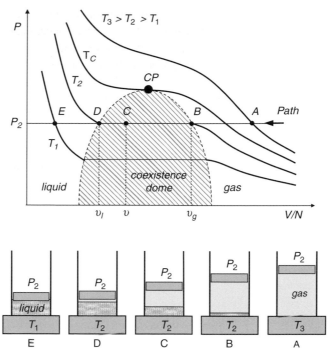

Figure 15.3   The *PV* diagram of a typical fluid is shown in the upper figure for a series of four isotherms. Only below the critical temperature is phase separation possible. During the first-order transition, the liquid phase separates from the gas with decreasing temperature at a fixed pressure, P_2, as illustrated in the lower set of figures.

piston on which a fixed load is maintained throughout the cooling process. Illustrations of the contents of the container at different points during the cooling process are provided in Fig. 15.3. Between T_3 and T_2, the system remains a gas, which contracts with cooling. At T_2, the liquid phase begins to condense into the bottom of the container and an interface (boundary between the two phases) forms. Between point B and point D, heat flows out of the system as the gas is transformed into liquid. The total heat leaving is the *latent heat of vaporization*. At point D, the entire system has transformed into liquid and additional cooling only results in a slight decrease in volume associated with the liquid's thermal expansion. The entire process can be reversed in vaporizing the liquid and will require an identical input of latent heat.

Lever rule

Between points B and D in Fig. 15.3, both the pressure and the temperature are fixed. These fixed values correspond to some point on the vaporization curve of Fig. 15.2, where both liquid and gas coexist. Because both P and T are fixed,

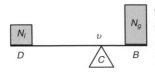

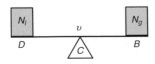

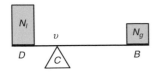

Figure 15.4

The lever rule for phase
transitions. As the transition
proceeds, the average specific
volume shifts from gas-like to
liquid-like. At each stage the
numbers of particles in each
phase form proportions that
maintain a balanced lever.

only the specific volume is changing during the transformation. However, the
specific volume of each phase separately is not changing: the specific volume
of the gas, v_g, is given by point B and that of the liquid, v_l, by point D in
Fig. 15.3. What is changing then are the *proportions* of gas and liquid, which
can be determined from the PV diagram using what is known as the *lever rule*.

The average specific volume of the system is,

$$v = \frac{V}{N} = \frac{N_l v_l + N_g v_g}{N} = X v_l + (1 - X) v_g, \qquad (15.4)$$

where N_l and N_g are the numbers of liquid and gas particles at any point during
the transition and the fraction of liquid particles is,

$$X = N_l/N. \qquad (15.5)$$

Now v must be located somewhere between points B and D, say at point C, and
the two line segments BC and BD have a ratio,

$$\frac{BC}{BD} = \frac{v_g - v}{v_g - v_l} = X \qquad (15.6)$$

equal to the fractional amount of liquid. During the transition from gas to
liquid, the position of v moves continuously from point B to point D. As
illustrated in Fig. 15.4, the point acts as a moving fulcrum that is always
positioned so as to properly balance the relative amounts of liquid and gas
particles present. At any point in the journey, the relative length of the "lever
arm" BC, extending from the fulcrum to the gas end is a measure of the liquid
present, while the other lever arm CD is a measure of the gas present.

15.2.3 TV diagram

Our final projection of the phase diagram is that onto the TV-plane, of which an
illustration is shown in Fig. 15.5 for a series of fixed pressures (isobars)
increasing from P_1 to P_3. Again, we see that at low densities, the isobars
approximate the linear dependence expressed by the ideal gas law, but that the
curves become progressively distorted as the density increases. We now repeat
our experiment with the gas in a cylinder with a movable piston, but this time
we fix the temperature at T_2 and compress the gas by increasing the pressure.
We find, as shown in Fig. 15.5, the same progression of the system from a gas
into a liquid. Again, an amount of latent heat is released as particles of gas
condense into the liquid phase in moving from point B to point D in Fig. 15.5,
and the fraction of liquid can be determined using the lever rule.

15.2.4 Order parameter

Let us step back and consider what is happening during the above transitions in terms of changing levels of order. In the gas at point B, each

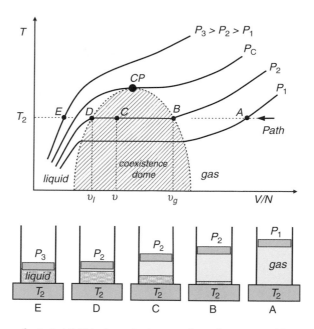

The *TV* diagram of a typical fluid is shown in the upper figure for a series of four isobars. Only below the critical pressure is phase separation possible. During the first-order transition, the liquid phase separates from the gas with increasing pressure at a fixed temperature, T_2, as illustrated in the lower set of figures.

particle has a rather large volume of space, v_g, all to itself. This free-range existence implies a large degree of disorder in the sense that it is difficult to predict where the particle is actually located. By comparison, the particles in the liquid at point D have far less available volume ($v_l \ll v_g$), and like chickens in a henhouse, are confined to a limited region of space where we can better predict their location. Although the liquid and gas remain disordered, in the sense that both are rotationally invariant, the confinement of the particles of the liquid causes short range correlations to appear and amounts to a level of ordering. To quantify this level of ordering, we introduce an appropriate *order parameter*, ϕ, such as,

$$\phi = X\left(\frac{n_l - n_g}{n_C}\right) \tag{15.7}$$

to monitor the developing order. Here, n_C is the density of the system at the critical point. To see how this order parameter works, consider either of the condensation experiments illustrated in Fig. 15.3 or Fig. 15.5. As we move from gas to liquid, X changes from zero (no order) to $X = 1$ (complete order). Meanwhile, the magnitude of this ordering is reflected by the density difference between the endpoints ($n_l - n_g$) which, in turn, depends on how far below the critical point the transition occurs.

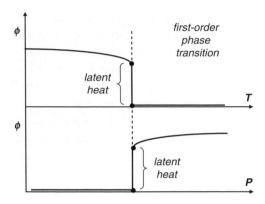

Figure 15.6 Variation of the order parameter for the first-order phase transitions along the paths illustrated in Figures 15.3 and 15.5.

The variation of the above order parameter for the gas–liquid transition is sketched in Fig. 15.6, both in the case when P is increased or T is decreased to the transition point (P_2, T_2) of Fig. 15.3 and Fig. 15.5. Here, the order parameter changes discontinuously at the transition and is an example of what is known as a *first-order transition*. First-order transitions are characterized by (1) a discontinuity in the order parameter and (2) a finite amount of latent heat entering or exiting during the transition. Shortly, we will discuss features of *second-order transitions* in which the order parameter increases continuously and for which there is no latent heat involved.

15.3 Supercooling/heating and nucleation

Although the transition point, where both phases coexist, is well defined, the little experiment with cooling a gas in a cylinder that we described above is not entirely accurate. In many instances, the gas can be cooled *beyond* the transition point without the development of any liquid. That is, it can be *supercooled*. Similarly, the liquid can often be *superheated* above the transition without the development of any gas. How is this possible when minimization of free energy would dictate that the phase transition should occur precisely at the transition point?

Classical nucleation theory can help us understand the phenomenon of supercooling (and of superheating). Let us consider the first tiny drop of liquid that initially forms from the gas in our little experiment. Where did it come from? Why is located where it is? The answers to these questions lie in the density fluctuations present in the gas. As a result of some random chance, a small number of gas particles, aided by the reduced thermal agitation present below T_c, came into close proximity and stuck together to become a small amount of liquid. The N_l particles of this drop contain an amount of free energy, $G_{liq} = N_l\mu_l$, where μ_l is the free energy per liquid particle, also known as the *chemical potential*. Because the system is super-cooled, the chemical potential of the liquid is lower than that of the gas, and we would expect the drop to be energetically favorable. However, the formation of this drop comes with an additional energy penalty. The drop is formed *within* the surrounding gas and so its existence necessitates the formation of an interface whose surface tension adds to the free energy of the drop by an amount,

$$G_{surface} = 4\pi R^2\gamma, \tag{15.8}$$

where R is the radius of the drop and γ is the surface tension (energy per area). Together, the total free energy of the drop is,

$$G_{drop} = N_l\mu_l + 4\pi R^2\gamma, \tag{15.9}$$

while the free energy of the remaining gas is,

$$G_{gas} = (N - N_l)\mu_g. \tag{15.10}$$

So in the process of forming the drop, the change in free energy of the system as a whole is,

$$\Delta G_{drop} = G_{final} - G_{initial} = \left[N_l\mu_l + 4\pi R^2\gamma + (N - N_l)\mu_g\right] - \left[N\mu_g\right]$$
$$= 4\pi R^2\gamma - \left(\frac{4\pi}{3}R^3 n_l\right)\Delta\mu, \tag{15.11}$$

where $\Delta\mu = \mu_g - \mu_l$.

A plot of ΔG_{drop} as a function of the drop size is presented in Fig. 15.7, which displays a maximum for a *critical nucleation size*, R_c, given by

$$R_c = \frac{2\gamma}{n_l\Delta\mu}. \tag{15.12}$$

For small drops with $R < R_c$, the droplet is unstable and vaporizes back into the gas phase. For drops larger than R_c, the free energy is reduced by increasing the drop size and the droplet begins to grow in size by condensing additional particles from the gas. Now we can understand the phenomenon of supercooling. Just at the coexistence point, $\Delta\mu = 0$ and so, according to Eq. (15.12), the critical nucleation size is infinite. At this point, there is no density fluctuation likely to produce such an infinitely large drop and so in spite of the possibility of both phases coexisting, only the gas phase is present. As cooling proceeds below the transition point, $\Delta\mu$ increases (see Fig. 15.1b) and the critical drop size decreases. Eventually, the critical size becomes comparable to the average size of density fluctuations present in the gas and condensation of a liquid phase begins. Similar

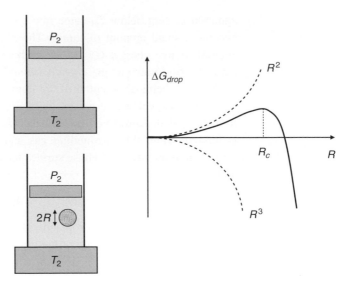

Density fluctuations drive the spontaneous development of a drop of liquid of radius R in a gas. If the size of the drop is less than the critical nucleation size, R_c, the drop is unstable and will dissolve. If the size is larger than R_c, the drop will grow and the entire system will condense into the liquid phase.

analysis explains the phenomenon of superheating, where a tiny bubble of gas must eventually form with a size larger than a corresponding critical size.

15.4 Critical phenomena

So far we have examined first-order phase transitions, for which there is a finite amount of latent heat entering or exiting the system, and for which the order parameter changes discontinuously. We now turn our attention to an example of a second-order phase transition, which for fluids occurs near the critical point. We again imagine performing an experiment, but now with the system confined to a box whose volume per particle is fixed at the critical value, $v_C = n_C^{-1} = V_C/N$, of the critical point. As illustrated in Fig. 15.8, we imagine decreasing temperature from some point in the supercritical fluid region, such that our system proceeds towards the critical point along a vertical path that strikes the coexistence dome head-on.

In Fig. 15.8, we sketch the condition of the system at each point along the path, as well as the corresponding lever arm analysis. At point A, far above the critical point, the system is a supercritical fluid of fixed density that is decreasing in temperature (and in pressure) with cooling. As it cools the particles become ever more prone to their mutual attraction, but as yet remain

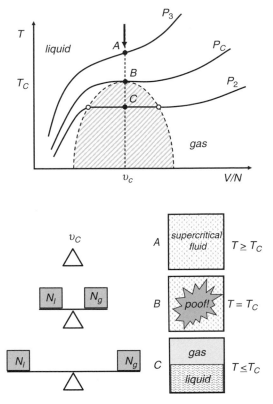

Figure 15.8 Illustration of the second-order phase transition occurring near a critical point in fluids. Here we imagine approaching the critical point by cooling the contents of a box whose volume and particle number is fixed to match the critical density, n_C. As we cool further below T_c, equal portions of liquid and gas appear.

unable to condense. Conversely, at point C far below the critical point, we find that the system is separated into a liquid and a gas phase with a well-defined boundary or meniscus between the two phases. By virtue of the lever rule, each phase contains approximately equal numbers of particles. What then happens at the critical point itself? It is as though the particles of the gas were *suddenly* able to take advantage of their attractive interaction and condense with roughly half of the particles instantly forming into a liquid! Just above the critical point, we had a disordered phase and, without any latent heat flow, the system spontaneously developed a more ordered phase.

Some appreciation for how this spontaneous transformation occurs is seen in the shape of both the critical isotherm and critical isobar in Fig. 15.3 and Fig. 15.5, just at the critical point. In both instances, the curve just flattens to a slope of zero at the critical point. This means that quantities like the isothermal compressibility,

$$\chi_T \equiv -\frac{1}{V}\left(\frac{\partial V}{\partial P}\right)_T, \tag{15.13}$$

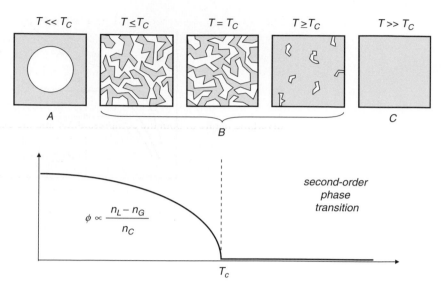

Figure 15.9 An illustration showing the developing pattern of large-sized density fluctuations near the critical point. As T_c is approached from above, small clusters of liquid phase sporadically begin to develop and grow in size with cooling. Just at the critical point, these liquid clusters are substantially interconnected and together comprise roughly half the particles in the box (via the lever rule). Below T_c, the particles of the liquid phase rapidly densify while the particles of the gas quickly expand and the order parameter, shown in the lower figure, develops continuously below T_c in a form characteristic of a second-order phase transition.

and the isobaric expansivity,

$$\alpha_P \equiv \frac{1}{V}\left(\frac{\partial V}{\partial T}\right)_P, \tag{15.14}$$

both *diverge* near the critical point. For the case of cooling, this means that the density of the gas is unstable and that any small decrease in temperature can result in an infinitely large density increase. Similarly, any infinitesimally small increase in pressure can cause an infinitely large density increase. In both situations, these large density increases represent the nearly spontaneous formation of a liquid-like phase (the high density and more ordered phase).

15.4.1 A closer look: density fluctuations

Although the transition at the critical point occurs rather spontaneously, in reality it is a continuous transition developing over a very narrow interval of temperature or pressure change. What actually happens in passing through the transition region is illustrated by the series of sketches shown in Fig. 15.9. In this panel of sketches, however, we illustrate the evolution of the system as it might appear if our experiment were conducted in a zero-gravity environment (such that the denser phase would not sink in the container). Far above the critical point, the system is entirely gas with a range of small density fluctuations randomly occurring. Because the attractive forces are ineffective at these temperatures, the fluctuations quickly dissipate. As the system approaches near to, but still just above the critical point, these random fluctuations in density are *amplified* by the diverging nature of both the compressibility and the expansivity and they lead to a segregation of the system into small regions of liquid (caused by a positive density fluctuation) and small regions of gas (promoted by a negative density fluctuation). Since the system is now closer to T_c, the attractive forces are more effective in allowing these regions to persist for longer periods of time.

Imagine what happens when a small region of high density forms and persists for some time. Because this region is more ordered, it suffers a drop in entropy that necessitates a small outflow of heat to its surroundings. The region cools and grows, as the surroundings heat and expand, in a sort of feedback process that is the source for the divergences in Eq. (15.13) and Eq. (15.14). Here, above the critical point, the surface tension is negligibly small and, instead of forming droplets, the liquid regions tend to assume dendritic patterns of a self-similar, fractal nature. Just at the critical point, these randomly forming liquid "clusters" persist for very long times and develop branches that span the entire container. Here the system can be viewed as either dendritic liquid clusters in a gas or, equivalently, as dendritic gas clusters in a liquid with roughly equal numbers of particles devoted to either phase. Below the transition, the system separates completely into a liquid drop and a surrounding gas. With additional cooling, the liquid drop densifies while the surrounding gas becomes less dense. A plot of the order parameter, shown in Fig. 15.9, displays a rapid, but *continuous* development near the transition point that is the signature of a *second-order transition*.

On approaching the critical point from either direction, we see a developing structure in the form of large-sized density fluctuations. At some point, these fluctuations grow to a size, ξ, comparable to that of the wavelength of visible light and result in intense scattering known as *critical opalescence*. Away from the transition, where the density fluctuations are small, little scattering occurs and the majority of incident light exits in the forward direction – the system appears transparent (provided that there is no absorption). However, as the critical point is approached, the system becomes turbid or opalescent (i.e. cloudy or milky) to an extent where almost all of the incident light is scattered away.

Critical opalescence: the Ornstein–Zernike theory

An early attempt to interpret the scattering of light near the critical point was advanced by Ornstein and Zernike (1914), and it is fruitful to review their approach here. Recall that the structure factor for an isotropic medium is,

$$S(q) = 1 + \langle n \rangle \int g(r) e^{-i\vec{q} \cdot \vec{r}} d^3 \vec{r} = 1 + \langle n \rangle \Gamma(q),$$

where $\Gamma(q)$ is just introduced as shorthand for the Fourier transform of the pair distribution function. If we are interested in the angular variation of *visible* scattered light, for which q is small, we might consider attempting an expansion of $\Gamma(q)$. However, since *g(r)* develops large fluctuations near the critical point, its Fourier components at small q will tend to diverge and it is unlikely that the expansion will properly converge. Instead, Ornstein and Zernike consider the quantity,

$$C(q) \equiv \frac{\Gamma(q)}{1 + \langle n \rangle \Gamma(q)} \propto \begin{cases} \Gamma(q) \approx 0, & \text{for } T \gg T_C \\ \langle n \rangle^{-1}, & \text{for } T \approx T_C, \end{cases} \tag{15.15}$$

which remains finite on approach to the critical point. This can now be expanded in a Taylor series,

$$C(q) \approx c_o + \left(\frac{\partial C}{\partial q} \right)_{q \approx o} q + \frac{1}{2!} \left(\frac{\partial^2 C}{\partial q^2} \right)_{q \approx o} q^2 + \cdots$$
$$= c_o + c_1 q + c_2 q^2 + c_3 q^3 + \cdots,$$

where the coefficients are given by,

$$c_\ell = \frac{1}{l!} \left(\frac{\partial^l C}{\partial q^l} \right)_{q=0} = \frac{(-i)^l}{l!} \int C(r) r^l (\cos \theta)^l r^2 dr \sin \theta d\theta d\phi. \tag{15.16}$$

From Eq. (15.16), one notes that the angular integration will cause all coefficients of odd l to vanish such that, to leading order in q,

$$C(q) \approx c_o + c_2 q^2. \tag{15.17}$$

From Eq. (15.15) and Eq. (15.17), we find that the *inverse* of the scattered intensity increases with increasing q as,

$$S^{-1}(q) = 1 - \langle n \rangle C(q) \approx -\langle n \rangle c_2 \left\{ \frac{1 - \langle n \rangle c_o}{-\langle n \rangle c_2} + q^2 \right\} \approx R^2 \{ \kappa^2 + q^2 \}, \tag{15.18}$$

where,

$$R^2 = -\langle n \rangle c_2 = \frac{2\pi \langle n \rangle}{3} \int r^4 C(r) dr,$$

and

$$\kappa^2 R^2 = 1 - \langle n \rangle c_o = S^{-1}(0). \tag{15.19}$$

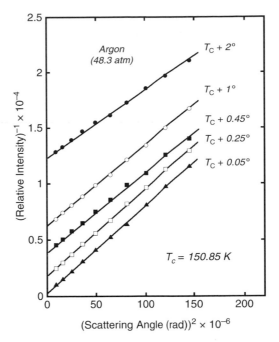

Figure 15.10

Small angle X-ray scattering from argon near its critical point ($T_c = 150.85$ K), plotted in the manner of the Ornstein–Zernike result of Eq. (15.18). Note that the x-axis is proportional to q^2. With cooling toward T_c, the y-intercept of each curve is seen to approach zero, implying a divergence in the compressibility, χ_T, in accordance with Eq. (15.20). (Adapted from Thomas and Schmidt, 1963.)

An example of this behavior is shown in Fig. 15.10, where the inverse of the scattered intensity from a container of argon is plotted against the square of the scattering wave vector so as to highlight the linear relation with q^2 in Eq. (15.18). Here the slope is proportional to R^2 and the intercept is proportional to

$$S^{-1}(0) = \kappa^2 R^2 \propto \chi_T^{-1} \propto (T - T_c)^\gamma, \qquad (15.20)$$

where, as was pointed out in Chapter 9, the scattered intensity of a liquid or gas at small q is proportional to the compressibility. In the Ornstein–Zernike theory, the pair distribution function (obtained by an inverse Fourier transform of Eq. (15.18)) has the form,

$$g_{OZ}(r) \propto \frac{1}{R^2}\frac{e^{-\kappa r}}{r}, \qquad (15.21)$$

which, curiously, is seen to match that for the structure of a random walk (see Eq. (8.8)). The exponential term thus represents a cutoff for the self-similarity and, in the light of how κ vanishes on approach to the critical point (see

Eq. (15.19)), we can identify its inverse as that of the diverging size or correlation length of the clusters,

$$\xi = \kappa^{-1} \propto (T - T_c)^{-\nu}. \tag{15.22}$$

In more careful measurements, it is found that the structure factor displays a slightly different q-dependence, of the form $S(q) \approx q^{-2+\eta}$, such that more generally,

$$g(r) \propto \frac{1}{R^2} \frac{e^{-r/\xi}}{r^{(d-2+\eta)}}. \tag{15.23}$$

15.5 Magnetic phase transitions

There is remarkable similarity in the features of phase transitions in both fluids and magnetic materials. Like fluids, a magnetic material consists of a large number of interacting objects, and its properties are influenced by various external fields. However, unlike the fluid phase transitions we have discussed thus far, the magnetic phase transitions we consider here take place while the material remains in the solid phase. In this case the particles remain fixed and the density of the crystal is largely impervious to changes in external pressure. Instead, the property of interest for a magnetic system is its magnetization, M, under the influence of temperature and an external magnetic field, H.

As described in some detail in Chapter 4, magnetic materials are composed of particles which, as a result of an imbalance in the atomic electron configuration, possess a net angular momentum and, in turn, a net magnetic moment, $\vec{\mu}$. There are certain quantum mechanical restrictions regarding the direction that the spin of an atom can take in space; but aside from these, the spin is otherwise unrestricted and may point in any random direction, as illustrated in Fig. 15.11.

When the spins are randomly oriented, the magnetization,

$$\vec{M} = n\langle\vec{\mu}\rangle \tag{15.24}$$

vanishes. But if an external magnetic field is applied, each spin experiences a torque,

$$\vec{\tau} = \vec{\mu} \times \vec{H}, \tag{15.25}$$

which acts to align the magnetic moment in the direction of the external field.

We now begin to see the parallels emerging between magnetic and fluid materials. For both, temperature increases the thermal agitation of the particles and, as a field, tends to randomize or disorder the system. For magnets, temperature tends to randomize the spins, causing the magnetization to vanish. Conversely, the external field, H, acts to order the system by aligning the spins through a mechanical process involving work. This mimics the effect of

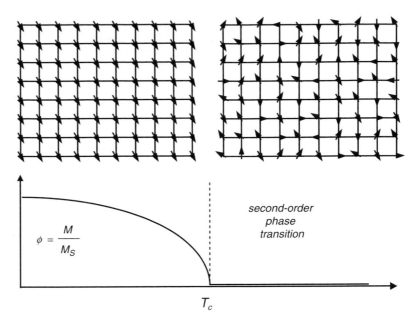

Illustration of the second-order phase transition in magnets. In the presence of a weak external field above T_c, magnetic spins are randomly oriented and produce no net magnetization. Just below T_c, small clusters of aligned spins sporadically form and grow with cooling until all spins are aligned. Consequently, the order parameter (i.e. the magnetization) increases in a continuous fashion.

pressure in the fluid case, which orders the system by performing work to increase the particle density. The fields P and H are thus counterparts. For the fluid system, ordering was measured by the development of a higher density (liquid) phase in place of the lower density (gas) phase. For the magnetic system, ordering is characterized by the magnetization, M, which serves as an appropriate order parameter,

$$\phi = \frac{M}{M_S},$$ (15.26)

where $M_S = n\mu$ is the maximum value of the magnetization obtained when all the spins are aligned.

15.5.1 Exchange interaction

In the case of fluids, we required an attractive interaction between the particles that would encourage them to stick together. The source of this attraction was the pair interaction (i.e. the van der Waals attraction), which took hold only for $T < T_c$. In the magnetic case it is the *exchange interaction* introduced in Chapter 4 (see Eq. (4.27)). In this interaction, a lowering of energy is achieved

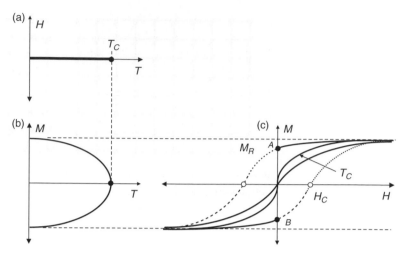

Figure 15.12

Phase diagrams for the paramagnetic–ferromagnetic phase transition. (a) The *HT* diagram shows that the transition occurs in the absence of an external field. (b) The *MT* diagram displays a shape similar to the coexistence dome of the fluid systems. However the two branches do not represent two phases, but rather two possible directions for the magnetization depending on the direction of an external field. (c) The *MH* diagram is shown for three isotherms (solid curves). Note that for the critical isotherm, the slope $\chi_m = \partial M/\partial H$, diverges for $H \rightarrow 0$. A residual magnetization, M_R, is present in the system at weak fields below T_c and can only be removed by applying a reverse field of magnitude H_c.

whenever two spins are aligned and, provided that the thermal energy is low enough, the aligned state remains stable. The boundary between where the exchange interaction is ineffective and where it begins to take hold, occurs at the *Curie temperature* (or critical temperature), T_c. Above T_c, spins are randomized and can only be aligned by application of an external magnetic field. When that magnetic field is removed the spins return to a randomized condition. This is much like compressing a gas above T_c – all we achieve is a packing of the particles without any cohesion and return to a less ordered condition on removal of the pressure. Above T_c, the magnetic system is referred to as *paramagnetic*. Below T_c, the material enters into the *ferromagnetic* phase. Here the exchange interaction dominates over the thermal agitation and when the external field is removed, the magnetization does not vanish.

15.5.2 Magnetic phase diagrams

Like the fluids, we can discuss the paramagnetic–ferromagnetic phase transition in terms of various projections of the three-dimensional (H, M and T) phase diagram. The first of these is the *HT* diagram (corresponding to the *PT* diagram of fluids), shown in Fig. 15.12a. It is rather simple and consists of a single, horizontal line segment extending below T_c at $H = 0$. The diagram simply indicates that the randomized (paramagnetic) phase is separated from the ordered (ferromagnetic) phase at that temperature. The external field, as mentioned above, only serves to bias the direction of the magnetization.

The *TM* diagram (corresponding to the *TV* diagram of fluids) is shown in Fig. 15.11b and exhibits a dome-like shape, comparable with the coexistence dome in fluids. Here, however, there are not *two* coexisting phases indicated, but rather two options for the direction of the magnetization of a single ordered phase, depending on the bias provided by a vanishingly small external field *H*. In the light of Eq. (15.26), we see that the paramagnetic–ferromagnetic transition is a second-order phase transition in which the magnetization develops continuously, without the action of any external work (performed by *H*, which can be zero) or equivalent latent heat.

The last projection is the *MH* diagram (corresponding to the *PV* diagram of fluids), which is shown in Fig. 15.11c. Three individual isotherms are displayed. The first of these is above T_c, in the paramagnetic phase, where application of the external field increases the magnetization towards its saturation value, M_S. As the field is removed, thermal energy randomizes the spins and the magnetization returns to zero along the same curve. Below T_c, in the limit of a vanishingly small external field, the magnetization is finite as a result of the spontaneous ordering. When the field is changed below T_c, the magnetization traces out the sort of hysteresis loop that was discussed previously in Chapter 4.

What happens just at T_c? Note here how the T_c isotherm in Fig. 15.12c just becomes vertical in passing through the origin and mimics how the T_c isotherm of a fluid just becomes flat near the critical point (see Fig. 15.3). For fluids, this flattening of the isotherm indicated a diverging compressibility wherein small pressure fluctuations created large density variations and the formation of dendritic structures. Here, it is the magnetic susceptibility,

$$\chi_m \equiv \frac{\partial M}{\partial H},\tag{15.27}$$

that is diverging near the critical point and similarly implies that any small fluctuation in the local field will motivate a large fluctuation of magnetization. Like the fluid case, the divergence has its source in a feedback process that leads to the formation of dendritic patterns near T_c.

15.6 Universality: the law of corresponding states

In our discussion above, we have witnessed considerable similarity between the fluid and magnetic phase transitions near their respective critical points. Although each addresses different properties (particle density versus spin alignment), and each involves a different interaction (van der Waals versus exchange), each exhibits a diverging response just near the critical point.

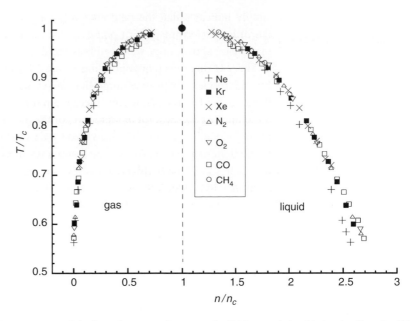

Figure 15.13 Demonstration of the law of corresponding states for fluids near their critical point. Here the *TV* coexistence curves of a diverse set of fluid systems are seen to collapse to a common (universal) curve upon scaling the temperatures and densities to those at the critical point. (Adapted from Guggenheim, 1945.)

Indeed, in careful experiments, these divergences are seen to exhibit power law dependences of the form,

$$\chi_T \approx (T - T_C)^{-\gamma}$$
$$\chi_m \approx (T - T_C)^{-\gamma}, \tag{15.28}$$

where the exponent, $\gamma \approx 1.3$, displays a similarity that is remarkably unexpected for what would seem to be very different systems. Furthermore, many second-order phase transitions display an interesting *law of corresponding states*, such that properties (like the coexistence curves shown in Fig. 15.13 for a variety of fluids) behave identically when the state variables (P, V and T) are scaled to values (P_c, V_c and T_c) at the critical point. This is only a foreshadowing of the many similarities we will encounter between the behaviors of unlike systems when they approach a critical point. As it turns out, many of the similarities arise from the geometry of the isotherms very near T_c, and we often speak about such different systems as belonging to the same *universality class*, because of how they share similar *critical exponents*. We will examine this universality in more detail in Chapter 17, but first, we examine an important model regarding the structure of the density fluctuations, known as percolation theory, which captures much of the underlying physics behind critical behavior.

Summary

- Phase transitions are a consequence of the thermodynamic requirement for minimization of free energy of a system of particles, and generally involve a change in the degree of order as characterized by some appropriate order parameter.

- Phase transitions are classified into two types. First-order transitions exhibit a discontinuous change in the order parameter and an exchange of latent heat. Second-order transitions exhibit a continuous change in the order parameter and occur in the absence of any latent heat exchange.

- Diverging susceptibilities near a critical point drive the production of large, spatial fluctuations in the order parameter. In the case of fluids, these density fluctuations cause substantial scattering, known as critical opalescence.

- Laws of corresponding states, wherein the behaviors of different systems appear identical when thermodynamic fields are appropriately scaled to those at the critical point, are characteristic of second-order phase transitions.

Exercises

15.1. Some people have reported being scalded by water that suddenly began boiling vigorously when removed from a microwave oven. Oven manufacturers now recommend that water be heated with a non-metallic stirrer or toothpick inserted into the liquid to reduce the chances of this happening. Explain why the sudden boiling occurs and how the stirrer aids in reducing the effect.

15.2. (a) Derive Eq. (15.12) for the critical nucleation size. (b) Show also that the barrier for homogeneous nucleation is $\Delta G|_{R_c} = \frac{16\pi\gamma^3}{3n_l^2\Delta\mu^2}$. (c) Show that for small undercoolings of amount ΔT, the barrier can be expressed as $\Delta G|_{R_c} = \frac{16\pi}{3}\left(\frac{\gamma^3}{n_l^2}\right)\left(\frac{T_b}{L_V}\right)^2(\Delta T)^{-2}$, where T_b is the boiling point and L_V is the latent heat of vaporization.

15.3. In *heterogeneous* nucleation, an additional surface is introduced as a catalyst for the growth of nucleation sites. Imagine then a droplet of liquid forming on the surface, as illustrated in Fig. 15.14. The liquid drop takes the form of a spherical section and meets the surface with a contact angle θ. (a) Show that the contact angle is given by

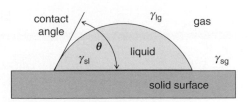

Figure 15.14

$\cos\theta = (\gamma_{sg} - \gamma_{sl})/\gamma_{lg}$. (b) Show that the critical nucleation radius is unchanged and equal to that for homogeneous nucleation (Eq. (15.12)). (c) Show, however, that the nucleation barrier is reduced over that of homogeneous nucleation (see Exercise 15.2) by a factor, $f(\theta)$ such that

$$\Delta G|_{R_c} = \left[\frac{16\pi\gamma^3}{3n_l^2\Delta\mu^2}\right](2 - 3\cos\theta + \cos^3\theta)/4.$$

15.4. Show that the slope of the liquid–vapor coexistence curve is given by, $\frac{dP}{dT} = \frac{S_g - S_l}{V_g - V_l}$.

15.5. A liquid crystal undergoes a transition from the isotropic phase, where the particles are randomly oriented, to the nematic phase where the particles are oriented in a common direction. A common choice for the order parameter is $\phi = \left\langle \frac{3\cos^2\theta - 1}{2} \right\rangle$, where θ is the angle between the long axis of the particle and the director. Verify that this is an appropriate order parameter by showing that it vanishes in the disordered phase.

Suggested reading

I much recommend Schroeder's thermodynamics textbook. The text by Stanley is devoted almost entirely to the subject of critical phenomena in both fluids and magnetic systems.

C. Kittel and H. Kroemer, *Thermal Physics*, 2nd Ed. (W. H. Freeman and Co., San Fransico, 1980).

D. V. Schroeder, *An Introduction to Thermal Physics* (Addison Wesley Longman, New York, 2000).

H. Eugene Stanley, *Introduction to Phase Transitions and Critical Phenomena* (Oxford University Press, New York, 1971).

16 Percolation theory

Introduction

Percolation theory refers to properties of a simple experiment in which random events produce features common to second-order transitions; namely a continuously developing order parameter and self-similar, critical-like fluctuations. The model itself is quite simple, yet as we will see, it has been used extensively to interpret many phenomena found in nature, including not only the conditions under which liquids percolate through sand (from which the theory obtains its name), but also the manner in which stars form in spiral galaxies.

In this chapter, we investigate the percolation process in some rigorous detail to demonstrate how percolation clusters develop in a self-similar, power law manner near the percolation threshold. We also take this opportunity to introduce both the finite-sized scaling and renormalization techniques. Both of these techniques exploit the inherent self-similarity to gain insight into the critical exponents that characterize a second-order phase transition, and will prove useful to us in the next chapter.

16.1 The percolation scenario

At the heart of percolation theory is the question of how long-range connections develop through a random process. Consider a geometrical lattice of some arbitrary dimension such as the two-dimensional networks of pipes shown in the form of a square lattice in Fig. 16.1a. Here, the pipes are fully connected and fluid is free to flow from one edge of the network to the other. Suppose we now insert valves throughout this arrangement of pipes in one or the other of two ways. In the first instance, which corresponds to *bond percolation*, the valves are placed inside the pipes (i.e. inside the "bonds" between intersections), as shown in Fig. 16.1b. In the alternate case, referred to as *site percolation*, the valves are placed at the intersection of the pipes. Again, when all the valves are opened, the network is fully connected and fluid can flow readily from one side to the other. But, if all the valves are closed, the network is fully unconnected and fluid is unable to flow anywhere.

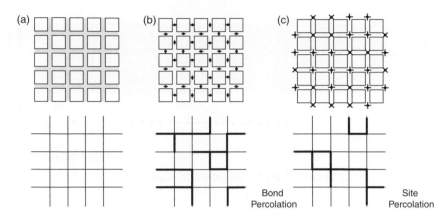

(a) A network of interconnected pipes form a square lattice. (b) In the case of bond percolation, the valves are inserted within the connecting pipe. (c) For site percolation, the valves are inserted at points where pipes connect. In either situation, valves are randomly opened and we inquire as to the degree of connectivity that develops.

Percolation theory asks, "how does the connectivity of the network develop as valves are opened at random?" Suppose a fraction, p, of the closed valves are opened. Obviously, if $p = 0$, the connectivity remains absent and if $p = 1$, the connectivity is maximum. However, at some intermediate value of p, as is the case for Fig. 16.1c, there will occur a dendritic network of connections for which the fluid can just manage to flow from one side to the other.

In Fig. 16.2, site percolation is carried out on a square lattice with each panel illustrating the evolution of the connectivity as a function of the fraction of open valves. In the figure we represent an opened valve by a black dot which indicates that the valve is activated or that the site is "occupied". Initially, all the sites are unoccupied (corresponding to closed valves and a complete absence of any connections). In Fig. 16.2b, one-quarter of the sites have been randomly chosen for occupancy and we see that the lattice consists mainly of a large number of isolated occupied sites (monomers) with only some occasional occurrences of pair and triplet combinations (dimers and trimers), for which limited connectivity is present. In the figure, dark lines represent these connections and the number of sites, s, that participate in such a "cluster" provides a measure of its mass. Clearly, the distribution of these s-mers decreases with mass, as illustrated in Fig. 16.2b, where we count some 21 monomers, eight dimers, two trimers and one each of a 4-mer and a 6-mer.

As p is increased further, such as in Fig. 16.2c, where $p = 0.50$, we find that the distribution of cluster masses has evolved. The average mass of the largest-sized clusters, s_{max}, has increased and the numbers of very small-sized clusters have decreased. Upon reflection, we see that two mechanisms of cluster growth are at work. The first of these is a rather innocuous self-growth mechanism, in which the random introduction of a newly occupied site

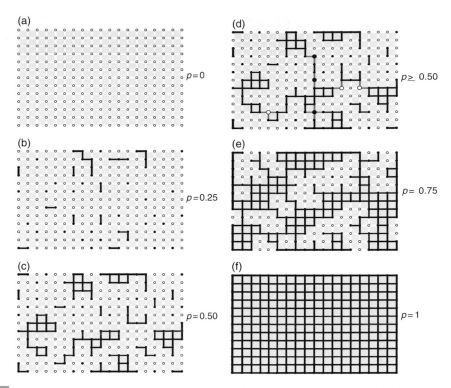

Figure 16.2   Site percolation is carried out on a 13 × 19 square lattice. (a) Sites on an unoccupied lattice become randomly occupied. (b) At 25% occupancy, a small number of other-than-monomer clusters have developed and (c) by 50% a well-defined distribution of cluster sizes is present. (d) At this point we see that the introduction of as few as three additional (but well-chosen) occupied sites will produce a spanning cluster that connects one side of the system to the other. Two examples are shown in (d): when the three white circles are added to the panel in (c), a spanning cluster forms horizontally across the system and when the three black circles are instead added, a spanning cluster forms vertically. (e) Additional sites above the percolation threshold rapidly incorporate remaining finite clusters into the mass of the spanning cluster. (f) At 100% occupancy, connectivity reaches its maximum.

happens to occur beside an existing cluster, causing its mass to increase by one. The second of these is a more severe aggregation process, in which the random introduction of a newly occupied site happens to link together two pre-existing clusters, causing the mass of the resulting cluster to far exceed that of either of the two from which it formed.

The aggregation mechanism is a key feature to the sharpness of the connectivity transition. To see this, consider again the case of $p = 0.50$ as it is reproduced in Fig. 16.2d. In this figure, two situations are shown in which just three additional sites (corresponding to a minor increase to $p \approx 0.51$) have been (non-randomly) added. In one instance, the added sites produce a very large cluster that spans from left to right across the network, while in the other

Table 16.1 Percolation thresholds for both site and bond percolation conducted on various lattices (data obtained from Stauffer (1985).)

Dimension, d	Lattice type	p_c (site)	p_c (bond)	Coordination
1	Chain	1	1	2
2	Honeycomb	0.696	0.653	3
2	Square	0.593	0.500	4
2	Diamond	0.428	0.388	4
2	Triangular	0.500	0.347	6
3	SC	0.312	0.249	6
3	BCC	0.245	0.179	8
3	FCC	0.198	0.119	12

case, the cluster spans from top to bottom. When this "spanning" cluster forms, a percolating pathway is produced and long-range connectivity of the network is first established.

16.1.1 Percolation threshold: the spanning cluster

In the example of Fig. 16.2d, the three added sites were not randomly chosen, to make a point. Nevertheless, they could have occurred randomly to produce a spanning cluster at $p \approx 0.51$. Had other sites been randomly selected, the percolation path might not yet have formed. In fact, the precise fraction of occupied sites at which the pathway forms would vary from any one simulation to the next, depending upon what series of random choices were actually made. To circumvent this, we need to study the scenario multiple times to produce an appropriate ensemble average. If this is done, one finds there is a well-defined critical fraction, p_c, at which a spanning cluster forms to produce a complete connection from one side of the lattice to the other. At this *percolation threshold*, the connectivity "percolates" across the entire network. Studies of the percolation scenario indicate that the specific value of p_c depends not only on the dimensionality of the network, but also on the type of lattice (square, triangular, hexagonal, etc.) and on the percolation variety (site or bond percolation), as summarized in Table 16.1.

What now happens above this percolation threshold? As illustrated in Fig. 16.2e, the aggregation process remains active and, now that a spanning cluster exists, the largest of the remaining finite clusters are quickly incorporated into it. As a result, the mass of the spanning cluster grows quite dramatically just above p_c, while the remaining population of finite clusters diminishes rapidly. If we define P to represent the probability that any randomly selected site is connected with, and thus a member of the spanning cluster, we would find (with appropriate ensemble averaging) a dependence for P on p, like that illustrated in Fig. 16.3. Below p_c, a spanning cluster does not yet exist and so

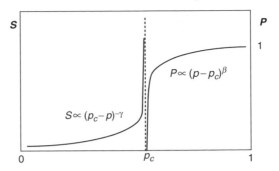

Figure 16.3 An illustration of how the average cluster mass diverges in a power law manner on approach to the percolation threshold, and how the mass of the spanning cluster develops rapidly (in a power law fashion) just above the threshold.

the probability that a site belongs to a spanning cluster is zero. At p_c, the spanning cluster just forms and grows quickly as a result of further aggregation. The probability P becomes finite here and increases rapidly. As p approaches 1, all of the finite clusters are eventually incorporated into the spanning cluster (see Fig. 16.2f) and P approaches unity.

Look closely at Fig. 16.3. What does it remind you of? Hopefully, you will recognize its strong similarity with that of an order parameter for a second-order phase transition occurring at $p = p_c$! Indeed, percolation is an example of a second-order transition and provides an excellent introduction to some of the phenomena that arise near such transitions.

16.1.2 A closer look: cluster statistics

Now that we have outlined the concepts of the percolation scenario, let us look at the properties of these clusters in some greater detail. Although the series of figures in Fig. 16.2 is illustrative of the process, we really imagine that the lattice is far more extensive than the 13×19 grid shown (infinite in fact), and therefore the distribution of cluster masses is well-defined and forms a continuous function of the mass s. In this case, we could define a cluster mass distribution function, n_s, to represent the number of s-mers per lattice site,

$$n_s = (\# \, s\text{-mers})/N_{\text{TOT}}, \tag{16.1}$$

where $N_{\text{TOT}} \approx L^d$ is the total number of sites on a d-dimensional lattice of length L. One consequence of this normalization scheme for the distribution is that the first moment of the distribution just equals the fraction of occupied sites p,

$$\sum_{s=1}^{\infty} s n_s = \frac{1}{N_{\text{TOT}}} \sum_{s=1}^{\infty} (\# \, s\text{-mers}) \times s = \frac{N_{\text{Occupied}}}{N_{\text{TOT}}} = p. \tag{16.2}$$

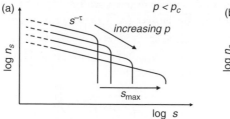
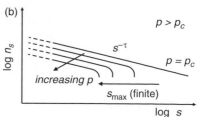

Figure 16.4   The evolution of the cluster mass distribution of finite-sized clusters during the percolation process. (a) Below the percolation threshold, the average cluster mass (as well as the maximum cluster mass, s_{max}) increases with increasing occupancy and smaller-sized clusters develop a self-similar distribution described by a power law, $n_s \approx s^{-\tau}$. As the occupancy increases, cluster–cluster aggregation forces a reduction in the population of smaller clusters. (b) Above the threshold, a spanning cluster is present which rapidly accumulates mass by the incorporation of the remaining finite-sized clusters. Both the maximum cluster mass and population of finite-sized clusters decrease as a result.

Cluster mass distribution: homogeneous functions

What is the functional form of n_s? Clearly it must have certain properties. Firstly, it most likely is a decreasing function of the cluster mass because at any $p < p_c$ there are always more small clusters than there are large clusters (see for example Fig. 16.2c). Secondly, for $p < p_c$, it must drop to zero rather rapidly for masses $s \approx s_{max}$, where s_{max} is the average mass of the *largest* clusters present. Lastly, we note that the aggregation process has a built-in self-similarity: the aggregation occurs similarly at all scales. For example, a newly occupied site is just as likely to merge two 5-mers into a 10-mer as it is to merge two 50-mers into a 100-mer. Consequently, n_s should reflect this self-similarity and the distribution (at least for $s < s_{max}$) should appear the same at different scales. A function that displays this sort of self-similarity is a *homogeneous* function of the form $g(\lambda x) = \lambda^a g(x)$, of which the most common example is a power law, $g(x) \approx x^{-\tau}$. For these reasons, we assume the cluster mass distribution to be of the form,

$$n_s \approx s^{-\tau} f(s/s_{max}) \approx s^{-\tau} e^{-s/s_{max}}, \tag{16.3}$$

where the function $f(z)$ produces the desired cutoff for $s > s_{max}$. A plausible choice for this cutoff function is an exponential as suggested in Eq. (16.3).

An illustration of the cluster mass distribution is shown for $p < p_c$ in Fig. 16.4a in a double logarithmic scale for which the cutoff appears quite dramatic. Note that as p increases, the distribution spreads outwards as a result of the aggregation of smaller clusters into larger ones and downwards as a result of the depopulation of smaller clusters. Aside from the cutoff, the distribution is self-similar in that it appears the same for all mass scales less than s_{max}. Indeed, the cutoff itself provides the only evidence of a relevant or special mass scale, as it marks a bend in the distribution function.

How does this relevant mass scale (s_{max}) depend on p? The growth process occurring for the largest clusters is clearly non-linear and must lead to a divergence of s_{max} near p_c, where s_{max} produces the spanning cluster extending across the infinitely large lattice. To a good approximation, this divergence is well described by another power law,

$$s_{max} \approx |p_c - p|^{-1/\sigma}. \tag{16.4}$$

Average cluster mass

The largest clusters have a mass, s_{max}, but what is the *average* mass of any arbitrarily chosen cluster? To determine this we need to employ the cluster mass distribution Eq. (16.3) that was just developed, and recall that n_s is the number of s-mers per lattice site. The quantity sn_s is the probability that any arbitrarily chosen site belongs to an s-mer, and the fraction of clusters with mass s is then,

$$w_s = \frac{sn_s}{\sum\limits_{s} sn_s} = \frac{sn_s}{p}. \tag{16.5}$$

The average cluster mass we seek is just the weighted average,

$$S = \sum_s s w_s = \frac{\sum\limits_{s} s^2 n_s}{p}. \tag{16.6}$$

Since we are assuming a large lattice and thus a continuous distribution of masses, we can employ an integral,

$$S \approx \int_0^\infty s^2 n_s \mathrm{d}s = \int_0^\infty s^{2-\tau} \mathrm{e}^{-s/s_{max}} \mathrm{d}s, \tag{16.7}$$

where n_s from Eq. (16.3) has been incorporated. To carry out the integration, we choose the change of variables $z = s/s_{max}$, such that,

$$S \approx (s_{max})^{3-\tau} \int_0^\infty z^{2-\tau} \mathrm{e}^{-z} \mathrm{d}z \approx (s_{max})^{3-\tau}. \tag{16.8}$$

Here, the integral over z contains the cutoff function (which rapidly falls to zero) and produces an unknown, but finite, constant. Incorporating the power law divergence for s_{max} from Eq. (16.4), we find (not surprisingly) that, like s_{max}, the average cluster mass also diverges on approach to p_c as another power law:

$$S \approx |p_c - p|^{-\gamma}, \quad \gamma = \frac{3-\tau}{\sigma} \quad (\tau < 3), \tag{16.9}$$

as is illustrated in Fig. 16.3.

The spanning cluster

As discussed earlier, a spanning cluster forms at p_c and accretes mass rapidly thereafter. As it grows, remaining finite-sized clusters that are not part of the original spanning cluster are "swallowed up" by the spanning cluster, causing the cluster mass distribution of the finite clusters to recede and eventually vanish, as illustrated in Fig. 16.4b. To determine the variation of P just above the threshold, we recognize that the occupancy, p, varies continuously across the transition. Just before the spanning cluster forms, $P = 0$ and $p = \sum_{s=1}^{\infty} sn_s \, (p \approx p_c)$, as given by Eq. (16.2). Just above the threshold, p can be separated into those sites belonging to the spanning cluster, P, and those remaining occupied sites found in other, finite-sized clusters,

$$p = P + \sum_{\text{finite } s} sn_s \, (p \geq p_c), \qquad (16.10)$$

For p to remain continuous across the transition, we thus require

$$\sum_{s} sn_s \, (p \approx p_c) \approx P + \sum_{\text{finite } s} sn_s \, (p \geq p_c), \qquad (16.11)$$

or, upon rearranging and switching again to an integral form,

$$P \approx \int_{0}^{\infty} \left[s^{-\tau} - s^{-\tau} e^{-s/s_{\max}} \right] s \, ds = \int_{0}^{\infty} s^{1-\tau} \left[1 - e^{-s/s_{\max}} \right] ds. \qquad (16.12)$$

Integration by parts results in,

$$P \approx \frac{s^{2-\tau}}{2-\tau} \left[1 - e^{-s/s_{\max}} \right] \Bigg|_{0}^{\infty} + \int_{0}^{\infty} \frac{s^{2-\tau}}{\tau - 2} \frac{e^{-s/s_{\max}}}{s_{\max}} \, ds, \qquad (16.13)$$

and, since P must remain finite, $\tau > 2$ and the first term vanishes. As for the remaining integral, we again apply the change of variables $z = s/s_{\max}$ to obtain,

$$P \approx \frac{(s_{\max})^{2-\tau}}{\tau - 2} \int_{0}^{\infty} z^{2-\tau} e^{-z} \, dz \approx (s_{\max})^{2-\tau} \approx |p - p_c|^{\beta}, \quad \beta = \frac{\tau - 2}{\sigma} \quad (\tau > 2),$$

$$(16.14)$$

and find that P vanishes near p_c in a power law fashion, as is illustrated in Fig. 16.3.

Fractal clusters

Examine again the clusters in Fig. 16.2c, just near the percolation threshold. Their shape is determined by a random process and one sees that they possess a rather open structure containing voids and filaments. Indeed, if this figure were reproduced using a much larger sized lattice that approaches that of an infinite space, we would find that the clusters are another example of *fractal* structures (like those introduced in Chapter 8), whose mass is related to their spatial size by a fractal dimension,

$$s \propto R_s^{D_f}. \tag{16.15}$$

Here, R_s is some appropriate measure of the effective diameter or radius of the cluster, such as its radius of gyration.

The fractal structure is evidenced in all reasonably large clusters, including those largest clusters, $s = s_{\max}$, for which we can define an important relevant length scale, the *correlation length*,

$$\xi \propto \left(s_{\max}\right)^{1/D_f}. \tag{16.16}$$

The correlation length is significant as it establishes the one (and only) relevant length scale controlling the entire percolation process. To see this, just review Eq. (16.3), Eq. (16.8) and Eq. (16.14) where S, P and the cutoff are ultimately determined by the divergence of $s_{\max}$, itself a consequence of the growing correlation length,

$$\xi \approx \left(s_{\max}\right)^{1/D_f} \approx \left(|p_c - p|^{-1/\sigma}\right)^{1/D_f} \approx |p_c - p|^{-v}, \quad v = \frac{1}{\sigma D_f}. \tag{16.17}$$

The correlation length can be viewed as a measure of the granularity of the emerging sea of clusters, which coarsens appreciably with increasing p, culminating, at the percolation threshold, in the appearance of a correlation length comparable to the finite size, L, of the system itself.

16.2 Scaling relations

We have thus far characterized the percolation process by the divergence of several quantities described by a host of yet unknown critical exponents $(\tau, v, \beta, \gamma, \sigma, D_f)$. What can we say about the actual values these exponents assume? In this section we show how the necessity that the percolation transition proceeds in a continuous fashion through the threshold region places constraints on what value these exponents can assume. These constraints appear in the form of relationships between the exponents that limit us to just *two*, truly unique, exponents for any given dimension.

We have seen that the percolation threshold marks a unique situation in which the system is delicately balanced between two states. One of these is just below p_c, where a spanning cluster is not yet present but a self-similar distribution of large finite clusters (with fractal shapes) is poised to complete the long-range connectivity of the lattice. The other is just above p_c, where a spanning cluster has now appeared and is rapidly consolidating other clusters into its structure. Above the threshold, the mass of this spanning cluster is given as,

$$s_{\text{max}}^{\text{above } p_c} \propto PL^d, \tag{16.18}$$

for a lattice containing L^d sites. Just below the threshold, $P = 0$, and, although the spanning cluster does not "officially" exist yet, there is at least one very large cluster whose mass is given by Eq. (16.16) as, $s_{\text{max}}^{\text{below } p_c} \propto \xi^{D_f}$. Just at the threshold,

$$s_{\text{max}}^{\text{below } p_c} = s_{\text{max}}^{\text{above } p_c},$$

and, because $\xi \to L$ in this limit,

$$\xi^{D_f} \approx P\xi^d. \tag{16.19}$$

Since $P \approx \Delta p^\beta$ and $\xi \approx \Delta p^{-\nu}$ (where $\Delta p \equiv |p - p_c|$), this matching requires that

$$[\Delta p^{-\nu}]^{D_f} \approx \Delta p^\beta [\Delta p^{-\nu}]^d,$$

or,

$$D_f = d - \beta/\nu. \tag{16.20}$$

This last equation is an example of a *scaling relation*: an expression relating different critical exponents. Other scaling relations can be determined (Ex. 3) where one finds that, aside from the dimension of the lattice, it suffices to know only two critical exponents in order to determine all of the rest.

16.2.1 Finite-sized scaling

We need only determine two of the critical exponents. But how might these unknown exponents be determined? Naively, one would think that we could just measure them experimentally by performing the percolation repeatedly on a computer-generated lattice. However, the exponents determined in this way turn out to be sensitive to the size of lattice we use, and will not in general reflect the "true" exponents associated with an infinite-sized lattice. Instead, *finite-sized scaling* provides a mechanism for extrapolating the features of percolation found on a finite lattice to that which would occur on an infinite lattice.

Suppose, as illustrated in Fig. 16.5, we conducted the percolation scenario on an infinite lattice to some fraction $p < p_c$, but chose to partition this space into sub-lattices of size $L = \xi$. For the original lattice ($L = \infty$) at $p < p_c$, $L < \xi$, and so a spanning cluster is not present and $P(L = \infty) = 0$. However, for the partitioned lattices ($L \approx \xi$), half will have a spanning cluster and half will not. For these, $P(L \approx \xi) \approx 0$ such as would occur when $p \approx p_c$. If the sub-lattices are further partitioned to ones in which $L < \xi$, almost all will contain a spanning cluster that is reasonably well developed with $P(L < \xi) > 0$, as occurs when $p > p_c$. What we have achieved by this partitioning is a percolation transition caused, not by the growth of the correlation length, but by the shrinking of the

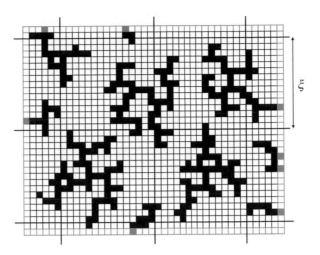

Figure 16.5 Percolation conducted on an infinite lattice is arbitrarily partitioned into very many smaller systems of length L. The presence or absence of a spanning cluster in these smaller systems will depend on the size of the partitioning relative to the correlation length, ξ.

system size in relation to a *fixed* correlation length. All of this suggests that, when dealing with a finite system, we must generalize $P(p)$ to include an implicit dependence on the finite size of the system in relation to the correlation length:

$$P(p) \Rightarrow P(p, L). \tag{16.21}$$

What is the likely form for $P(p,L)$? Because of the inherent self-similarity, we might anticipate a power law dependence and, since the percolation illustrated in Fig. 16.5 transpires as L decreases, an appropriate form would be $P(L) \approx L^{-A}$, where A is some, yet undetermined, exponent. But, as we saw for the cluster mass distribution, there are limitations to self-similarity whenever the length scales exceed the correlation length. These limitations are incorporated by introducing a cutoff function (also referred to as a scaling function) and so we anticipate,

$$P(p, L) \approx L^{-A} f(z), \quad z = L/\xi = L \Delta p^{\nu}. \tag{16.22}$$

The exponent A must now be set in such a way that $\lim_{L \to \infty} P(p, L) \approx L^0 \Delta p^{\beta}$, and becomes independent of L. For this we need $f(z) = z^{\beta/\nu}$ or $A = \beta/\nu$. Then,

$$P(p, L) \approx L^{-\beta/\nu} f(L \Delta p^{\nu}). \tag{16.23}$$

The utility of this result lies in its applicability to computer simulations conducted on lattices of sequentially increasing size L near p_c. By plotting $P(L)$ against L on a double logarithmic scale, the resulting slope provides a direct determination of the ratio of two "true" (i.e. $L = \infty$ lattice)

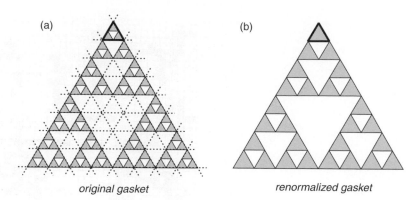

Figure 16.6 An illustration of the renormalization approach for a self-similar Sierpinski gasket. The left-hand gasket (a) is partitioned (dashed lines) into cells containing four tiny triangles, as emphasized in the topmost triangle. The cell emphasized contains three solid triangles and one (inverted) vacant triangle and is condensed to appear in the renormalized gasket (b) as a single solid triangle. Some of the cells in the left gasket contain four vacant triangles and are condensed to produce a corresponding vacant triangle in the renormalized gasket.

critical exponents. By similar reasoning, the L-dependence of other quantities, such as the average cluster mass,

$$S(p, L) \approx L^{\gamma/\nu} f(L\Delta p^{\nu}), \tag{16.24}$$

can be employed to obtain other exponents.

16.2.2 Renormalization

Clearly an experimental route to determining the critical exponents is available in the form of finite-sized scaling. But are there any theoretical approaches? One that appears to work well is known as *renormalization group theory*. While we only develop it here at a conceptual level, renormalization group theory has been widely applied to a variety of phenomena, well removed from condensed matter theory, including nuclear physics.

Again, the approach centers on exploiting the inherent self-similarity present. As a familiar example, consider once more the geometrical fractal shown in Fig. 16.6, known as a Sierpinski gasket. Suppose we "renormalize" the gasket on the left of the figure by replacing or "condensing" a cell of three smallest triangles to form a single larger triangle. The result (shown on the right in the figure) appears with its self-similar structure intact. Indeed, if we took the gasket on the right and rescaled it down by a factor of two in size, it would identically replace any of the three large triangles that make up the original gasket. This is the nature of self-similar structures.

Now consider a percolation lattice, like that illustrated in Fig. 16.7, in which a group of $b^d = 9$ sites (or cells) are combined to form a single super cell. If we

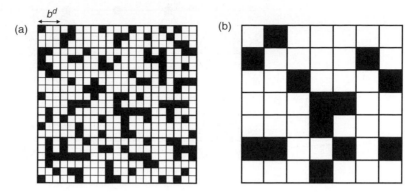

Figure 16.7 (a) Percolation on a d-dimensional lattice is partitioned into supercells of length b, each containing b^d subcells. (b) Under a renormalization operation, each supercell in the left-hand figure is assigned an occupancy. In this example, a supercell is occupied if its subcells contain a cluster that spans either vertically or horizontally across the supercell.

are at $p = p_c$, where self-similarity is present at all length scales ($n_S \approx s^{-\tau}$), then we anticipate that condensing the cells of the original lattice on the left to form the supercells of the lattice on the right should (like the Sierpinski example) retain the inherent structure.

The rules under which a set of cells in the original lattice is condensed are somewhat arbitrary, but here we might take it to mean that, if the set of b^d cells contains a cluster that spans the set, its corresponding super cell will be an occupied site. However, if the set of original cells does not contain a spanning cluster, the super cell will be labeled unoccupied.

For the original lattice, $\xi \propto \Delta p^{-\nu}$. For the renormalized lattice,

$$\xi \propto b(\Delta p')^{-\nu}, \qquad (16.25)$$

where p' is generally different from p except just in the limit that p approaches p_c, where $p = p' = p_c$. Because we are near $p = p_c$, we demand that in either lattice the correlation length should still be comparable to the overall lattice size and hence identical for each lattice,

$$\Delta p^{-\nu} = b(\Delta p')^{-\nu}. \qquad (16.26)$$

This then implies that the exponent is given by,

$$\nu = \frac{\log b}{\log(\Delta p'/\Delta p)}. \qquad (16.27)$$

As an example of applying the renormalization technique to percolation, we consider site percolation on the triangular lattice shown in Fig. 16.8. Sites of the original lattice are depicted by circles, three of which (those at the corners of the shaded triangles) are to be condensed to form a super lattice depicted by

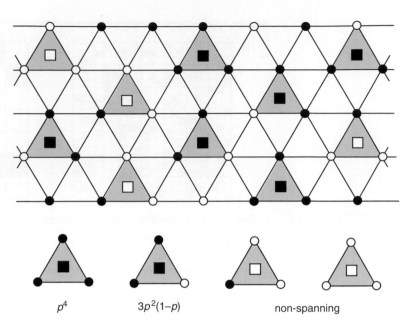

p^4 $3p^2(1-p)$ non-spanning

Figure 16.8 A triangular lattice of circles is renormalized to the larger triangular lattice of squares. In performing this renormalization, a cell of the new lattice is occupied if the three surrounding sites of the original lattice produce a spanning cluster either by containing two or three occupied sites (lower figure). The probability for such clusters to occur in the original lattice is indicated in each case.

the squares. For the super lattice, p' is the probability that any arbitrary site of the super lattice is occupied. Likewise, p is the probability that any arbitrary site of the original lattice is occupied, while $1-p$ is the probability that the site is vacant. We consider a triplet of original sites to contain a spanning cluster if at least two of the sites are occupied. The probability for all three being occupied is p^3 and for any two it is $p^2(1-p)$ times the number of possible permutations. Hence the probability that the triplet produces an occupied super cell is,

$$p' = p^3 + 3p^2(1 - p). \tag{16.28}$$

At $p = p_c$, $p = p'$ and there are three solutions for Eq. (16.28) at $p^* = 0, 0.5$ and 1. Two of these solutions (also known as fixed points) are trivial and reflect the obvious result that, in the case of $p = 0$ or $p = 1$, the condensing will lead to all supercells empty or occupied, respectively. The unique solution occurs at $p^* = \frac{1}{2}$, which happens to coincide with the value of p_c obtained experimentally (see Table 16.1).

To determine the exponent for our triangular example using Eq. (16.27), we first express p' as a Taylor expansion about the unique solution $p^* = p_c$,

$$p' \approx p_c + \frac{dp'}{dp}\bigg|_{p_c} (p - p_c) = p_c + [6p_c(1 - p_c)](p - p_c). \tag{16.29}$$

Quantity	Exponent	$d = 2$	$d = 3$	$d = 4$	$d = 5$	Bethe
$P \approx \Delta p^{\beta}$	β	0.14	0.39	0.56	0.67	1
$S \approx \Delta p^{-\gamma}$	γ	2.39	1.8	1.43	1.22	1
$\xi \approx \Delta p^{-\nu}$	ν	1.33	0.9	0.64	0.51	1/2
$n_s \approx s^{-\tau}$	τ	2.05	2.2	–	–	5/2
$R_s \approx s^{1/D_f}$	D_f	1.9	2.5	3.12	3.69	4
$D \approx \Delta p^{\mu}$	μ	1.3	2.0	–	–	3

Table 16.2 Critical exponents for percolation in several dimensions (data obtained from Jan, Hong and Stanley (1985) and Zallen (1983).)

For $p_c = \frac{1}{2}$, we find that

$$\Delta p' / \Delta p \approx [6 p_c (1 - p_c)] = 3/2. \tag{16.30}$$

Including this result in Eq. (16.27) together with our renormalization factor $b^2 = 3$, we obtain,

$$\nu = \frac{\log(\sqrt{3})}{\log(3/2)} = 1.355, \tag{16.31}$$

which is favorably close to the experimental value of 4/3 (see Table 16.2). In general, the renormalization approach improves when a larger number of original cells are condensed (e.g. $b^2 > 3$ in our case of a triangular lattice), but of course the development of p' corresponding to Eq. (16.28) then becomes more cumbersome.

16.2.3 Universality and the mean field limit

A summary of the critical exponents for percolation is provided in Table 16.2. It is interesting to note that the exponents only depend on the dimension of the lattice. They do not depend upon the type of lattice or whether the percolation is of the site or bond variety. This invariance is referred to as *universality* and is a recurrent feature of phase transitions in condensed matter physics.

Also included in Table 16.2 are the exponents determined (analytically) for the so-called Bethe lattice. A Bethe lattice is formed by repeated branching of a site to z adjacent sites, as illustrated in Fig. 16.9 for $z = 3$. The Bethe lattice is not a conventional sort of lattice because none of the branches ever intersect (as they would in say a square lattice), but rather extend indefinitely without crossing each other. Such a lattice is clearly impractical as it would require sites to become infinitely crowded. Nevertheless, it has an inherent connectivity that can be analyzed in a percolation scenario.

Particularly significant is that the Bethe lattice mimics the behavior of a conventional lattice of very high (infinite) dimension. For conventional lattices of size L, the volume increases as,

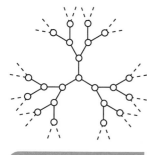

Figure 16.9

The Bethe lattice for coordination number $z = 3$.

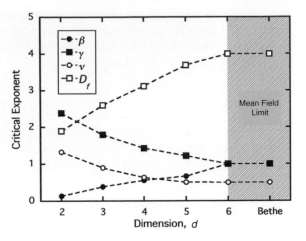

Figure 16.10 Critical exponents of percolation plotted against the lattice dimension to illustrate the approach to so-called mean field values above a critical dimension $d \geq 6$. (Data from Jan, Hong, and Stanley (1985) and Zallen, 1983.)

$$\text{Volume} \propto L^d,$$

while the surface increases as,

$$\text{Surface} \propto L^{d-1}.$$

Thus the surface to volume ratio of a conventional lattice decreases as,

$$\frac{\text{Surface}}{\text{Volume}} \propto L^{-1} \propto (\text{Volume})^{-1/d}. \tag{16.32}$$

However, for the Bethe lattice we see that each new generation of branches produces more sites than the previous generation. So after any generation, the majority of the total number of sites resides on the surface and the surface to volume ratio approaches a constant. According to Eq. (16.32), this would correspond to a conventional lattice of infinite dimension ($d = \infty$).

In Fig. 16.10 we have plotted the exponents of Table 16.2 as a function of the dimension. Two things are apparent. The exponents exhibit a systematic variation with dimension and all approach some fixed, d-independent, value for $d \geq 6$. It would seem that there is some sort of "homogenization" of the percolation process occurring as the dimension reaches $d = 6$, in which the divergent features of the transition assume some simple, innocuous limit. This limit is often referred to as the *mean field limit* in connection with a similar "homogenization" of the critical exponents associated with thermal and magnetic phase transitions, to be discussed in more detail in the next chapter. There, sites are associated with objects (e.g. magnetic spins that are either up or down) that interact with one another in a pairwise fashion. However, in the mean field limit, these discrete, site-to-site, interactions are approximated by a homogeneous "average" interaction similar to the interaction of the object with an external field (a mean field).

16.3 Applications of percolation theory

16.3.1 Orchard blight and forest fires

As mentioned in the introduction, percolation concepts are relevant to a wide field of processes found in science. As a first example, illustrating the practical applications of percolation theory, we consider the predicament faced by a fruit farmer who wishes to plant an orchard of fruit-bearing trees. We suppose the trees are planted on a square grid with a lattice spacing a. Farmland is expensive and the farmer strives to maximize his land use by planting the trees close together. However, he also recognizes that there is a risk that if one tree becomes diseased, the disease is likely to spread to all the other trees and destroy the entire orchard. Obviously, if the branches of the trees touch each other, the likelihood of disease transfer is very high but this probability for disease transfer should fall off in some exponential fashion at larger separations, as shown in Fig. 16.11.

This is a bond percolation problem on a square lattice, since the probability for disease transfer is equivalent to the probability that a bond has formed between two trees (i.e. sites). From Table 16.1, we see that $p_c = 0.50$ for this bond percolation scenario. Thus, if the farmer plants the trees with a spacing larger than the critical spacing a_c for which disease transfer probability is 50%, the likelihood of a devastating wipeout of some large fraction of the orchard is reduced, while valuable farmland is best utilized. In a fully analogous manner, this same problem applies to other transfer processes such as the spread of fire in a forest or the spread of disease among a population.

16.3.2 Gelation

Percolation also provides a suitable model for the process of gelation. No doubt we are all familiar with the edible gelatin ("Jell-O"), found in many school cafeterias, that is able to sustain its shape without being entirely rigid. Gelatin such as this forms when long polymer fibers dissolved in a solvent begin to form crosslinks between one another to produce an extended network. Where and when the crosslinks form is a random process and hence the connectivity of the polymer strands is well accounted for by percolation theory.

In gelatin, the crosslinking occurs with cooling and can be reversed by heating. For this reason, the edible gelatin we are most familiar with is known

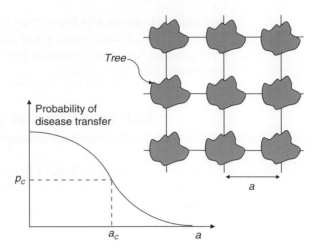

Upper right-hand figure shows trees planted in an orchard in the form of a square lattice of spacing a. Lower left-hand figure illustrates how the likelihood of disease transferring from one tree to another might depend on their separation. For separations greater than a_c, this probability resides below the corresponding bond percolation threshold and limits the spread of disease to a small number of finite clusters of trees.

as a reversible or *physical gel*. On the other hand, a similar gelation process occurs in epoxy resins, where two components (the epoxy and the hardening agent) are mixed and chemically react to produce permanent crosslinks over time at a fixed temperature. These crosslinks are thermally stable and epoxy resins are an example of what are known as irreversible or *chemical gels*.

In either case, crosslinks form randomly below the percolation threshold (also known as the gel point) and the solution, known as a sol, consists of small clusters of crosslinked material. As a result of the crosslinking the sol becomes progressively less fluid and the viscosity increases. The increase in viscosity is most acute near the threshold, where it is often observed to diverge in a power law manner of the form,

$$\eta \approx (p_c - p)^{-k}. \tag{16.33}$$

Just above the gel point, the system is referred to as a gel. Here, low frequency shear waves can be supported by the extensive network of crosslinked strands and the material begins to assume the properties of a solid. With increasing reaction above the gel point, the network becomes stronger and the shear modulus increases towards a limiting maximum value in a power law of the form,

$$G \approx (p - p_c)^{t}. \tag{16.34}$$

Although much of the process of gelation can be understood conceptually in terms of percolation theory, this cannot predict the values of the two exponents (k, t) that characterize the divergent trends in viscosity and shear modulus. These two quantities represent *dynamical* aspects that are not directly addressed by percolation theory, which is itself only a model for structural connectivity.

16.3.3 Fractal dynamics: anomalous diffusion

Because percolation theory addresses mainly the structural features of a percolating system, our interpretations of dynamical processes associated with a percolating system must involve the introduction of additional assumptions regarding time-dependent parameters. New exponents, like k and t, discussed for the sol–gel transition above, arise and are of a dynamical origin. Although values for these new exponents are often not predicted, scaling arguments can be applied, much like the finite-sized scaling discussed in connection with Eq. (16.22), to forge insightful scaling relationships between the various relevant exponents. We anticipate then that we could express any dynamical quantity, $Q(p,t)$, in a scaling form analogous with Eq. (16.22),

$$Q(p,t) = t^A f(z), \quad z = t/\tau = t\Delta p^x, \tag{16.35}$$

in which time scales are placed in reference to a characteristic time scale (τ) instead of length scales in reference to a characteristic length (ξ). A good example of this is the analysis of the *anomalous diffusion* occurring on a percolating lattice.

Diffusion is a dynamical process and as we have learned in Chapter 9, can be understood in terms of the motion of a walker performing a random walk. For an unrestricted space, we saw that the mean squared displacement of the walker evolves linearly in time:

$$\langle r^2(t) \rangle = Dt, \tag{16.36}$$

where D is the diffusivity. But what happens if the walk is restricted to those occupied sites of the percolating lattice?

Imagine yourself as the walker who parachutes down onto any one of many large clusters that have formed just below p_c. With dice in hand, you begin walking randomly. However, the space you have available to walk in is a fractal filled with branches upon branches forming a tenuous pathway, not unlike a labyrinth. You make some headway at first, but before too long you wander down a dead end that requires many rolls of the dice in order to randomly retrace your steps. Alternatively, you may find that you are located in a smaller cluster connected to other clusters only by a narrow passage. Again, to migrate through this bottleneck between the two regions requires many rolls of the dice and curtails growth of the mean squared displacement at long times. Instead of advancing linearly with time, as in Eq. (16.36), we find that the walk proceeds in an anomalous, sub-diffusive manner described by a power law of the form,

$$\langle r^2(t) \rangle \propto t^{1-a}. \tag{16.37}$$

Of course, this advance of the mean squared displacement with time is limited by the finite size of the cluster, R_s, you happen to land on. While at short times the mean squared displacement grows according to Eq. (16.37), it is limited at long times by the extent of the cluster itself,

$$\langle r^2(t \rightarrow \infty) \rangle_s = R_s^2. \tag{16.38}$$

To obtain a robust and meaningful measure of the diffusion present below p_c, we would need to imagine that an army of walkers is parachuted to every cluster and average their mean square displacement. We would find that each performs a similar sub-diffusive walk at short times, but differ in the limiting mean squared displacement they ultimately attain, depending on the size of the cluster they happen to inhabit. The average of this limiting mean square displacement is given by,

$$\langle R_s^2 \rangle = \sum_s s n_s R_s^2 \approx \int_0^\infty s \left(s^{-\tau} e^{-s/s_{\max}} \right) \left(s^{2/D_f} \right) ds, \tag{16.39}$$

where again, $s n_s$ is the probability that a given lattice site belongs to a cluster of mass s. Using techniques we employed previously, one finds that

$$\langle R_s^2 \rangle \approx \Delta p^{\beta - 2v}. \tag{16.40}$$

The average mean squared displacement of the walker is illustrated in Fig. 16.12 for clusters below the percolation threshold. At short times, the walker explores a fractal landscape and the mean squared displacement evolves in a sub-diffusive manner. However, depending on the level of occupied sites p, there is a crossover from this anomalous diffusion to a fixed, time-independent mean squared displacement, $\langle R_s^2 \rangle \approx \Delta p^{\beta - 2v}$, found at very long times. That is,

$$\langle r^2(t) \rangle = \begin{cases} t^{1-a}, & t < \tau_- \\ \langle R_s^2 \rangle \approx \Delta p^{\beta - 2v}, & t > \tau_-. \end{cases} \tag{16.41}$$

As can be seen from the figure, this fixed value diverges on approach to p_c and the crossover point occurs at a characteristic time,

$$\tau_- \approx \Delta p^{(\beta - 2v)/(1-a)}, \tag{16.42}$$

which also diverges on approach to the transition. Thus, just at p_c, the anomalous diffusion of Eq. (16.37) is unlimited.

Alright, now what happens above p_c? Imagine we parachute a walker who (mostly likely) lands on the spanning cluster. Very near p_c, this cluster still retains much of its fractal character, but as p increases, more of the dead ends become interconnected and the bottlenecks become less constricting due to the

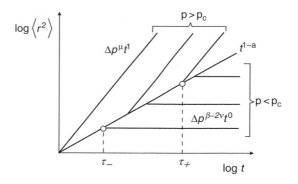

Diffusion on a percolating lattice displays anomalous diffusion at short times. Below the threshold, the mean squared displacement of a random walker is limited at long times (greater than τ_-) by the finite cluster size. Above the threshold, the random walk returns to normal diffusion at long times (greater than τ_+).

accretion of mass onto the spanning cluster. For short distances, the walker still explores a fractal landscape and the mean squared displacement starts off as before (Eq. (16.37)). However, at much longer distances, the reduction of dead ends and bottlenecks allows the mean squared displacement to increase in a normal, linear manner, but with a diffusivity that decreases as p approaches p_c from above,

$$\langle r^2(t) \rangle = \begin{cases} t^{1-a}, & t < \tau_+ \\ D(p)t \approx [\Delta p^\mu]t, & t > \tau_+. \end{cases} \qquad (16.43)$$

Again, from consideration of Fig. 16.12, we can conclude that

$$\tau_+ \approx \Delta p^{-\mu/a}, \qquad (16.44)$$

which again diverges on approach to p_c from above the threshold.

We see above that the dynamics are underscored by two time scales, given by Eq. (16.42) and Eq. (16.44). As we approach the percolation threshold, we need both time scales to be coincident so that the dynamic pattern (Eq. (16.41)) seen below p_c will join smoothly with that (Eq. (16.43)) above p_c. Thus the two exponents describing each time scale should match, and we can obtain the scaling relationship,

$$\mu = \left(\frac{a}{1-a} \right)(2v - \beta). \qquad (16.45)$$

An alternative route to this same scaling relationship is to apply the scaling approach introduced at the start of this section (see Eq. (16.35)). In this approach, we assume a scaling form for the mean squared displacement,

$$\langle r^2(p,t) \rangle_\pm = t^{1-a} f_\pm(z), \quad z = t\Delta p^x, \qquad (16.46)$$

where x is some, yet undetermined scaling exponent. At $p > p_c$, we seek a scaling function $f_+(z)$ that in the limit of large z will approach the desired $\langle r^2(p, t \to \infty)\rangle_+ = \Delta p^\mu t^1$. For this situation we must choose $f_+(z) \approx z^{\mu/x}$, so that

$$x = \mu/a. \tag{16.47}$$

At $p < p_c$, we seek a scaling function $f_-(z)$ that in the limit of large z will approach the desired $\langle r^2(p, t \to \infty)\rangle_- = \Delta p^{\beta-2\nu} t^0$. For this situation, we must choose $f_-(z) \approx z^{a-1}$, so that

$$x = (2\nu - \beta)/(1 - a). \tag{16.48}$$

Again, we find in combining Eq. (16.47) with Eq. (16.48), the same scaling relation of Eq. (16.45).

Summary

- Percolation theory considers the connectivity of a lattice with a fraction, p, of lattice sites randomly occupied. Bond percolation and site percolation are two common variants of the problem.

- As the occupancy increases, clusters develop with a self-similar distribution.

- At the percolation threshold, p_c, a continuous pathway (spanning cluster) forms that connects one side of the lattice to the other.

- The percolation threshold depends upon the lattice geometry and type (bond or site), but the critical exponents are universal and only depend on the dimension.

- Finite-sized scaling and renormalization are two approaches to determining critical exponents.

- Anomalous diffusion refers to the properties of a random walk conducted on a self-similar percolating lattice, and allows percolation theory to be extended to dynamical processes.

Exercises

16.1. Make a histogram of $n_s(s)$ for the situation shown in Fig. 16.2c. How does this histogram compare to $n_s(s)$ given by Eq. (16.3)?

16.2. Show how Eq. (16.39) becomes Eq. (16.40).

16.3. Show that all of the scaling exponents can be expressed in terms of just two together with the dimension, d, by expressing $v = f(\sigma, \tau, d)$ and $D_f = g(\sigma, \tau, d)$.

16.4. Apply renormalization to the square lattice with $b^d = 4$. In condensing, take a set of four original cells to contain a spanning cluster only if it produces a connection from left to right (not top to bottom). Show that the percolation threshold occurs at $p_c = 0.62$ with $v = 1.635$. Comment on this result in the light of the true values in Table 16.1.

16.5. Consider a Bethe lattice with branching z. Show that at the end of n branching generations,

(a) the volume (equal to the number of sites) is given by,

$$\text{Volume} = 1 + z \left\{ \frac{(z-1)^n - 1}{(z-2)} \right\},$$

(b) the surface (equal to the number of sites on the outermost layer) is given by,

$$\text{Surface} = z(z-1)^{n-1},$$

(c) and that, for large numbers of generation (n), the surface to volume ratio is given approximately by,

$$\frac{\text{Surface}}{\text{Volume}} \approx \frac{z-2}{z-1}.$$

16.6. A forest is modeled by pine trees (with branches spanning a diameter of $D = 10$ feet each) laid out on a 2D hexagonal (i.e. triangular) lattice of spacing $a > D$. The probability that an ember originating from any given tree will be sufficiently hot to ignite branches of another tree a distance r away (measured from tree center to tree center), is given as, $P(r) = \left\{ 1 - \tanh^2(r/4D) \right\}$. For this situation, determine the minimum spacing a_c for which percolation of a forest fire would be likely to occur.

16.7. As mentioned in the text, both the viscosity and shear modulus display power law dependencies on Δp, during the sol–gel transition. However, these two quantities are also frequency dependent and near the p_c, $G \approx \omega \eta \approx \omega^u$. The characteristic relaxation rate for a viscoelastic material is, $\omega_C = G_\infty / \eta_o \approx \Delta p^{t_\infty + k_o}$, and is seen to vanish at the percolation threshold. Let us attempt to introduce the frequency dependence by a scaling ansatz of the form:

$$G^*(\omega) = G(\omega) + i\omega\eta(\omega) = G_\infty \omega^u f_\pm(\omega/\omega_C) \text{ where at } p > p_c,$$

$$G(\omega) = G_\infty \omega^u f_+(\omega/\omega_C) \approx \begin{cases} G_\infty \omega^{-1} \Delta p^{t_o}, & \omega < \omega_C \\ G_\infty \omega^0 \Delta p^{t_\infty}, & \omega > \omega_C, \end{cases}$$

and at $p < p_c$,

$$\eta(\omega) = G_\infty \omega^{u-1} f_-(\omega/\omega_C) \approx \begin{cases} G_\infty \omega^0 \Delta p^{k_o}, & \omega < \omega_C \\ G_\infty \omega^1 \Delta p^{k_\infty}, & \omega > \omega_C. \end{cases}$$

(a) Develop appropriate forms for the piecewise scaling functions $f_\pm(z)$ required to match the above limiting frequency dependencies.

(b) Use these scaling functions to obtain the following scaling relations:
$$k_o = t_\infty(u^{-1} - 1), t_o = t_\infty(u^{-1} + 1), \text{ and } k_\infty = t_\infty(2u^{-1} - 1).$$

Suggested reading

Stauffer's introduction to percolation is an enjoyable read for anyone at any level. The last three articles are significant works on the topic of anomalous diffusion.

D. Stauffer, *Introduction to Percolation Theory* (Taylor and Francis, Philadelphia, 1985).

R. Zallen, *The Physics of Amorphous Materials* (John Wiley and Sons, New York, 1983).

S. Havlin and D. Ben-Avraham, *Adv. Phys.* **36**(6), 695 (1987).

Y. Gefen, A. Aharony and S. Alexander, *Phys. Rev. Lett.* **50**, 77 (1983).

A. L. Efros and B. I. Shklovskii, *Phys. Status Solidi B* **76**, 475 (1976).

Mean field theory and renormalization

Introduction

Our discussion of critical phenomena surrounding a second-order phase transition has thus far focused only on qualitative features. We have now examined three systems, fluids, magnets and random percolation, each of which displays an abruptly sharp transition from a less ordered to a more ordered phase as the respective transition point is encountered. Each shows a similar pattern of developing structure just in advance of the critical point. Fluctuations in the respective order parameter display a self-similar structure that is limited only by a single relevant length scale, the correlation length ξ, which diverges on approach to the transition point. For fluids and magnets, this structure arises from inherent fluctuations that are amplified by either a diverging compressibility or susceptibility, respectively.

In this chapter, we explore more quantitative, theoretical approaches taken to understand the features of second-order phase transitions. The simplest of these are the *mean field theories*, in which the pairwise interaction (needed to produce a phase transition) is introduced in the form of an average field. In this approach, the effects of the growing fluctuations of the order parameter near the critical point are ignored. Although the mean field approach does meet many of the requirements and does predict divergences of certain quantities near the transition, the critical exponents predicted by the theory do not match those seen experimentally. Obtaining correct exponents requires a more advanced approach involving renormalization techniques that exploit the self-similar structure of the fluctuations near the critical point and allow all of the various critical exponents to be inter-related, such that knowledge of any two yields all the others.

17.1 Mean field theory

17.1.1 The mean field approximation

In this chapter we want to examine theoretical models that might account for the second-order phase transitions of both the fluids and magnetic materials which we introduced in Chapter 15. We begin with what are known as mean field theories and consider (in parallel) two primary examples of the mean field approach: the van der Waals model for fluids and the Ising model for magnetic materials.

In order for a phase transition to exist, we must have attractive forces between the particles of the system that will allow them to stick together. At a microscopic level, these attractive forces are described by pairwise interactions between any two particles and provide for a lowering of the free energy, which as we learned in Chapter 15, is fundamentally responsible for phase changes. For fluids of the van der Waals type, the fundamental interaction is the van der Waals or Lennard-Jones potential,

$$u(r_{ij}) = 4\varepsilon \left\{ \left(\frac{\sigma}{r_{ij}} \right)^{12} - \left(\frac{\sigma}{r_{ij}} \right)^{6} \right\},$$

introduced in Chapter 3, which describes the potential energy stored by two particles separated by a distance r_{ij}. For magnetic materials, the fundamental interaction is the exchange interaction introduced in Chapter 4,

$$u(r_{ij}) = -2J_{ex}(r_{ij})\, \vec{S}_i(0) \cdot \vec{S}_j(r_{ij}), \tag{17.1}$$

which describes the stored energy associated with two adjacent electron spins separated by a distance r_{ij}. For both cases, the total ensemble-averaged energy stored in the interaction is obtained by summing over all pair combinations. For isotropic systems, we found in Chapter 3 that this energy can be expressed as,

$$\langle U_{\text{tot}} \rangle = \left\langle \sum_{i,j} u(r_{ij}) \right\rangle = \frac{N}{2} \sum_{j=2,N} \langle u(r_{1j}) \rangle. \tag{17.2}$$

The *Ising model* is a simplified version of a ferromagnet that incorporates two major restrictions. Firstly, the magnetic moments are assumed to arise only from the electron spin (i.e. $J = S$ and $L = 0$), so that the moment is proportional to the electron spin. In this case, the exchange interaction in Eq. (17.1) can then be expressed alternatively as,

$$u_{\text{Ising}}(r_{ij}) = -2\tilde{J}_{ex}(r_{ij})\, \vec{\mu}_i(0) \cdot \vec{\mu}_j(r_{ij}), \tag{17.3}$$

where the proportionality constant has been incorporated into $\tilde{J}_{ex}(r_{ij})$. Secondly, the magnetic moments (or the spins) are restricted to point in only one of two directions: up or down. For the Ising model containing N spins, we can thus express the total energy as,

$$\langle U_{\text{tot}} \rangle = -\frac{N}{2} \sum_{j=2,N} 2\tilde{J}_{ex}(r_{1j}) \langle \vec{\mu}_1(0) \cdot \vec{\mu}_j(r_{1j}) \rangle = -N \int_b^{\infty} \tilde{J}_{ex}(r)\Gamma(r)\, n\, d^3\vec{r},$$

$$\tag{17.4}$$

where n is the number density and the quantity in brackets is recognized as the *moment–moment* (or *spin–spin*) *correlation function*, $\Gamma(r) = \langle \vec{\mu}(0) \cdot \vec{\mu}(r) \rangle$, first encountered in Chapter 4.

In a similar manner, the total energy for a fluid is given (see Eq. (3.23)) as,

$$\langle U_{\text{tot}} \rangle = \frac{N}{2} \int_b^\infty g(r)u(r)\, n\, \mathrm{d}^3 \vec{r}, \tag{17.5}$$

where $u(r)$ is the van der Waals interaction and $g(r)$ is the pair distribution function which describes the average spatial variation of the density from one point to another.

Now herein lies our dilemma. To obtain the total internal energy, which we need to evaluate the free energy, we must integrate a quantity over all space that is inherently unstable just near the critical point. As we learned in Chapter 15, strong fluctuations develop near the critical point and so both $\Gamma(r)$ and $g(r)$ will be difficult to evaluate analytically there. Mean field theory then enters as an approximation scheme for dealing with this problem, by ignoring the fluctuations altogether and replacing them by average (mean) values. As an example, for the Ising model, the interaction between the central spin $\vec{\mu}_1(0)$ and another spin $\vec{\mu}_j(r)$ is replaced by a separation-independent average magnetic moment, $\vec{\mu}_j(r) = \langle \vec{\mu} \rangle = \vec{M}/n$, proportional to the magnetization. This allows the internal energy of Eq. (17.4) to be expressed as,

$$\langle U_{\text{tot}} \rangle / N = -\vec{\mu}_1(0) \cdot \lambda \vec{M} = -\vec{\mu}_1(0) \cdot \vec{H}_{\text{int}}, \tag{17.6}$$

where,

$$\lambda = \int_b^\infty \tilde{J}_{ex}(r)\mathrm{d}^3\vec{r}, \tag{17.7}$$

is a material-dependent parameter that provides a measure of the interaction strength. In this approach, the interaction of a moment with its neighbors is replaced by the interaction of the moment with an internal field, $\vec{H}_{\text{int}} = \lambda \vec{M}$, and this explains why the approximation is termed "mean field".

In a similar fashion, the developing density fluctuations for a fluid near the critical point are ignored. In the mean field approach, the density is treated as spatially uniform so that $g(r) = 1$ and

$$\langle U_{\text{tot}} \rangle = -N\,n\,a = -\frac{N^2}{V}a, \tag{17.8}$$

where,

$$a = \frac{1}{2}\left| \int_b^\infty u(r)\mathrm{d}^3\vec{r} \right| \tag{17.9}$$

provides a similar measure of the attractive interaction.

17.2 The mean field equation of state

17.2.1 Fluids: the van der Waals model

Following the thesis work of van der Waals, we now apply the mean field approximation to a fluid. Beginning with the non-interacting equation of state (i.e. the ideal gas law), we imagine what happens to a gas of given entropy when the interaction is instantly introduced. When the particles are suddenly allowed to attract one another, their impacts with the walls of the container are reduced. Using the first law of thermodynamics, together with Eq. (17.8), one can show (Ex. 1) that the pressure decreases by an amount,

$$\Delta P = -\frac{N^2}{V^2}a. \tag{17.10}$$

In addition to this reduction in pressure, van der Waals also modified the ideal gas law to adapt it to real particles with a finite volume b. Together, these two changes result in the *van der Waals equation of state*,

$$P = \frac{Nk_BT}{(V - Nb)} - \frac{N^2}{V^2}a. \tag{17.11}$$

Examples of the isotherms produced by Eq. (17.11) are shown in Fig. 17.1. At high temperatures, the van der Waals model produces behavior consistent with

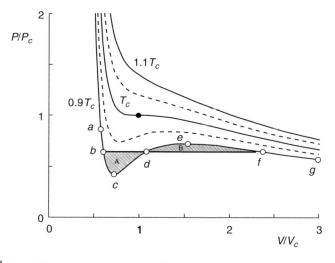

Figure 17.1   The *PV* diagram for five isotherms of the van der Waals model ($T/T_c = 0.9$, 0.95, 1, 1.05 and 1.1). Above T_c, each isotherm is single valued but becomes multivalued below the critical temperature. As an example, the isotherm for 0.9 T_c exhibits three values of volume (points labeled b, d and f) associated with a single pressure.

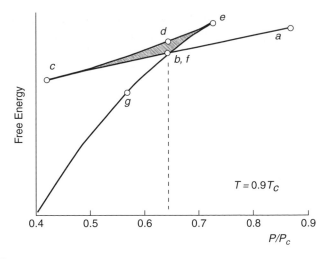

Figure 17.2 A plot of the Gibbs free energy obtained from the van der Waals model (Eq. (17.13)) plotted against the pressure for a fixed temperature ($T = 0.9T_c$) below the critical point. Because the total change in free energy in the shaded loop must vanish, the two shaded areas shown in Figure 17.1 must be equal. This then provides a convenient method for determining the coexistence curve.

the non-interacting ideal gas law, as expected at temperatures above T_c, where thermal agitation renders the attractive interaction ineffective. As the temperature approaches T_c, a distortion of the isotherm develops in which the slope tends toward zero at some special V_c. Below T_c, the isotherms assume a multivalued property wherein three volumes are associated with any $P < P_c$.

The region below T_c corresponds to the region of liquid–vapor coexistence and, to better evaluate the meaning behind these multivalued isotherms, we consider the free energy of the van der Waals model. Recalling that $dG = -SdT + VdP$, we have for a fixed temperature,

$$\left(\frac{\partial G}{\partial V}\right)_T = V\left(\frac{\partial P}{\partial V}\right)_T = -\frac{Nk_B TV}{(V - Nb)^2} + 2\frac{N^2}{V^2}a. \qquad (17.12)$$

Integrating this over volume for the same fixed temperature (Ex. 4) provides the free energy,

$$G = -Nk_B T \ln(V - Nb) + \frac{(Nk_B T)(Nb)}{(V - Nb)} - 2\frac{N^2}{V}a + c(T), \qquad (17.13)$$

where $c(T)$ is the integration constant, which could depend upon what fixed temperature is being considered. When the free energy is plotted against pressure (by evaluating G from Eq. (17.13) and P from Eq. (17.11) for a series of volumes at fixed temperature) we find, as shown in Fig. 17.2, that it produces an unusual triangular loop whenever $T < T_c$. Several points located in the figure are mapped to corresponding points on the isotherm in Fig. 17.1. Now because the second law dictates that the system should adopt the phase

(a)

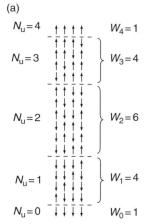

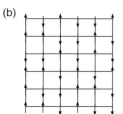

(b)

(a) A summary of all the possible configurations for an Ising system containing four spins. These 16 distinct configurations are divided into five macrostates according to the net number of upward spins. (b) Just one of the $2^{36} = 6.87 \times 10^{10}$ possible configurations of an Ising system containing 36 spins.

with the lowest free energy, we see that the system will in fact transition directly from the gas at point f to the liquid at point b (during which a first-order phase transition occurs). This transition occurs at a fixed pressure, as indicated by the points f and b in Fig. 17.2. Furthermore, because the net change in free energy around the triangular loop of Fig. 17.2 must vanish,

$$\int_{\text{loop}} dG = \int_{\text{loop}} V dP = 0; \qquad (17.14)$$

the critical pressure can alternatively be determined directly from the PV diagram as that fixed pressure for which the two areas (A and B in Fig. 17.1) are equal. Similar analysis of other isotherms thus allows the coexistence dome to be mapped out.

17.2.2 Magnets: the Ising model

In the Ising model, the spins are restricted to be directed either up or down and so too is the external field. Consequently, the total interaction between an external field and a system of N spins, configured with N_u spins up and N_d spins down, is simply,

$$U = \mu H(N_d - N_u) = \mu H(N - 2N_u), \qquad (17.15)$$

while the net magnetization is

$$M = n\langle \mu \rangle = n\frac{\mu(N_u - N_d)}{N} = -n\frac{U}{NH}. \qquad (17.16)$$

Before we can introduce the mean field interaction of Eq. (17.6), we must first determine the equation of state that describes the Ising model in the absence of particle–particle interactions. That is, we first need the corresponding "ideal gas law" for the Ising system. Our point of entry is the entropy of the spin system, given by statistical mechanics as $S = k_B \ln W$, where the multiplicity, W, represents the number of possible ways that the N spins can be configured such that a given situation of N_u and N_d is achieved.

The Ising model is an example of a two-state system. Like a coin that can assume either heads or tails, the magnetic moment can only assume up or down. For such a system, the number of microstates associated with a given macrostate of N_u and N_d is given by,

$$W = \frac{N!}{N_u! N_d!} = \frac{N!}{N_u!(N - N_u)!}. \qquad (17.17)$$

By way of a validation of this result, consider the small system of only $N = 4$ spins, illustrated in Fig. 17.3. For this small system, there are a total of $2^4 = 16$ possible arrangements of the spins that have been divided into five groups according to the values of N_u and N_d. For each group, one can see that Eq. (17.17) correctly predicts the number of configurations in each macrostate.

Given Eq. (17.17), the entropy is,

$$S = k_B \ln W = k\{\ln N! - \ln N_u! - \ln(N - N_u)!\}$$
$$\approx k_B\{N \ln N - N_u \ln N_u - (N - N_u) \ln(N - N_u)\}, \qquad (17.18)$$

where use has been made of Sterling's approximation, $\ln N! = N \ln N - N$, in this situation of very large N. Because the Ising model is constructed from a fixed lattice, its volume does not change and from the thermodynamic identity $(dU = TdS - PdV)$ we find

$$T^{-1} = \left(\frac{\partial S}{\partial U}\right) = \frac{\partial S}{\partial N_u}\frac{\partial N_u}{\partial U} = -\frac{1}{2\mu H}\frac{\partial S}{\partial N_u} = \frac{k_B}{2\mu H}\ln\left(\frac{N - U/\mu H}{N + U/\mu H}\right). \quad (17.19)$$

Rearranging this, the energy of the non-interacting system is,

$$U = N\mu H\left(\frac{1 - e^{2\mu H/k_B T}}{1 + e^{2\mu H/k_B T}}\right) = -N\mu H \tanh\left(\frac{\mu H}{k_B T}\right), \qquad (17.20)$$

and the magnetization, given by,

$$M = n\mu \tanh\left(\frac{\mu H}{k_B T}\right) = M_S \tanh\left(\frac{\mu H}{k_B T}\right), \qquad (17.21)$$

constitutes the Ising equation of state as it relates the three state variables, H, M and T.

The mean field interaction is easily added in the form of an effective field, $H_{int} = \lambda M$, and in the case of such an interacting Ising system, the equation of state becomes,

$$M = M_S \tanh\left(\frac{\mu(H + \lambda M)}{k_B T}\right). \qquad (17.22)$$

As illustrated in Fig. 17.4, solutions to the equation of state are obtained when the left-hand side, $y_L = M$, matches the right-hand side, $y_R = M_S \tanh(\mu\lambda M/k_B T)$, in the case of a vanishingly small external field. One sees that at high temperatures there is only one solution found at the origin where $M = 0$. However, below some critical temperature, there appear two additional solutions (one positive and the other negative), which correspond, to the partial ordering of spins either up or down. These two solutions correspond to the upper and lower branches of the MT– phase diagram (see Fig. (15.12)) discussed in Chapter 15.

17.3 Law of corresponding states

Thus far, both mean field models for fluid and magnetic phase transitions are performing well. Each produces a transition in which order spontaneously appears. Next, we inquire as to whether these models conform to a law of

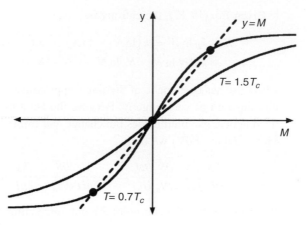

Figure 17.4 The equation of state for the Ising model. Solid curves represent the right-hand side of Eq. (17.22) for two cases above and below the critical temperature. The dashed line represents the left-hand side of Eq. (17.22). At temperatures above the critical temperature, the solutions to Eq. (17.22) are single-valued (equal to zero). Below the critical point, the solutions become multivalued with two non-zero values of the magnetization corresponding to the two allowed directions for aligned spins.

corresponding states, such that their respective equations of state may be expressed in an invariant form when scaled by representative state variables such as $p = P/P_c$, $v = V/V_c$ and $t = T/T_c$. For the van der Waals model, these variables are those at the critical point, which can be derived from the properties of the critical isotherm,

$$\left(\frac{\partial P}{\partial V}\right)_{\text{at C.P.}} = 0 \quad \text{and} \quad \left(\frac{\partial^2 P}{\partial V^2}\right)_{\text{at C.P.}} = 0. \tag{17.23}$$

These two conditions, together with the van der Waals equation of state, allow one to determine (Ex. 2) that

$$P_C = \frac{a}{27b^2}, \quad V_C = 3Nb, \quad \text{and} \quad k_B T_C = \frac{8a}{27b}. \tag{17.24}$$

When these values are substituted back into the equation of state, a scaled form,

$$\left\{\frac{P}{P_C} + 3\left(\frac{V_C}{V}\right)^2\right\}\left(3\left(\frac{V}{V_C}\right) - 1\right) = 8\frac{T}{T_C} \quad \text{or} \quad (p + 3/v^2)(3v - 1) = 8t, \tag{17.25}$$

is found and confirms that the van der Waals model exhibits the expected invariance and that the transition appears the same regardless of the size of the particles (b) or the degree of attraction (a).

For the magnetic case, two of the state variables (M and H) happen to be zero at the critical point. For these two, it is customary to scale instead by the saturation magnetization (M_S) and the corresponding saturated mean field, respectively,

$$m = M/M_S, \quad h = H/\lambda M_S. \tag{17.26}$$

From Fig. 17.4, we notice that the critical temperature is characterized by the isotherm whose slope ($(\partial y_R/\partial M)|_{M\to 0}$) just equals unity. One can then show (Ex. 5) that the Ising critical temperature is,

$$k_B T_c = \lambda \mu M_S = \lambda n \mu^2. \tag{17.27}$$

Making these substitutions into the equation of state in Eq. (17.22), one obtains a scaled form,

$$m = \tanh\left(\frac{h}{t} + \frac{m}{t}\right) = \frac{\tanh\left(\frac{h}{t}\right) + \tanh\left(\frac{m}{t}\right)}{1 + \tanh\left(\frac{h}{t}\right)\tanh\left(\frac{m}{t}\right)}. \tag{17.28}$$

Since our interest resides in features near the critical point that are present in a vanishingly small external field, we can use the following expansion,

$$\tanh(x) = x - \frac{1}{3}x^3 + \cdots, \tag{17.29}$$

to reduce Eq. (17.28) to,

$$\frac{h}{t} = \frac{m - \tanh(m/t)}{1 - m\tanh(m/t)}. \tag{17.30}$$

17.4 Critical exponents

Thus far, the mean field approach is quite successful. In each case, the introduction of a mean field has produced a phase transition involving a spontaneous ordering, and one in which the ordering appears identically for each case regardless of the details of the particles or the strength of the interaction. In both cases, we see that near the critical transition, the isotherms become distorted leading to divergences in the compressibility or susceptibility and the amplification of any order parameter fluctuations present in the system.

Near the critical point, there are a number of power law-dependent properties including: the compressibility and susceptibility,

$$\begin{aligned}
\chi_T &= -\frac{1}{V}\left(\frac{\partial V}{\partial P}\right)_T \propto (\Delta t)^{-\gamma_1} \\
\chi_m &= \left(\frac{\partial M}{\partial H}\right)_T \propto (\Delta t)^{-\gamma_2},
\end{aligned} \tag{17.31}$$

the order parameter,

$$\phi = \frac{1}{2}\frac{n_l - n_g}{n_C} \propto (-\Delta t)^{\beta_1}$$

$$\phi = \frac{M}{M_S} \propto (-\Delta t)^{\beta_2}$$

(17.32)

and the shape of the critical isotherm,

$$(\Delta p) \propto (\Delta v)^{\delta_1}$$

$$h \propto m^{\delta_2},$$

(17.33)

where

$$\Delta t = (T - T_C)/T_C = t - 1$$
$$\Delta p = (P - P_C)/P_C = p - 1$$
$$\Delta v = (V - V_C)/V_C = v - 1$$

are scaled variables that describe relative proximity of the critical point. We now ask, "What do these two mean field models predict for the values of their critical exponents?"

To evaluate these critical exponents we restrict our interest in the scaled equations of state to conditions very near the critical point (this corresponds to $p \approx 1$, $v \approx 1$ and $t \approx 1$ in the van der Waals model and $h \approx 0$, $m \approx 0$ and $t \approx 1$ in the Ising model) and so consider expansions of both Eq. (17.25) and Eq. (17.30) in these limits. For the van der Waals model, a Taylor expansion of Eq. (17.25) about $v = 1$ produces,

$$p = 4t - 3 - 6(t-1)(v-1) + 9(t-1)(v-1)^2 - \frac{3}{2}(9t-8)(v-1)^3 + \cdots$$

$$= 4t - 3 - 6\Delta t \Delta v + 9\Delta t (\Delta v)^2 - \frac{3}{2}(9t-8)(\Delta v)^3 + \cdots,$$

(17.34)

where only terms up to order v^3 have been retained. For the Ising model, we apply the expansion of Eq. (17.29) to Eq. (17.30) to obtain,

$$\frac{h}{t} = \left\{ m - \left[\frac{m}{t} - \frac{1}{3}\left(\frac{m}{t}\right)^3 + \cdots \right] \right\} \times \left\{ 1 - m\left[\frac{m}{t} - \frac{1}{3}\left(\frac{m}{t}\right)^3 + \cdots \right] \right\}^{-1},$$

(17.35)

and upon treating the inverted term as a truncated binomial expansion $((1 \pm x)^{-n} \approx 1 \mp nx)$, obtain,

$$\frac{h}{t} = \left\{ m - \left[\frac{m}{t} - \frac{1}{3}\left(\frac{m}{t}\right)^3 + \cdots \right] \right\} \times \left\{ 1 + m\left[\frac{m}{t} - \frac{1}{3}\left(\frac{m}{t}\right)^3 + \cdots \right] \right\},$$

(17.36)

which, after collecting the terms up to order $(m/t)^3$ reduces to,

$$
\begin{aligned}
h &= m(t-1) + m^3 \left[\frac{1}{3t^2} + \left(1 - \frac{1}{t} \right) \right] + \cdots \\
&= m\Delta t + m^3 \left[\frac{1}{3t^2} + \frac{\Delta t}{t} \right] + \cdots .
\end{aligned}
\tag{17.37}
$$

17.4.1 Compressibility and susceptibility

For the van der Waals model, the compressibility near the critical point is given from Eq. (17.34) as,

$$
(-V_C \chi_T)^{-1} = \frac{P_C}{V_C} \left(\frac{\partial p}{\partial v} \right) \approx \frac{P_C}{V_C} \left\{ -6\Delta t + 18\Delta t \Delta v - \frac{9}{2} (t + 8\Delta t)(\Delta v)^2 \right\}.
\tag{17.38}
$$

In the limit that Δv approaches zero, this yields

$$
\chi_T \approx \frac{1}{6P_C} (\Delta t)^{-1} = \frac{1}{6P_C} (\Delta t)^{-\gamma_1},
\tag{17.39}
$$

with the critical exponent, $\gamma_1 = 1$. For the Ising model, the susceptibility is similarly obtained from Eq. (17.37) as,

$$
(\chi_m)^{-1} = \left(\frac{\partial H}{\partial M} \right)_T \approx \left(\frac{\lambda M_S}{M_S} \right) \left(\frac{\partial h}{\partial m} \right) = \lambda \left\{ \Delta t + 3m^2 \left[\frac{1}{3t^2} + \frac{\Delta t}{t} \right] \right\}, \tag{17.40}
$$

which, if we eliminate m by taking its limit ($m \approx 0$) at the critical point, we find,

$$
\chi_m^{-1} = \lambda \Delta t,
\tag{17.41}
$$

or,

$$
\chi_m = \frac{1}{\lambda} (\Delta t)^{-\gamma_2},
\tag{17.42}
$$

with $\gamma_2 = 1$.

17.4.2 Order parameter

For the van der Waals order parameter (i.e. the coexistence curve), we inquire as to how Δv depends on Δt below T_c. To eliminate the pressure from Eq. (17.34), we note from the scaled equation of state in Eq. (17.25) that, near the critical point, $p = 4t - 3$ and that Eq. (17.34) reduces to,

$$
\frac{3}{2} (t + 8\Delta t)(\Delta v)^2 \approx \Delta t (9\Delta v - 6).
\tag{17.43}
$$

Again, in the limit that Δv and Δt approach zero while t approaches unity, we find,

$$\Delta v \approx \pm 2(-\Delta t)^{1/2}, \tag{17.44}$$

from which one can show (Ex. 8) that the order parameter varies as a power law,

$$\phi = \frac{1}{2}\frac{n_l - n_g}{n_C} \approx (-\Delta t)^{1/2} = (-\Delta t)^{\beta_1}, \tag{17.45}$$

with a critical exponent $\beta_1 = \frac{1}{2}$.

For the Ising model, we obtain the order parameter (m) by applying the limit $h \approx 0$ to Eq. (17.37) to find,

$$-\Delta t = m^2 \left[\frac{1}{3t^2} + \frac{\Delta t}{t}\right], \tag{17.46}$$

or,

$$m \approx \pm\sqrt{3}(-\Delta t)^{\beta_2}, \tag{17.47}$$

with $\beta_2 = \frac{1}{2}$.

In a similar fashion, one can show (Ex. 9) that the shape of the critical isotherm of both the van der Waals and Ising models is given by Eq. (17.33) with $\delta_1 = \delta_2 = 3$. This is remarkable. Not only do both the van der Waals and Ising models predict divergences of related quantities (like the compressibility and susceptibility) near the critical point, but both models (dealing with very different physical systems) also display an *identical value* for the critical exponent. It is as though each model, at least in its scaled form, produces an equivalent geometrical property in which the shapes of the isotherms and their respective derivatives develop identically in the vicinity of the critical point. This geometrical connection is reinforced by a generalized theory of phase transitions, known as the *Landau Theory*, which we will now examine.

17.5 Landau theory

You may have noticed that, in general, the equation of state can be derived from the free energy. For fluids, it can be derived from either the Gibbs free energy by,

$$V = \left(\frac{\partial G}{\partial P}\right)_T, \tag{17.48}$$

or via the Helmholtz free energy by,

$$P = -\left(\frac{\partial F}{\partial V}\right)_T. \tag{17.49}$$

In a similar fashion, for magnetic systems the equation of state can be obtained either by,

$$M = -\left(\frac{\partial G}{\partial H}\right)_T \quad \text{or} \quad H = \left(\frac{\partial F}{\partial M}\right)_T. \tag{17.50}$$

Furthermore, the compressibility and susceptibility are seen to involve second derivatives of the free energy and it reasons that the similarities between both the fluid and magnetic phase transitions boil down to inherent similarities *already* present in the functional form of the free energy. To explore this, the Landau theory supposes that the free energy (here the Helmholtz free energy) can be expanded as a power series of the respective order parameter of the form,

$$F(t, \phi) = g_o(t) + g_1(t)\phi + g_2(t)\phi^2 + g_3(t)\phi^3 + g_4(t)\phi^4 + g_5(t)\phi^5 + \cdots, \tag{17.51}$$

where the various temperature-dependent coefficients are assumed to be given by similar expansions about the critical point of the form,

$$\begin{aligned}
g_o(t) &= f_{00} + f_{01}\Delta t + f_{02}\Delta t^2 + f_{03}\Delta t^3 + \cdots \\
g_1(t) &= f_{10} + f_{11}\Delta t + f_{12}\Delta t^2 + f_{13}\Delta t^3 + \cdots \\
g_2(t) &= f_{20} + f_{21}\Delta t + f_{22}\Delta t^2 + f_{23}\Delta t^3 + \cdots
\end{aligned} \tag{17.52}$$

$$\vdots$$

Now in many instances, including the two cases we have discussed thus far, the free energy is unchanged on inversion of the order parameter. For example, in the magnetic case, the free energy depends only on the size of the magnetization and not on whether it is directed up or down. Thus, for this situation, the free energy must be an even function of the order parameter,

$$F(t, \phi) = g_o(t) + g_2(t)\phi^2 + g_4(t)\phi^4 + \cdots. \tag{17.53}$$

The general equation of state is then given (using the magnetic system as an example) as,

$$H = \left(\frac{\partial F}{\partial M}\right)_T \propto \left(\frac{\partial F}{\partial \phi}\right)_t, \tag{17.54}$$

and the generalized susceptibility is given by,

$$\chi^{-1} = \left(\frac{\partial H}{\partial M}\right)_T \propto \left(\frac{\partial^2 F}{\partial \phi^2}\right)_t. \tag{17.55}$$

Note that since the susceptibility diverges on approach to the critical point, its inverse must vanish and so the coefficient f_{20} in Eq. (17.52) must be zero. To leading order, the equation of state is,

$$H \approx 2g_2(t)\phi + 4g_4(t)\phi^3, \tag{17.56}$$

which just below the critical point in a vanishing field becomes,

$$\phi^2 \approx -\frac{g_2(t)}{2g_4(t)} \approx \frac{f_{21}}{2f_{40}}(-\Delta t), \tag{17.57}$$

so that the order parameter again displays,

$$\phi \approx \pm\sqrt{\frac{f_{21}}{2f_{40}}}(-\Delta t)^{1/2}, \tag{17.58}$$

with the mean field exponent, $\beta = \frac{1}{2}$.

For the susceptibility above the critical point (where $\phi = 0$),

$$\chi^{-1} \propto \left(\frac{\partial^2 F}{\partial \phi^2}\right)_t \approx 2g_2(t) \approx 2f_{21}(\Delta t)^1, \tag{17.59}$$

implying the mean field exponent $\gamma = 1$. The shape of the critical isotherm itself is obtained from the equation of state in Eq. (17.56) in the limit that Δt approaches zero. In this limit (with $f_{20} = 0$), $g_2(t)$ *vanishes and,*

$$H \approx 4f_{40}\phi^3 \quad \text{or} \quad \phi \approx \left(\frac{1}{4f_{40}}\right)^{1/\delta} H^{1/\delta}, \tag{17.60}$$

with the mean field exponent $\delta = 3$.

17.6 Renormalization theory

The mean field approach, in which the growing fluctuations in order parameter occurring near the critical point are ignored in favor of an average field, is largely successful in explaining the critical phenomena including the universality seen between both the van der Waals and Ising models. However, the various critical exponents obtained in the mean field approximation, as summarized in Table 17.1, are clearly at odds with those values obtained

Table 17.1 Critical exponents observed experimentally for both fluids and magnet are compared with those from mean field theory (data obtained from Stanley (1971)).			
Quantity	Critical exponent	Experimental range	Mean field value
Specific heat	α	0 to 0.2	0
Order parameter	β	0.3 to 0.4	0.50
Susceptibility	γ	1.1 to 1.4	1.0
Critical isotherm	δ	4.2 to 4.4	3.0

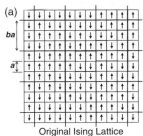

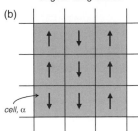

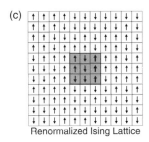

Figure 17.5

In the process of renormalization, an Ising spin lattice is (a) partitioned into cells of size $ba \ll \xi$. (b) The contents of each cell are then replaced by either a single up or down spin based on its average or net spin, according to Eq. (17.63). (c) Finally, the lattice is shrunken back to its original perspective according to Eq. (17.61).

by careful experiments conducted near the critical point. It seems then that the neglected fluctuations do have an effect on the transition after all. But how would we incorporate this effect for fluctuations that, by design, are unpredictable? There is one property of these fluctuations that can guide us: their structure is self-similar for length scales smaller than the correlation length.

17.6.1 A matter of perspective

Again, by presuming that the fluctuations are self-similar, we are assuming that the spatial patterns they form will appear unchanged under a process of renormalization. For example, the Sierpinski gasket that we encountered in Chapter 8 is a geometrical fractal that is self-similar. Viewed far away, we see the overall pattern of three solid triangles with one inverted vacancy. If we advance closer, our visual perspective effectively expands one of these three solid triangles. At some point of approach, this single triangle reveals itself to contain three solid triangles and an inverted vacancy, and appears to our eyes to be identical with the pattern we observed when we were farther away.

In the renormalization approach, we essentially seek a mathematical framework by which to describe how the self-similarity is maintained during transformations that function much like the alteration of our visual perspective when we move closer and farther away. In this way we can exploit the self-similarity to place restrictions (scaling relations) on the various exponents such that each can be predicted from knowledge of just two fundamental exponents: one (ω) that characterizes the self-similarity associated with the order parameter and another (v) that characterizes how the correlation length diverges on approach to the critical point.

17.6.2 Kadanoff spin renormalization

Following the example of Kadanoff, we will consider our system to be an Ising lattice whose spins are either up or down. In order to achieve the visual effect we desire, we will need to perform two operations on this lattice, as illustrated in Fig. 17.5. Firstly, we need to partition the system into cells of size $ba \ll \xi$, and follow some recipe to condense the contents of each cell into a single, Ising spin. This results in a lattice that appears magnified, as though we had moved closer. To counter this, we need to perform a second operation that amounts to our backing away from the lattice until it reappears as it did at the start. Thus, after condensing the spins, we need to shrink all the distances by a factor of b, such that a distance in the renormalized system becomes,

$$r' = r/b, \tag{17.61}$$

where r is the corresponding distance in the original lattice.

Now, how do we condense the spins of an Ising lattice so as to achieve a cell that contains a single Ising spin? An obvious recipe would begin by replacing the contents of the cell by the average of the spins contained inside. For a given cell labeled α, this average is,

$$\mu_\alpha^{avg} = \frac{1}{b^d} \sum_{i=1}^{N_\alpha} \mu_i^\alpha, \tag{17.62}$$

where the sum runs only over the contents of the cell. However, this average value alone is unlikely to serve as a renormalized Ising spin, whose value must be $\mu_\alpha = \pm 1$, because the average could be anywhere between these limits. In order to properly renormalize our original lattice into a *self-similar* Ising lattice, the average spin of a cell will need to be adjusted so that its corresponding Ising spin is either *all* up or *all* down. To do this we assume that there is an appropriate adjustment factor such that,

$$\mu_\alpha = b^\omega \mu_\alpha^{avg}. \tag{17.63}$$

Validity of the adjustment factor

Our choice for b is restricted to be less than the correlation length. When b is small compared to ξ, we will typically find that the contents of the cell are almost entirely up or entirely down, and so little or no adjustment is required ($b^\omega \approx 1$). But as b increases, the average of the contents will tend to decrease from ± 1 towards zero and the adjustment will need to increase to compensate. It reasons then that the exponent ω must be greater than zero. This adjustment may seem an unrealistic fabrication, since the average spin of any cell will vary and would require different adjustment factors. However, when we incorporate the necessary ensemble averaging, the notion of a common adjustment factor takes on a proper meaning.

Renormalized correlation function

Now that we have our recipe for carrying out the renormalization, let us look at how the spin correlation function remains self-similar during the transformation. The spin correlation function for the original system can be expressed as,

$$\Gamma(r) = \langle \mu(0)\mu(r) \rangle = \frac{1}{N} \sum_i^N \mu_i(0) \frac{1}{N_r} \sum_{j \neq i}^{N_r} \mu_j(r), \tag{17.64}$$

where N is the total number of spins and N_r the number of spins that are a distance r away from a central spin. The spin correlation function of the renormalized lattice is,

$$\Gamma(r') = \langle \mu_\alpha(0)\mu_\beta(r') \rangle = \frac{1}{N_C}\sum_\alpha^{N_C} \mu_\alpha(0)\frac{1}{N_{C,r'}}\sum_{\beta \neq \alpha}^{N_{C,r'}} \mu_\beta(r'), \qquad (17.65)$$

where $N_C = Nb^{-d}$ is the total number of cells created in the partitioning and $N_{C,r'} = N_r b^{-d}$ is the number of cells a distance r' away from a central cell. When expressed using our adjustment factor, this becomes,

$$\Gamma(r') = \frac{b^{2\omega}}{N_C N_{C,r'}}\sum_\alpha^{N_C} \mu_\alpha^{avg}(0)\sum_{\beta \neq \alpha}^{N_{C,r'}} \mu_\beta^{avg}(r') = \frac{b^{2\omega}b^{2d}}{NN_r}\sum_\alpha^{N_C} \mu_\alpha^{avg}(0)\sum_{\beta \neq \alpha}^{N_{C,r'}} \mu_\beta^{avg}(r')$$

$$= \frac{b^{2\omega}b^{2d}}{NN_r}\sum_\alpha^{N_C}\left(\frac{1}{b^d}\sum_i^{N_\alpha} \mu_i^\alpha(0)\right)\sum_{\beta \neq \alpha}^{N_{C,r'}}\left(\frac{1}{b^d}\sum_j^{N_\beta} \mu_j^\beta(r)\right)$$

$$= b^{2\omega}\frac{1}{N}\sum_i^{N} \mu_i(0)\frac{1}{N_r}\sum_{j \neq i}^{N_r} \mu_j(r) = b^{2\omega}\Gamma(r),$$

$$(17.66)$$

and we find that the spin correlation function undergoes renormalization with a scaling exponent 2ω.

In the renormalized lattice, distances of the original lattice are rescaled to $r' = b^{-1}r$. This includes the correlation length,

$$\xi' = b^{-1}\xi \approx b^{-1}(\Delta t)^{-\nu}, \qquad (17.67)$$

and implies that the renormalization has the effect of creating a new lattice that is farther away from the critical point,

$$\xi' \approx (\Delta t')^{-\nu} \Rightarrow \Delta t' \approx b^{1/\nu}\Delta t. \qquad (17.68)$$

Thus we can consider the development of the correlation function on approach to the critical point to be given by,

$$\Gamma(r, \Delta t) = b^{-2\omega}\Gamma(r', \Delta t') = b^{-2\omega}\Gamma(b^{-1}r, b^{1/\nu}\Delta t). \qquad (17.69)$$

The choice for the partitioning is arbitrary, so let us consider choosing it such that $b = r$. Then,

$$\Gamma(r, \Delta t) = r^{-2\omega}\Gamma(1, r^{1/\nu}\Delta t) = r^{-2\omega}f(z = r\Delta t^\nu) = r^{-2\omega}f(z = r/\xi). \quad (17.70)$$

If we compare this with the form obtained from Ornstein–Zernike in Chapter 15, or that for the pair distribution function of randomly generated fractals with fractal dimension D_f in Chapter 8,

$$\Gamma(r) = \frac{e^{-r/\xi}}{r^{(d-2+\eta)}} \propto \frac{e^{-r/\xi}}{r^{d-D_f}}, \quad D_f = 2 - \eta, \qquad (17.71)$$

we see that the exponent ω is merely a measure of the dendritic or fractal nature of the fluctuation patterns emerging near the critical point and that

$$\omega = \frac{1}{2}(d - 2 + \eta) \approx \frac{d - D_f}{2}. \tag{17.72}$$

The renormalized field

Odd though it may seem, in the process of renormalizing the lattice, we have also caused an effective alteration of the external field. Since energy can neither be created nor destroyed, it must remain unchanged during the renormalization process. Because we have redefined our Ising spins, we must accordingly adjust the external field so that

$$U = -H \sum_i^N \mu_i = -H' \sum_\alpha^{N_C} \mu_\alpha = -H' b^\omega \sum_\alpha^{N_C} \mu_\alpha^{\text{avg}} = -H' b^{\omega - d} \sum_\alpha^{N_C} \sum_i^{N_\alpha} \mu_i^\alpha. \tag{17.73}$$

remains unchanged. For this to be, the external field must be correspondingly renormalized as,

$$H' = b^{d-\omega} H. \tag{17.74}$$

17.6.3 Scaling relations

The free energy is likewise conserved on transforming the lattice and is only redistributed in a new way. In particular, the Gibbs free energy per particle must remain unaltered if both lattices are to represent an Ising system and, since the free energy of a condensed cell is b^d times that of a single site,

$$G(\Delta t, H) = b^{-d} G(\Delta t', H') = b^{-d} G(b^{1/\nu} \Delta t, b^{d-\omega} H). \tag{17.75}$$

Note that this homogeneous function contains only two undetermined (fundamental) exponents (ω and ν) aside from the dimension. Since all the diverging quantities present near the critical point can be obtained by partial derivatives of the free energy, renormalization offers the means to forge relationships between the various exponents such that, once any two are known, all others can be determined. To see how this works, let us start with the equation of state,

$$M = \left(\frac{\partial G}{\partial H}\right)_T.$$

This, from Eq. (17.75), undergoes renormalization as,

$$M(\Delta t, H) = b^{-d}\left(\frac{\partial G(\Delta t', H')}{\partial H}\right)_T = b^{-\omega}\left(\frac{\partial G(\Delta t', H')}{\partial H'}\right)_T \qquad (17.76)$$
$$= b^{-\omega}M(b^{1/\nu}\Delta t, b^{d-\omega}H),$$

where we have made use of Eq. (17.74). The magnetization M is also the order parameter and, to examine its temperature dependence in a vanishing field ($H = 0$), we arbitrarily choose our scaling factor such that $b = (-\Delta t)^{-\nu}$. Then,

$$M(\Delta t, 0) = [(-\Delta t)^{-\nu}]^{-\omega}M((-\Delta t)^{-1}\Delta t, 0) = (-\Delta t)^{\nu\omega}M(-1, 0) \propto (-\Delta t)^{\beta}, \qquad (17.77)$$

and we obtain the following scaling relation:

$$\beta = \nu\omega. \qquad (17.78)$$

A similar approach to the shape of the critical isotherm ($\Delta t = 0$) is obtained by arbitrarily choosing $b = H^{-1/(d-\omega)}$. From Eq. (17.76), we find,

$$M(0, H) = (H^{-1/(d-\omega)})^{-\omega}M(0, H^{-1}H) = H^{\omega/(d-\omega)}M(0, 1) \propto (H)^{1/\delta}, \qquad (17.79)$$

where,

$$\delta = \frac{d}{\omega} - 1. \qquad (17.80)$$

Continuing on to the susceptibility,

$$\chi_T(\Delta t, H) = \left(\frac{\partial M(\Delta t, H)}{\partial H}\right)_T = b^{-\omega}\left(\frac{\partial M(\Delta t', H')}{\partial H}\right)_T$$
$$= b^{d-2\omega}\left(\frac{\partial M(\Delta t', H')}{\partial H'}\right)_T \qquad (17.81)$$
$$= b^{d-2\omega}\chi_T(b^{1/\nu}\Delta t, b^{d-\omega}H).$$

Again, choosing $b = (\Delta t)^{-\nu}$, the susceptibility (for $H = 0$) can be expressed as,

$$\chi_T(\Delta t, 0) = [(\Delta t)^{-\nu}]^{d-2\omega}\chi_T((\Delta t)^{-1}\Delta t, 0) = (\Delta t)^{-\nu(d-2\omega)}\chi_T(1, 0) \propto (\Delta t)^{-\gamma}, \qquad (17.82)$$

where,

$$\gamma = \nu(d - 2\omega). \qquad (17.83)$$

As a final example, we also consider the specific heat, which is defined by,

$$C_H(\Delta t, H) = -T\left(\frac{\partial^2 G(\Delta t, H)}{\partial T^2}\right)_H = -Tb^{(2/\nu)-d}\left(\frac{\partial^2 G(\Delta t', H')}{\partial T'^2}\right)_H$$
$$= b^{(2/\nu)-d}C_H(b^{1/\nu}\Delta t, b^{d-\omega}H). \qquad (17.84)$$

Again, one can show that,

$$C_H(\Delta t, 0) = [(-\Delta t)^{-v}]^{(2/v)-d} C_H(-1, 0) \propto (-\Delta t)^{-\alpha}, \qquad (17.85)$$

where,

$$\alpha = 2 - vd. \qquad (17.86)$$

From Eqs. (17.78), (17.80), (17.83) and (17.86), one can determine a wide variety of scaling relations, such as:

$$\begin{aligned} \alpha + 2\beta + \gamma &= 2 \\ \alpha + \beta(\delta + 1) &= 2 \\ \gamma(\delta + 1) &= (2 - \alpha)(\delta - 1) \\ \gamma &= \beta(\delta - 1). \end{aligned} \qquad (17.87)$$

that restrict the critical exponents. Although renormalization has reduced our problem to the determination of two exponents, it still does not yet provide a recipe for determining the exponents v or ω. For this, one would need to proceed along lines much as we did in the renormalization example in Chapter 16, to develop an accurate recipe for how compacting the contents of each cell specifically relates $\Delta t'$ to Δt and $\Gamma(r')$ to $\Gamma(r)$. Then, exponents like v and ω could, in principle, be determined by,

$$v = \frac{\log b}{\log(\Delta t/\Delta t')} \quad \text{and} \quad 2\omega = \frac{\log b}{\log(\Gamma(r)/\Gamma(r'))}. \qquad (17.88)$$

Summary

- Mean field theories are those in which the fluctuations in the order parameter near the critical point are ignored and replaced by an order parameter that is spatially uniform.

- Mean field theories, including the generalized Landau theory, are successful in accounting for both the diverging behavior of certain quantities near the critical point and for laws of corresponding states. However, mean field theory predicts critical exponents that are often not in agreement with experiment.

- The determination of critical exponents is improved by incorporating fluctuations in the order parameter through a renormalization technique which assumes that the fluctuations are self-similar.

- The assumption of self-similar fluctuations leads to a set of scaling relations that constrict the various critical exponents to just two, independent, exponents.

Exercises

17.1. Show how Eq. (17.10) can be obtained from Eq. (17.8) using the thermodynamic identity ($dU = TdS - PdV$) and the ideal gas law, assuming that the interaction is turned on without disturbing the entropy.

17.2. Derive the expressions given in Eq. (17.24) for the state variables of the critical point in the van der Waals model.

17.3. Verify the law of corresponding states given in Eq. (17.25). (Hint: do this by substituting the critical point state variables of Eq. (17.24) into Eq. (17.25) to obtain the van der Waals equation of state.)

17.4. Derive the Gibbs free energy of the van der Waals model given in Eq. (17.13) starting from Eq. (17.12).

17.5. Derive the expression given in Eq. (17.27) for the Ising model Curie temperature.

17.6. Verify the law of corresponding states given in Eq. (17.28) for the Ising model. (Hint: do this by substituting the critical point state variables of Eq. (17.26) and Eq. (17.27) into Eq. (17.28) to obtain the Ising equation of state given by Eq. (17.22).)

17.7. Consider the scaled Ising equation of state given by Eq. (17.28) in the limit of a vanishing magnetic field. Find solutions for this equation both above and below T_c (e.g. at $T/T_C = 1.3, 1.2, 1.1, 0.9, 0.8, 0.7$ etc.) and plot these solutions as a function of T/T_C. How does your result compare with the MT phase diagram shown in Fig. 15.12?

17.8. Derive Eq. (17.45) from Eq. (17.44) by converting from specific volumes to densities.

17.9. Show that the shapes of the critical isotherms in the mean field approximation are given by Eq. (17.33) with $\delta_1 = \delta_2 = 3$.

17.10. Verify the scaling relations presented in Eq. (17.87).

Suggested reading

For more on the subject of renormalization, I recommend the first two texts. The basic principles of the Landau theory are presented in most standard thermodynamics texts. The two listed below are just personal favorites.

H. Eugene Stanley, *Introduction to Phase Transitions and Critical Phenomena* (Oxford University Press, New York, 1971).

P. M. Chaikin and T. C. Lubensky, *Principles of Condensed Matter Physics* (Cambridge University Press, New York, 2003).

C. Kittel and H. Kroemer, *Thermal Physics*, 2nd Ed. (W. H. Freeman and Co., San Francisco, 1980).

D. V. Schroeder, *An Introduction to Thermal Physics* (Addison Wesley Longman, New York, 2000).

18   Superconductivity

Introduction

We have now witnessed the similar patterns associated with second-order phase transitions in both fluid and magnetic systems. These patterns include laws of corresponding states and similarity in critical exponents, that govern how properties evolve near the transition point. Furthermore, the Landau theory provides a framework for understanding the commonality of these second-order phase transitions, in terms of similarity in the functional dependence of the free energy on an appropriately chosen order parameter, and a simple expansion that can be performed near the critical point. In this chapter, we examine yet another significant phase transition found in condensed matter: the transition of a material to a state of virtually infinite conductivity or superconductivity. Here the transition involves the sticking together of two electrons into a boson-like, superconducting charge carrier known as a Cooper pair, and we again find evidence of a second-order transition consistent with mean field theory.

18.1   Superconducting phenomena

18.1.1   Discovery

In 1908, H. K. Onnes perfected the technique for cooling helium gas to its condensation point and soon after began using this new technology to investigate the properties of various elements at ultra low temperatures. In one instance, Onnes was curious about the ultimate demise of the resistivity of an electronic conductor. As we saw in Chapter 12, the resistivity of most conductors decreases linearly with temperature at high temperatures, due to the scattering of electrons by lattice phonons, but approaches a limiting value at low temperatures, associated with a mean free path determined by macroscopic imperfections of the crystal lattice. In a series of studies, Onnes measured the resistance of gold and platinum and observed an approach to a limiting resistance at low temperatures. In an effort to eliminate the effects of imperfections,

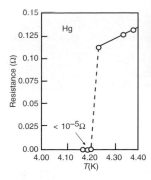

Figure 18.1

The resistance of a purified Hg specimen measured by Onnes demonstrating the transition to a state of ultralow resistance below 4.2 K. (Adapted from Onnes (1911).)

Table 18.1 Characteristic values for the critical temperature (T_c), critical field (H_c), specific heat increase ($(C_S - C_N)/C_N|_{T_c}$)) and energy gap (E_g) of several superconducting elements. (Data obtained from *Handbook of Chemistry and Physics* (1983) and from Aschroft and Mermin (1976).)

Element	T_c (K)	$H_c(0)$ (Gauss)	$(C_S-C_N)/C_N|_{T_c}$	$E_g(0)/k_B T_c$
Al	1.175	105	1.4	3.4
Cd	0.517	28	1.4	3.2
Hg (α)	4.154	411	2.4	4.6
In	3.408	282	1.7	3.6
Nb	9.25	2060	1.9	3.8
Pb	7.196	803	2.7	4.3
Sn	3.722	305	1.6	3.5
Ta	4.47	829	1.6	3.6
Tl	2.38	178	1.5	3.6
V	5.40	1408	1.5	3.4
Zn	0.85	54	1.3	3.2

he extended the study in 1911 to include mercury, which at that time could be refined to a highly pure form. The results of this study, shown in Fig. 18.1, are quite dramatic. A roughly linear temperature dependence was observed above about 4.2 K, which decreased abruptly to an immeasurably small resistance at lower temperatures. On reheating, the resistance was identically retraced, and Onnes concluded that mercury had undergone a unique phase transition to a new state characterized by virtually zero resistance – a "superconducting" phase.

Since then, many other elements of the periodic table have been shown to exhibit a similar superconducting transition at various transition temperatures, indicating that the transition is not unique to mercury. Values of the transition temperatures, T_c, as well as other characteristics of the transition are presented in Table 18.1.

18.1.2 Meissner effect

The next major development occurred in 1933 when W. Meissner and R. Ochsenfeld discovered another unique characteristic of the superconducting phase. They observed that, when a superconducting material is cooled below T_c in the presence of a weak magnetic field ($\vec{B} = \mu_o \vec{H}$), the field lines are totally expelled from the interior of the sample. In other words, below T_c the specimen behaves like a perfect diamagnetic material with magnetic susceptibility $\chi_m = -1$ and exhibits an effective magnetization that exactly counters the external field,

$$\vec{B} = \mu_o \vec{H} + \mu_o \vec{M} = 0, \quad \text{or} \quad \vec{M} = -\vec{H}. \tag{18.1}$$

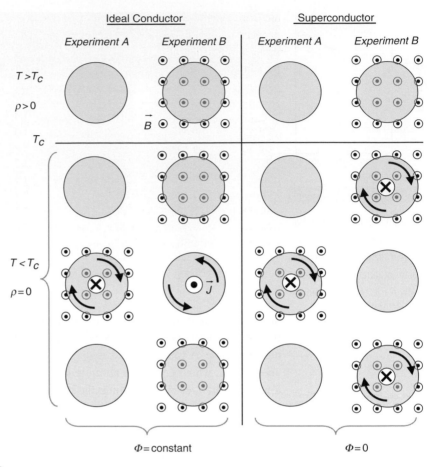

Figure 18.2 Two experiments (*A* and *B*) are conducted on both a normal, but ideal, conductor and on a superconductor. In experiment *A*, the specimen is cooled in zero-field and the field is switched on and then off only below T_c. In experiment *B*, the specimen is cooled in a non-zero field that is switched off and then back on below T_c. For the ideal conductor, the total flux inside the specimen (due to the external field and the induced currents) remains constant. For the superconductor, the total flux inside is always zero below T_c regardless of which experiment is performed.

The Meissner effect, which is often demonstrated by the levitation that occurs when a specimen is cooled below the transition temperature, is not merely a consequence of the vanishing resistivity. Rather, it is an altogether new property of the superconducting state. To see this, consider the set of experiments conducted both for an ideal ($\rho = 0$) conductor and for a super-conductor, that are illustrated in Fig. 18.2. For both materials, we imagine that the resistivity is finite above some critical temperature T_c and zero below that temperature. In experiment *A*, we imagine the specimens start off above T_c and have been left in the absence of any magnetic field for a sufficient length of time for any currents present to have decayed to zero. Both samples are then cooled below T_c into the regime where $\rho = 0$, at which point nothing happens. Now a magnetic field is introduced. The changing flux through both samples induces a circular current flow that, in this instance of zero resistivity, remains stable even after the magnetic flux has stopped changing. If the field is then switched off, the reverse change in flux brings both currents to zero again.

We can understand this first experiment rather easily from classical electomagnetic theory as a consequence of Faraday's law,

$$\text{emf} = \oint \vec{E} \cdot d\vec{l} = -\frac{\partial \Phi_B}{\partial t}. \tag{18.2}$$

Turning on or off the magnetic field causes an emf to appear which drives free charges in a circular fashion. For a material with $\rho = 0$, one can show (Ex. 1) that the current density, $\vec{J}$, increases at a rate given by,

$$\frac{\partial \vec{J}}{\partial t} = \frac{q_S^2 n_S}{m_S} \vec{E}, \tag{18.3}$$

where n_S, m_S and q_S are the concentration, mass and charge of the superconducting charge carriers, respectively. With a little manipulation, we can then express Faraday's law as,

$$\frac{\partial}{\partial t} \left\{ \frac{m_S}{q_S^2 n_S} \oint \vec{J} \cdot d\vec{l} \right\} = -\frac{\partial \Phi_B}{\partial t}, \tag{18.4}$$

or,

$$\frac{\partial}{\partial t} \left\{ \frac{m_S}{q_S^2 n_S} \oint \vec{J} \cdot d\vec{l} + \Phi_B \right\} = \frac{\partial}{\partial t} \{\Phi_J + \Phi_B\} = 0. \tag{18.5}$$

This last expression is really just a statement of Lenz' law pertaining to how Nature abhors a changing flux. As we see in the panel of illustrations of Fig. 18.2, the currents in an ideal conductor respond in whatever manner is needed to maintain the same net flux through the sample, regardless of the particular situation.

Returning again to Fig. 18.2, we now consider experiment B, in which both samples are prepared above T_c in the presence of a fixed magnetic field that has been applied for sufficient time that all induced currents have dissipated. Again, we cool both specimens into the region below T_c where $\rho = 0$. This time, something remarkable happens to the superconductor. The moment we enter the superconducting phase, we find that a current appears without any change of the magnetic flux! When we later turn off the field, this current disappears. In considering the behavior of the superconducting sample, it appears that current appears and disappears, not because of the *change* in magnetic flux, but rather in response to the presence or absence of the flux itself.

London equation

Regardless of whether the field is switched on or off, the total flux in the superconducting specimen always remains zero below T_c. It was this finding that motivated F. and W. London to propose that not only is the total flux in the superconductor constant, but that the constant value is always zero. From Eq. (18.5), this unique constraint implies that

$$\frac{m_S}{q_S^2 n_S} \oint \vec{J} \cdot d\vec{l} = -\Phi_B, \qquad (18.6)$$

which can be expressed alternatively using Stokes' theorem of vector calculus as,

$$\frac{m_S}{q_S^2 n_S} \int \left(\vec{\nabla} \times \vec{J} \right) \cdot d\vec{a} = -\int \vec{B} \cdot d\vec{a}. \qquad (18.7)$$

Since this condition must remain true regardless of the smallness of the specimen, we obtain the *London equation*,

$$\vec{\nabla} \times \vec{J} = -\frac{q_S^2 n_S}{m_S} \vec{B}, \qquad (18.8)$$

which expresses how the introduction of any weak external magnetic field in a superconductor is exactly countered by a rotating current of superconducting charge carriers near the surface.

Penetration length

Although the Meissner effect demonstrates that external magnetic fields are expelled from the interior of a superconducting sample, it is possible for some magnetic fields to just squeeze into the sample near the surface. If we recall Ampere's law,

$$\vec{\nabla} \times \vec{B} = \mu_o \vec{J}, \qquad (18.9)$$

for time-independent, static fields, we can apply a curl (and another Maxwell equation, $\vec{\nabla} \cdot \vec{B} = 0$) to obtain,

$$-\nabla^2 \vec{B} = \mu_o \vec{\nabla} \times \vec{J}. \qquad (18.10)$$

Upon introducing the London equation (Eq. (18.8)), this reduces to the following differential equation for the spatial variation of the magnetic field inside the sample:

$$\nabla^2 \vec{B} = \frac{\mu_o q_S^2 n_S}{m_S} \vec{B}. \qquad (18.11)$$

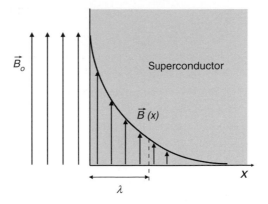

Figure 18.3 Although an external magnetic field is expelled from the interior of a superconductor, some field is permitted to penetrate near the surface. This field decreases to about 37% of its external intensity over a distance λ, known as the penetration length.

One can show (Ex. 2) that the current density is governed by a similar differential equation,

$$\nabla^2 \vec{J} = \frac{\mu_o q_S^2 n_S}{m_S} \vec{J}. \tag{18.12}$$

Imagine then a surface of the superconductor in the presence of a weak magnetic field, as presented in Fig. 18.3. Outside, the magnetic field has some value $\vec{B}_o$. In this one-dimensional situation, the variation of $\vec{B}(x)$ moving into the interior is governed by,

$$\frac{\partial^2 \vec{B}}{\partial x^2} = \frac{1}{\lambda^2} \vec{B}, \tag{18.13}$$

whose solution is,

$$\vec{B}(x) = \vec{B}_o e^{-x/\lambda}, \tag{18.14}$$

where,

$$\lambda = \sqrt{\frac{m_S}{\mu_o q_S^2 n_S}}. \tag{18.15}$$

In this instance, some magnetic field is allowed inside the superconductor, but the magnitude of the penetrating field decreases exponentially. The parameter λ is the *penetration length* and is a measure of how far the magnetic field (and the current density) can penetrate into a superconducting region. Using values of n_e from Table 12.1 for electrons in a typical conductor, one finds that $\lambda \approx 100$ nm.

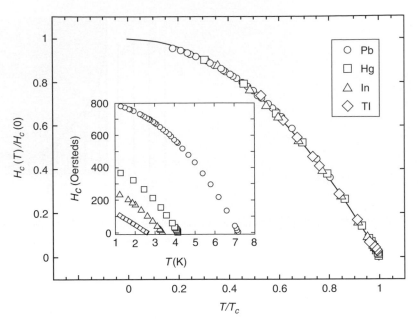

Figure 18.4 The critical external field required to destroy the superconducting state increases with decreasing temperature below T_c, and exhibits a law of corresponding states given by Eq. (18.16). Inset shows the critical fields prior to scaling. (Data from E. Maxwell and O. S Lutes, 1954; Decker et al., 1958; and Shaw et al., 1960.)

18.1.3 Critical field

In the above discussion, we specifically considered the expulsion of a "weak" magnetic field. There are, in fact, limits to the Meissner effect. If the applied field in a superconductor exceeds some temperature-dependent value, $H_c(T)$, then the field is no longer expelled and the resistivity returns to a non-zero value. That is, above the critical field H_c, the superconducting phase is *destroyed* and material returns to its "normal" (non-superconducting) phase. The size of the required critical field increases with decreasing temperature and, as shown in Fig. 18.4, is well described in terms of scaled variables by an empirical relation,

$$H_c(T) = H_c(0)\left\{ 1 - \left(\frac{T}{T_c}\right)^2 \right\} \approx (T_c - T) \text{ near } T_c. \tag{18.16}$$

Indeed, we see in Fig. 18.4 evidence for a law of corresponding states for the critical field that is suggestive of an underlying second-order phase transition.

18.1.4 Specific heat

When a superconducting material is cooled in a field that exceeds the maximum critical field ($H_c(0)$), it remains in the normal phase and its specific heat

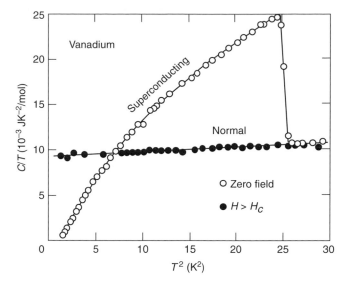

Figure 18.5 The specific heat of vanadium in both the normal ($H > H_c(0)$) and superconducting states. Note the quantity plotted on each axis is designed to highlight conformity to specific heat dependence for normal conductors. (Adapted from Bondi *et al.*, 1958.)

displays the usual temperature dependence of conventional electron conductors. Namely, there are two contributions: the Debye contribution due to the lattice vibrations which varies as T^3, and the weaker electronic contribution that decreases linearly with temperature. Both contributions are shown for vanadium in Fig. 18.5, where C/T is plotted against the square of the temperature to produce a linear dependence.

However, when the same specimen is cooled in the absence of an external magnetic field, thereby allowing the superconducting phase to form, the specific heat displays a discontinuous increase at T_c and a radically different temperature dependence below T_c. The size of the discontinuity at the critical temperature can be characterized by the dimensionless quantity,

$$\left. \frac{C_S - C_N}{C_N} \right|_{T_c}, \tag{18.17}$$

where C_S and C_N are the specific heats of the superconducting and normal phases, respectively. As seen in Table 18.1, this relative change in specific heat is curiously similar for a wide variety of materials.

From the specific heat, we can use a little thermodynamics to demonstrate that the superconducting state is one of lower entropy than the normal state. Recall that the specific heat (at constant volume) is given by $C = T(\partial S/\partial T)$. Since there is a discontinuous upward jump in C/T occurring at T_c for the superconducting phase, there must occur a discontinuous increase in the slope of $S(T)$ for this phase at T_c. As illustrated in Fig. 18.6, this implies that the entropy of the superconducting phase becomes less than that of the normal

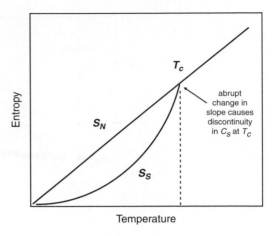

Figure 18.6 An illustration of how the entropy of both the normal and superconducting phases evolves differently below T_c. Because the entropy of the superconductor phase is lower than that of the normal phase, the superconductor is more highly ordered.

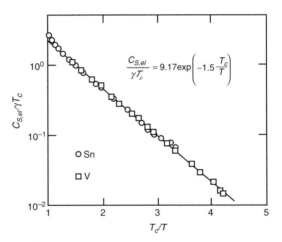

Figure 18.7 The electronic contributions to the specific heat of superconducting tin and vanadium are plotted on a logarithmic scale to illustrate the exponential dependence presented in Eq. (18.18) with $E_c = 1.5k_BT_c$. (Adapted from Bondi et al., 1958.)

phase below T_c. It is clear then that the superconducting transition involves some unknown ordering process.

At lower temperatures, the electronic contribution to the specific heat of the superconductor, obtained by subtracting off the Debye contribution, varies exponentially,

$$C_{S,el} \approx e^{-E_C/k_BT}, \qquad (18.18)$$

as illustrated in Fig. 18.7. Such an exponential temperature dependence is a hallmark of two-level systems (see for example, Eq. (11.32) in the limit of

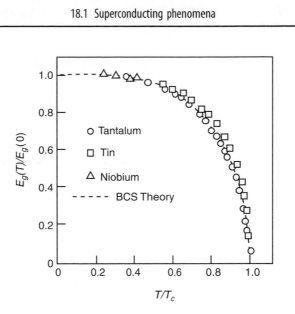

Figure 18.8 A small energy gap develops below T_c in superconductors and increases in size with cooling. A law of corresponding states is evident whose temperature dependence is well accounted for by BCS theory. (Adapted from Townsend and Sutton, 1962.)

small T) and suggests that the ordering which occurs at T_c produces a simple system of two energy levels with one excited state having an energy E_c above the ground state.

18.1.5 Energy gap

Experiments conducted in the 1950s revealed yet another characteristic of the superconducting phase: an energy gap. In the normal phase, above T_c, electrons of a conductor occupy energy levels up to the Fermi level. As we saw in Chapter 12, only those electrons near the Fermi level participate in the contributions to the specific heat, by being thermally excited into higher allowed levels of the conduction band. However, in a superconducting material, a new energy gap begins to form at T_c which is centered at the Fermi level. The energy gap that forms is quite small, only about 1 meV, and much smaller than the energy gap (≈ 1 eV) separating the valence band from the conduction band. This new energy gap increases with decreasing temperature, and as shown in Fig. 18.8, is well-described by an empirical law, similar to that of Eq. (18.16) for the critical field,

$$E_g(T) = E_g(0)\left[\frac{T_c - T}{T_c}\right]^{1/2}. \tag{18.19}$$

Once again, we find a law of corresponding states and further evidence for an underlying second-order phase transition.

Quest for an attractive interaction

In our survey of superconducting phenomena above, we have seen much evidence for an underlying second-order phase transition at work in the superconductivity transition. We see laws of corresponding states at work and evidence for a transition to a more ordered phase with lower entropy. However, these changes are not accompanied by any change in the structure of the material. Instead, the zero resistivity, various electrodynamic phenomena and the developing energy gap all suggest that the transition mainly involves a change in the behavior of the conduction electrons, such that they become more ordered and insensitive to scattering from the lattice.

As we have stressed before, phase transitions rely on the presence of some element of attractive interaction that overcomes the thermal agitation below T_c and which drives the development of a more ordered phase. For fluids, this attraction was the van der Waals interaction. For the magnetic system, it was the exchange interaction that favored the alignment of magnetic moments. For the superconducting phase transition, we face a big predicament: electrons do not like to "stick" together because they are charged particles repelled by the Coulomb force. It was this predicament that impeded the development of a microscopic theory of the superconducting transition for many years and the puzzle was only resolved in the 1950s after the isotope effect was discovered.

18.1.6 Isotope effect

In 1950, Maxwell and Reynolds observed that the critical temperature of a given superconducting material exhibited a surprising dependence on the average mass of the lattice atoms. By measuring the critical temperature in a series of samples with differing atomic mass, M, but identical electronic structure, they observed that

$$T_c \propto M^{-\beta}, \tag{18.20}$$

with the exponent $\beta \approx 0.5$ (see Fig. 18.9). This finding, known as the *isotope effect*, had a profound influence on later development of microscopic theories of the superconducting transition. As we have seen, the critical temperature is a measure of the degree to which thermal energy ($k_B T_c$) must be reduced for attractive interactions to become effective. Materials with a lower T_c have weaker attraction. The isotope effect is, in a sense, the "smoking gun" that points to the source of attraction that develops between the conduction electrons in the superconducting phase. The approximate square root dependence of T_c upon the atomic mass of the lattice atoms indicates that the attractive force is related to the fundamental

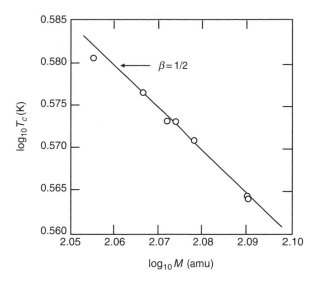

Figure 18.9 The critical temperatures for various isotopes of tin are plotted in a double logarithmic presentation. The critical temperature is a measure of the interaction energy between superconducting electrons and is seen to decrease as the square root of the mass of lattice atoms. (Adapted from Maxwell, 1952.)

frequencies of oscillation of the lattice of ion cores whose frequency decreases with increasing mass as,

$$\omega \approx \sqrt{\frac{4C_1}{M}} \propto M^{-1/2}. \tag{18.21}$$

Somehow, conduction electrons are able to achieve an attractive interaction at temperatures below T_c through some process that involves the phonon properties of the lattice.

18.2 Cooper pairs and the BCS theory

In 1957, some four and a half decades after superconductivity was first discovered, Bardeen, Cooper and Schrieffer ("BCS") developed a successful, microscopic theory of the superconducting state (see Bardeen, Cooper and Schrieffer (1957)). Unfortunately for us, the theory itself is constructed using a quantum mechanical framework that is beyond the scope of our current treatment of the subject. Nevertheless, we review here some of the salient features at a conceptual level.

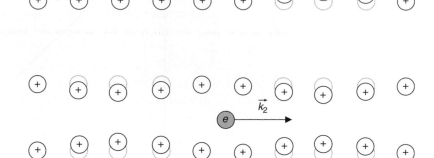

Figure 18.10 A conceptual illustration of the electron–phonon interaction that results in the formation of a Cooper pair. At the top, an electron traversing the lattice creates a minor distortion of the lattice. At the bottom, a second electron, traveling in the other direction, couples with the lattice distortion in such a manner as to reduce the overall energy of the electron pair.

18.2.1 Cooper pairs

The key ingredient of BCS theory is the notion that a pair of electrons with equal but opposite momenta ($\vec{k}_2 = -\vec{k}_1$) and opposite electron spin can be attracted through coupling with a lattice phonon. The formation of these so-called *Cooper pairs* results in a new quasi-particle characterized by a quantum mechanical wave function $\psi(\vec{r})$ that resides in a ground state of lowered energy.

Although the details of the attractive interaction are complicated, we can illustrate it conceptually, as shown in Fig. 18.10. Imagine a first electron with momentum $\vec{k}_1$ traveling through the lattice. At low temperatures, the lattice is fairly quiescent as there is little thermal energy to stimulate large numbers of phonons or any appreciable vibration of the ion cores. As the electron passes, ion cores are attracted to it by the Coulomb force and a slight distortion develops. The distortion is not instantaneous because the ion cores have inertia and, because the electron is really a wave, the induced distortion is impatterned in a manner specific to the $\vec{k}_1$ of the passing electron. In effect, a phonon is stimulated into existence by the passing electron. Now a second electron of opposite electron spin happens to travel in the opposite direction. If its wave vector, $\vec{k}_2$, has the same magnitude as $\vec{k}_1$, then it will encounter the phonon as one already preconfigured to its own wavelength. Like a key to its lock, the second electron can exploit the pre-existing lattice distortions and move along with less resistance. In reality, neither electron is first nor second, but rather both are coupled via the phonon as though an attractive force were active between them. In the BCS theory, this attractive force lowers the energy of the two-electron system and the two electrons, now coupled into a Cooper pair, enter into a new ground state (the BCS ground state) described by a superconducting wave function $\psi(\vec{r})$.

Treated as a quasi-particle, the Cooper pair has a net spin of zero. Therefore, the Cooper pair can behave as a *boson*. Like other bosons, such as phonons and photons, the Cooper pair is not restricted by the Pauli exclusion principle and an unlimited number of Cooper pairs can occupy the same quantum mechanical state described by the same wave function, $\psi(\vec{r})$. We can now begin to see how a superconductor loses its electrical resistivity. What were once two uncorrelated electrons that independently suffered scattering events on passing through the crystal are now two correlated electrons "glued together" into a single entity. One cannot be scattered unless the other is also identically scattered and so neither can be scattered without disrupting the coupling that holds them together. The binding of the two electrons in a Cooper pair requires energy input to separate them. This energy requirement is what creates the formation of the energy gap in the superconducting state and separates the BCS ground state from the uncoupled upper energy level. With cooling, this energy gap widens, implying that Cooper pairs become more tightly bound with decreasing temperature.

The overall lowering of energy in the system promotes the formation of more Cooper pairs and a self-accelerating condensation of Cooper pairs into the BCS ground state ensues, which is much like the condensation of a gas into a liquid or a paramagnet into a ferromagnet near a second-order phase transition. Each new Cooper pair enters into the BCS ground state and the superconducting wave function assumes the form of a macroscopic, coherent, wave function whose modulus is a measure of the density of Cooper pairs,

$$n_S = |\psi(\vec{r})|^2. \tag{18.22}$$

The superconducting current density associated with this wave function is

$$\vec{J} = \frac{q_S}{m_S} \langle \psi(\vec{r}) \| \vec{p} \| \psi^*(\vec{r}) \rangle. \tag{18.23}$$

Here the quantum mechanical operator for charged particles interacting with an electromagnetic field is given by,

$$\| \vec{p} \| = \left(-i\hbar \vec{\nabla} - q_S \vec{A} \right), \tag{18.24}$$

where $\vec{A}$ is the vector potential such that $\vec{B} = \vec{\nabla} \times \vec{A}$.

In situations where the density of Cooper pairs is uniform, any spatial variation of $\psi(\vec{r})$ can only appear in the phase of the wave,

$$\psi(\vec{r}) = |\psi| e^{i\theta(\vec{r})}. \tag{18.25}$$

One then finds,

$$\vec{J} = -\frac{q_S n_S}{m_S} \left(\hbar \vec{\nabla} \theta + q_S \vec{A} \right), \tag{18.26}$$

which, when curled,

$$\vec{\nabla} \times \vec{J} = -\frac{q_S n_S}{m_S}\left(\hbar \vec{\nabla} \times \vec{\nabla}\theta + q_S \vec{\nabla} \times \vec{A}\right) = -\frac{q_S^2 n_S}{m_S}\vec{B}, \qquad (18.27)$$

results in the London equation.

Without going into the details, we highlight two other significant predictions of the BCS theory that are well supported by experiment. BCS theory predicts that,

$$E_g(0) = 3.52 k_B T_c, \qquad (18.28)$$

and

$$\left.\frac{C_S - C_N}{C_N}\right|_{T_c} = 1.43. \qquad (18.29)$$

As can be seen from Table 18.1, both of these predictions are well supported by experiment.

18.2.2 Flux quantization

Another feature that emerges from the coherent nature of the superconducting wave function is the quantization of magnetic flux. Imagine, as illustrated in Fig. 18.11, a superconducting ring with a closed path C in its interior. Inside, deep beyond the penetration length of any magnetic field, both $\vec{B}$ and $\vec{J}$ are zero. Thus from Eq. (18.26) above,

$$\vec{\nabla}\theta = -\frac{q_S \vec{A}}{\hbar}. \qquad (18.30)$$

If we now integrate the gradient of the phase around the closed path,

$$\oint_C \vec{\nabla}\theta \cdot \mathrm{d}\vec{l} = \Delta\theta, \qquad (18.31)$$

we must obtain the total change in phase around the ring. But, because $\psi(\vec{r})$ is coherent, the wave function traveling around the ring must "meet up" with itself. The total change in phase must thus be some integer amount of 2π and so,

$$-\frac{q_S}{\hbar}\oint_C \vec{A} \cdot \mathrm{d}\vec{l} = -\frac{q_S}{\hbar}\int\left(\vec{\nabla} \times \vec{A}\right) \cdot \mathrm{d}\vec{a} = -\frac{q_S}{\hbar}\int \vec{B} \cdot \mathrm{d}\vec{a} = m2\pi. \qquad (18.32)$$

That is, the magnetic flux passing through the ring is *quantized* into units of,

$$\Phi_o = -\frac{h}{q_S} \qquad (18.33)$$

known as a *fluxon*. For Cooper pairs with $q_S = -2e$, a fluxon is only 2.07×10^{-15} Tm^2.

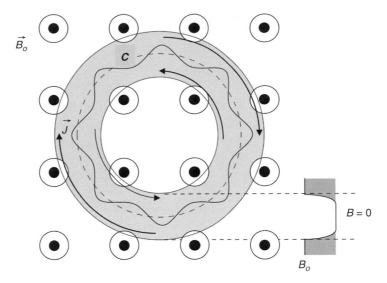

A superconducting ring placed in an external magnetic field has currents and a magnetic field only near the surface of the material. In the interior, the superconducting wave function, $\psi(\vec{r})$, is uniform aside from a varying phase. To maintain this uniformity, the phase change in any single circuit around the path C must appear in integers of 2π.

18.3 Thermodynamics: Ginzburg–Landau theory

18.3.1 Mean field theory

In our discussion above, we have seen many features of the superconducting phase transition that suggest, at least in the absence of any external magnetic field, that the transition is one of second order. Thus, as a starting point, we could examine this transition in terms of a mean field approach in which it is assumed that the order parameter is uniform. Returning to the Landau theory, we ask, "What is an appropriate order parameter for the superconducting phase transition?" The order parameter must be a measure of the continuously increasing order that develops below T_c and must be zero above T_c. Since the transition to the superconducting state arises from the spontaneous coupling of electrons into Cooper pairs, Ginzburg and Landau suggested using the macroscopic ground state wave function, ψ, as the order parameter. Then, from Eq. (17.53) of the previous chapter, the free energy density, $\tilde{F}_S(t, \psi) = F_S(t, \psi)/V$, is given for a second-order phase transition as,

$$\tilde{F}_S(t, \psi) = g_o(t) + g_2(t)|\psi|^2 + g_4(t)|\psi|^4 + \cdots, \tag{18.34}$$

where,

$$
\begin{aligned}
g_o(t) &= \tilde{F}_S(t, \psi = 0) = \tilde{F}_N(t) \\
g_2(t) &= f_{21}\Delta t \\
g_4(t) &= f_{40}.
\end{aligned}
\tag{18.35}
$$

To determine the variation of the order parameter, we seek the value that minimizes the free energy. This is obtained by taking a derivative of the free energy density with respect to ψ (actually ψ^*),

$$
\frac{\partial \tilde{F}_S(t, \psi)}{\partial \psi^*} = f_{21}\Delta t \psi_{eq} + 2 f_{40} |\psi_{eq}|^2 \psi_{eq} + \cdots = 0, \tag{18.36}
$$

from which we find that the equilibrium density of superconducting charges increases below T_c as,

$$
|\psi_{eq}|^2 = \frac{f_{21}(-\Delta t)}{2 f_{40}} = \frac{f_{21}}{2 f_{40}}(T_c - T), \tag{18.37}
$$

and see that the order parameter increases with the usual mean field critical exponent of ½. At equilibrium, the free energy density is then,

$$
\tilde{F}_S(t, \psi_{eq}) \approx \tilde{F}_N(t) + f_{21}\Delta t |\psi_{eq}|^2 + \cdots = \tilde{F}_N(t) - \frac{f_{21}^2}{2 f_{40}}(\Delta t)^2, \tag{18.38}
$$

which, as anticipated, is seen to be lower than that of the normal conducting phase whenever $T < T_c$.

Now that we know the free energy density in the superconducting state, we can assess how large an external magnetic field is required in order to destroy the phase. In general, application of an external field produces magnetization and reduces the free energy density by an amount,

$$
d\tilde{F} = -\vec{M} \cdot d\vec{B} = -\mu_o \vec{M} \cdot d\vec{H}. \tag{18.39}
$$

For a superconductor below the critical field, the magnetization is just $-\vec{H}$ (i.e. diamagnetic), and so the free energy density of the system increases with increasing applied field as,

$$
\tilde{F}_S(t, \psi_{eq}, H) = \tilde{F}_S(t, \psi_{eq}) - \int_0^H (-\mu_o H)dH = \tilde{F}_S(t, \psi_{eq}) + \frac{\mu_o H^2}{2}, \tag{18.40}
$$

as illustrated in Fig. 18.12. Assuming that the normal phase is not especially magnetic so that $\tilde{F}_N(t, H) \approx \tilde{F}_N(t)$, then the critical field corresponds to the point in Fig. 18.12 when the free energy of the superconducting phase exceeds that of the normal phase. From Eq. (18.38), this occurs when

$$
\tilde{F}_S(t, \psi_{eq}, H) = \tilde{F}_N(t) - \frac{f_{21}^2}{2 f_{40}}(\Delta t)^2 + \frac{\mu_o H_c^2}{2} = \tilde{F}_N(t), \tag{18.41}
$$

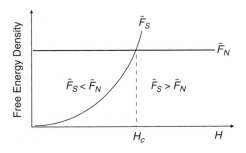

Figure 18.12 Variation of the free energy density for both the normal and superconducting phases in the presence of an external field. The critical field, H_c, is that field for which the free energy of the superconducting phase exceeds that of the normal phase.

or,

$$H_c = \frac{f_{21}}{\sqrt{\mu_o f_{40}}} |\Delta t|, \qquad (18.42)$$

and we find that, near T_c, the critical field increases linearly with decreasing temperature, in agreement with observations (see Eq. (18.16)).

18.3.2 Type II superconductors

We have waited until now to reveal a dirty little secret. Although all superconducting materials exhibit a vanishing of the resistivity, development of an energy gap, and a discontinuity in the specific heat at T_c, not all exhibit a complete Meissner effect below T_c. There is a class of superconducting materials, comprised mainly of alloys and low conducting materials, for which the external field is only *partly* expelled. Unlike those materials (Type I) that exhibit full expulsion of the magnetic field up to the thermodynamic critical field H_c, these other superconducting materials are referred to as Type II superconductors.

The differences between Type I and Type II superconductors are illustrated in Fig. 18.13. For Type I, the superconductor behaves as an ideal diamagnet whose magnetization exactly offsets the applied field until the critical field is reached and the normal phase returns (with a finite resistivity). In Type II materials, this diamagnetic behavior is seen only up to a lower critical field, H_{c1}. Above H_{c1}, the magnetization falls off and only some fraction of the external field is expelled. Note, however, that the resistivity remains zero above H_{c1} and only returns to the normal state value at some upper critical field, H_{c2}, where the magnetization is zero and the external field penetrates the entire sample.

Studies of Type II superconductors have revealed that a spatial pattern develops in the region between H_{c1} and H_{c2} in which the superconducting

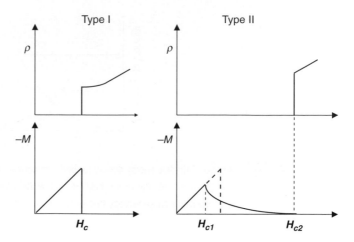

In Type I superconductors, the Meissner effect is maintained until the thermodynamic critical field H_c is reached, when an abrupt transition to the normal phase occurs. In Type II superconductors, the Meissner effect is maintained only to the lower critical field, H_{c1}. Between H_{c1} and the upper critical field, H_{c2}, an external magnetic field is only partially expelled.

phase (S) is punctuated by small regions of normal phase (N). As illustrated in Fig. 18.14, a view looking down along the direction of H finds isolated regions of normal phase comprised of vertical "tubes" that span from top to bottom through the sample. Because these tubes are composed of material in the normal phase, they do not expel magnetic field and it is through these tubes that some of the applied field is transmitted, giving the appearance of a reduced magnetization.

Because of the tubes, there is a spatial variation in the sample. Below H_{c1}, where the tubes are not present and the entire sample is superconducting, the wave function of the Cooper pairs is spatially uniform. Its amplitude is the same at all locations. When tubes form, this wave function can no longer be uniform and its amplitude must vary in space, especially in the region near a tube whose interior demands that $|\psi|^2$ be zero. When magnetic field penetrates through a tube, superconducting currents are set up, as shown in Fig. 18.14, whose function is to offset the penetration of the magnetic field into the superconducting region. These local "swirls" of superconducting current density are referred to as vortices, and the region of magnetization between H_{c1} and H_{c2} is often referred to as the *vortex* state.

18.3.3 The Ginzburg–Landau equations

To consider the features of the Type II superconductors, we now visit a significant extension of the Landau theory of phase transitions, first presented by Ginzburg and Landau in 1950. Our starting point is the realization that for Type II situations, the wave function (also known as the order parameter) is no

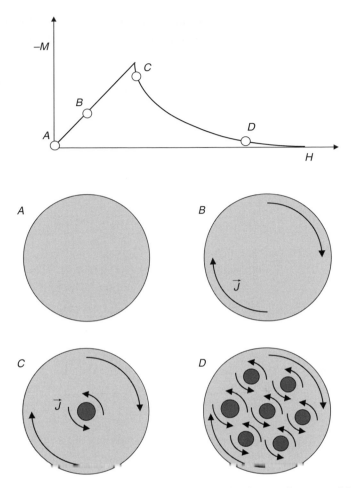

Figure 18.14 The developing structure of a Type II superconductor. (*A*) In the absence of an external field, no current is present. (*B*) Below H_{c1}, a current is produced in accordance with the London equation and the Meissner effect is observed in which the external field is fully expelled. (*C*) Near H_{c1}, a tube of normal phase nucleates into existence in the interior of the sample, creating a vortex of counter circulating current near its surface. (*D*) With increasing field, more tubes are formed until, at H_{c2}, they become too dense to sustain any superconducting current.

longer uniform and because tubes might exist, there can be regions in which the magnetic field is non-zero. To accommodate this, two new terms are introduced into the free energy density of Eq. (18.34) such that,

$$\tilde{F}_S(t, \psi, \vec{H}, \vec{A}) \approx \tilde{F}_N(t) + f_{21}\Delta t|\psi|^2 + f_{40}|\psi|^4 + \frac{\mu_o}{2}\left|\vec{H} - \vec{H}_i\right|^2$$
$$+ \frac{1}{2m_S}\left|\left(-i\hbar\vec{\nabla} - q_S\vec{A}\right)\psi\right|^2. \tag{18.43}$$

The first new term is pretty easy to understand. We saw above that when an applied field is fully expelled it increases the free energy density of a superconductor by $\mu_o H^2/2$. Consequently, if some of the field, $\vec{H}_i$, is allowed to penetrate through tubes, this will lower the free energy density. The last term derives from Eq. (18.24) and looks quite similar to a quantum mechanical kinetic energy ($E = (\hbar^2/2m)\nabla^2\psi$), but with a component derived from the local magnetic field, ($\vec{B} = \vec{\nabla} \times \vec{A}$). It is this last term, containing the gradient of the wave function, which will accommodate contributions to the free energy from a non-uniform order parameter.

Our approach parallels that for the case of a uniform order parameter, discussed earlier. We seek the order parameter that minimizes the free energy. But, since we are dealing with a spatially varying structure (i.e. regions of superconductor punctuated by normal tubes) we need to minimize the total free energy,

$$F_S = \int_V \tilde{F}_S(t, \psi, \vec{H}, \vec{A}) \mathrm{d}^3\vec{r}, \tag{18.44}$$

with respect to both $\psi(\vec{r})$ and $\vec{A}(\vec{r})$. This minimization is not trivial, but if done, one obtains two conditions. The first of these arises from minimization with respect to $\psi(\vec{r})$ and is,

$$\frac{1}{2m_S}\left|\left(-i\hbar\vec{\nabla} - q_S\vec{A}\right)\right|^2\psi + f_{21}\Delta t\psi + 2f_{40}|\psi|^2\psi = 0. \tag{18.45}$$

The second, arising from minimization with respect to $\vec{A}(\vec{r})$ is,

$$-\frac{iq_S\hbar}{2m_S}\left(\psi^*\vec{\nabla}\psi - \psi\vec{\nabla}\psi^*\right) - \frac{q_S^2}{m_S}|\psi|^2\vec{A} = \vec{J}. \tag{18.46}$$

Together, the two conditions are known as the *Ginzburg–Landau equations*.

We leave it as an exercise (Ex. 4) to show that in cases where the order parameter is uniform ($\psi(\vec{r}) = |\psi|e^{i\theta(\vec{r})}$), the second equation (Eq. (18.46)) just reduces to the London equation. Otherwise, in regions of superconducting phase where $\vec{A} = 0$, we find,

$$\frac{1}{2m_S}\left|\left(-i\hbar\vec{\nabla}\right)\right|^2\psi + f_{21}\Delta t\psi + 2f_{40}|\psi|^2\psi = 0, \tag{18.47}$$

or,

$$\frac{\hbar^2}{2m_S}\nabla^2\psi = -\left(f_{21}\Delta t + 2f_{40}|\psi|^2\right)\psi. \tag{18.48}$$

In the vicinity of T_c, where $|\psi|^2$ is small, this can be approximated as,

$$\nabla^2\psi = \frac{2m_S}{\hbar^2}f_{21}(-\Delta t)\psi, \tag{18.49}$$

whose solution in 1D is $\psi(x) = |\psi|e^{-x/\xi}$, where,

$$\xi = \frac{\hbar}{\sqrt{2m_S f_{21}(-\Delta t)}} \propto \left(\frac{T_c - T}{T_c}\right)^{-1/2}. \tag{18.50}$$

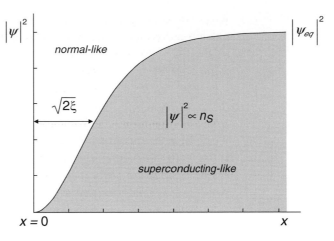

Figure 18.15 The Ginzburg–Landau equation puts limits on how rapidly the superconducting wave function can change spatially. Near a normal phase interface, this spatial change is limited to that of the coherence length, ξ.

More generally, if we retain the second, non-linear, term in Eq. (18.48), we find that the 1D solution for $\psi(x)$ near an interface at $x < 0$ is,

$$\psi(x) = \left(\frac{f_{21}(-\Delta t)}{2f_{40}} \right)^{1/2} \tanh\left(x/\sqrt{2}\xi \right), \qquad (18.51)$$

or,

$$n_S = |\psi(x)|^2 = |\psi_{eq}|^2 \tanh^2\left(x/\sqrt{2}\xi \right), \qquad (18.52)$$

provided the boundary conditions are such that $|\psi(x)|^2 = 0$ at $x \leq 0$ and approaches the equilibrium value $|\psi_{eq}|^2$ as $x \to \infty$. The variation of $|\psi(x)|^2$ traveling away from the boundary is shown in Fig. 18.15 and is illustrative of how the density of Cooper pairs vanishes on approach to an interface (i.e. a tube) of normal phase in a Type II superconductor over a distance comparable to ξ.

Coherence length

From the first Ginzburg–Landau equation, we find a new length scale, ξ, known as the Ginzburg–Landau *coherence length*. As seen from solutions to Eq. (18.48) and Eq. (18.49), the coherence length is a measure of how rapidly $\psi(\vec{r})$ can vary in space, and could be thought of as a measure of the "rigidity" of the macroscopic wave function of superconducting charges. A small coherence length means that the wave function is very pliable and can be bent around obstacles in a sharp manner, while a large coherence length implies that the wave function can only change gradually. In some respects, the coherence

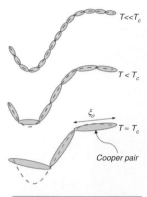

$T \ll T_c$

$T < T_c$

ξ_o

$T \approx T_c$

Cooper pair

Figure 18.16

The intrinsic coherence length is a measure of the extent over which Cooper pairs are correlated. This length scale diverges near T_c as the population of Cooper pairs vanishes. Since it is the Cooper pairs that make up the superconducting wave function, this wave function becomes increasingly more rigid near T_c.

length is similar to the correlation length discussed in Chapter 17. It too exhibits a divergence near T_c, with the same mean field exponent, $v = \frac{1}{2}$.

The BCS theory also provides a characteristic length scale known as the *intrinsic coherence length*,

$$\xi_o = \frac{2\hbar v_F}{\pi E_g}, \qquad (18.53)$$

where v_F is the speed of conduction of electrons near the Fermi surface and E_g is the superconducting energy gap. This intrinsic coherence length is, in essence, a measure of the mean extension of a Cooper pair; the distance over which two coupled electrons remain correlated. Since, as we observed in Eq. (18.19), the energy gap increases with decreasing temperature, ξ_o develops in a parallel fashion with the Ginzburg–Landau coherence length, and both diverge near T_c in the same power law manner.

We can now paint a simple conceptual picture for why the wave function becomes so rigid and inflexible near T_c. As illustrated in Fig. 18.16, at temperatures well below T_c there are a great many Cooper pairs and, according to Eq. (18.37), their density increases with decreasing temperature below T_c. Here, the intrinsic coherence length is short and we can think of the macroscopic wave function like a chain composed of a great many small links (each link a Cooper pair) that can readily bend on a small radius of curvature and easily navigate around obstacles. However, as temperature is increased towards T_c, the intrinsic coherence length increases and the density of Cooper pairs decreases. The wave function may still span the same region of space, but is now composed of a small number of large, bulky links that find it increasingly more difficult to bend or accommodate obstacles.

18.3.4 Type II critical fields

Type II superconductors are able to expel an external magnetic field up to values of H_{c1} where the vortex state develops. What then determines this lower critical field? Let us imagine that we are well below T_c where $H_c(T)$ is large. Here, the wave function is very pliable and can easily navigate about a tube of normal phase, without disturbing the equilibrium $|\psi_{eq}|^2$ in the bulk of the remaining superconducting phase. But we know that the free energy density of the normal phase is not thermodynamically favored below T_c, and the appearance of such a tube of normal phase should not occur unless it somehow lowers the overall free energy. The solution to this puzzle lies in the fact that the introduction of a tube produces an interface that can contribute either positive or negative *surface energy*. If this new surface energy is sufficiently negative, the overall free energy can be reduced, making the tube thermodynamically stable. If positive, then the tube would not be allowed to form and we would be stuck with a Type I behavior.

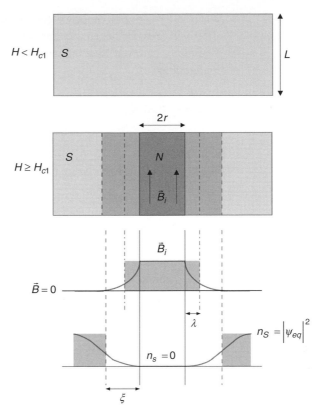

Figure 18.17 A superconducting slab experiences the nucleation of a vertical tube of normal phase in its interior. The tube itself has a nominal diameter of $2r$, but its magnetic field penetrates into the surrounding superconductor by a distance λ. Simultaneously, the density of Cooper pairs vanishes on approach to the new interface over a distance equal to the coherence length. The relative size of both λ and ξ determine whether the surface energy is positive or negative.

Lower critical field, H_{c1}

Consider then the nucleation of a single tube of normal phase in the center of a slab of superconducting phase of height L, as shown in Fig. 18.17. Shown near the lower portion of the figure is the spatial variation of both the magnetic field, which penetrates into the superconducting regions to a distance of λ, and the density of Cooper pairs, which become vanishingly small over a distance ξ near the vicinity of the tube. Although both $\vec{B}(\vec{r})$ and $\psi(\vec{r})$ vary smoothly near the tube, let us *approximate* the situation as though it were a sharp interface in which both $\vec{B}(\vec{r})$ and $\psi(\vec{r})$ change abruptly at λ and ξ, respectively, as shown by the shaded regions in Fig. 18.17.

Before the tube appears (at $H < H_{c1}$) the total free energy is initially given by Eq. (18.40) as,

$$F_{\text{init}} = \left(\tilde{F}_S(t, \psi_{eq}) + \frac{\mu_o H^2}{2} \right) V. \qquad (18.54)$$

When the tube first appears (at H just above H_{cI}), the magnetic field effectively extends out to a radius of $r + \lambda$, while the density of Cooper pairs is virtually zero out to a radius of $r + \xi$. The total free energy in this final configuration is,

$$F_{\text{final}} = \tilde{F}_S(t, \psi_{eq})\left(V - \pi(r + \xi)^2 L\right) + \frac{\mu_o H^2}{2}\left(V - \pi(r + \lambda)^2 L\right)$$
$$+ \tilde{F}_N(t)\left(\pi(r + \xi)^2 L\right). \tag{18.55}$$

The net change in free energy on forming the tube is then,

$$\Delta F = F_{\text{final}} - F_{\text{init}}$$
$$= \left[\tilde{F}_N(t) - \tilde{F}_S(t, \psi_{eq})\right]\pi(r + \xi)^2 L - \frac{\mu_o H^2}{2}\pi(r + \lambda)^2 L \tag{18.56}$$
$$= \left[\frac{\mu_o H_c^2}{2}\right]\pi(r + \xi)^2 L - \pi(r + \lambda)^2 L \frac{\mu_o H^2}{2},$$

where we recall from Eq. (18.41) that the difference in free energy density between the normal and superconducting phases, $\tilde{F}_N(t) - \tilde{F}_S(t, \psi_{eq})$, is just $\mu_o H_c^2(T)/2$.

Now, since the normal phase is not thermodynamically favored below T_c, we expect this tube will be as small a diameter as it can be, while still providing an interface. When we let $r \to 0$, we find,

$$\Delta F = \frac{\mu_o \pi L}{2}\left(H_c^2 \xi^2 - H^2 \lambda^2\right). \tag{18.57}$$

The tube will be stable only if $\Delta F < 0$ or,

$$H^2 \geq \frac{\xi^2}{\lambda^2} H_c^2, \tag{18.58}$$

and the threshold condition for a tube to form is then set by $\Delta F = 0$, where,

$$H_{c1} = \frac{\xi}{\lambda} H_c = \kappa^{-1} H_c. \tag{18.59}$$

For Type II behavior to occur, the nucleating field H_{cI} must be less than H_c, and thus for these materials, $\xi < \lambda$ and $\kappa > 1$.

One thing that we have neglected in allowing this tube to nucleate is that, in addition to its surface energy, it also creates magnetic flux passing through the surrounding superconducting phase. As we observed earlier, such flux is quantized and this places an additional constraint on the external field inside the tube,

$$\mu_o H(\pi \lambda^2) \geq \Phi_o. \tag{18.60}$$

In careful experiments, it has been shown that each tube in the mixed phase of the vortex state, no matter how many tubes might appear, individually

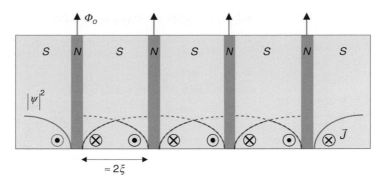

Figure 18.18 A cross-sectional view of the tubes shown in Fig. 18.14D near H_{c2}. When the tube density reaches a point that their individual coherence lengths overlap, the superconducting state becomes unstable. Note also, that in this same limit, the counter-propagating current densities of each vortex begin to cancel out.

allows only a single quantum (one fluxon) of magnetic flux. With this insight, we see that,

$$H_{c1} = \frac{\Phi_o}{\mu_o \pi \lambda^2}, \tag{18.61}$$

and, with Eq. (18.59),

$$H_c = \frac{\Phi_o}{\mu_o \pi \lambda \xi}. \tag{18.62}$$

Upper critical field, H_{c2}

As it turns out, the limitation of one quantum of flux per tube allows the upper critical field, H_{c2}, to be determined. As the external field is increased to H_{c1}, a first tube forms. Since each tube can handle only one fluxon of magnetic flux, further increases in H result in the formation of additional tubes (each carrying only one fluxon) and the density of tubes increases. At some point, the density of tubes will encounter a maximum where, as illustrated in Fig. 18.18, we can just maintain $n_C = |\psi_{eq}|^2$ in the surrounding superconducting regions. This maximum density corresponds to overlapping of the coherence length extending from neighboring tubes. At this point, the total number of tubes is,

$$N \approx \frac{V}{\pi \xi^2 L} = \frac{A}{\pi \xi^2}, \tag{18.63}$$

and the total magnetic flux passing through the system is

$$\Phi_{\text{tot}} = N\Phi_o \approx \mu_o H_{c2} A. \tag{18.64}$$

Solving, we find the upper critical field is given as,

$$H_{c2} \approx \frac{\Phi_o}{\mu_o \pi \xi^2},$$ (18.65)

or, taking into account Eq. (18.62),

$$H_{c2} \approx \frac{\mu_o \pi \lambda \xi H_c}{\mu_o \pi \xi^2} = \frac{\lambda}{\xi} H_c = \kappa H_c.$$ (18.66)

We see that the two critical fields of the Type II superconductor are each related to the thermodynamic critical field (H_c) of the underlying superconducting phase. In each case, the field is chiefly determined by κ, the ratio of the penetration length to the coherence length. By increasing κ, we can extend the superconducting state (with zero resistivity) to fields of very high strength, allowing for the commercial development of practical superconducting magnets.

18.3.5 High-T_c superconductors

For some years after the development of BCS theory, the highest T_c that could be obtained was thought to be about 30 K, and limited by the elasticity of the lattice and the density of electron states near the Fermi level. However, in 1986, researchers discovered superconductivity in certain cuprate–perovskite ceramic materials and shortly thereafter obtained T_cs in excess of the boiling point of liquid nitrogen, a cheaply available cooling medium. The first of these new "high-T_c superconductors" was $YBa_2Cu_3O_7$, whose T_c is between 90 and 95 K. Although these materials exhibit all the hallmarks of the superconducting phase, including the formation of Cooper pairs, the presence of such an enhanced attractive coupling between electrons poses significant challenges to the BCS theory. It seems clear that the binding of Cooper pairs in these new materials may be derived from a source of interaction other than that of lattice phonons.

Summary

- Superconductors exhibit zero electrical resistivity.

- In weak magnetic fields (below $H_c(T)$), magnetic field lines are expelled from the interior of a superconductor (Meissner effect), except near the surface, where they may penetrate to a depth of $\lambda = \sqrt{\frac{m_S}{\mu_o q_S^2 n_S}}$.

- In the absence of an external magnetic field, the superconducting phase transition is one of second order and displays both a diverging characteristic length scale and laws of corresponding states.

- In the BCS theory, superconducting charge carriers arise from an electron–phonon interaction that allows pairs of electrons with opposed momenta and spin to form into Cooper pairs. The Cooper pairs act much like bosons and form a coherent, macroscopic superconducting wave function, $\psi(\vec{r})$.

- There are two types of superconductors. Type I superconductors exhibit a complete Meissner effect, expelling magnetic field up to the thermodynamic critical field, $H_c(T)$. Type II superconductors exhibit an incomplete Meissner effect due to the nucleation of regions of normal phase. For Type II superconductivity to occur, $\kappa = \frac{\lambda}{\xi} > 1$.

Exercises

18.1. Derive Eq. (18.3) relating the current density and electric field in an ideal conductor.

18.2. Derive Eq. (18.12) for the current density near the surface of a superconductor in an external field.

18.3. Show that near T_c, the critical field given by quadratic form in Eq. (18.16) varies in a linear manner as $H_c(T) \approx 2H_c(0)(T_c - T)/T_c$.

18.4. Show that the second Ginzburg–Landau equation (Eq. (18.46)) just reduces to the London equation in cases where the order parameter is uniform ($\psi(\vec{r}) = |\psi|e^{i\theta(\vec{r})}$).

18.5. Consider an infinite 2D slab of superconducting material of thickness $2d$ aligned perpendicular to the y axis as shown in Fig. 18.19 with a uniform magnetic field of H_o applied along the z-axis.

(a) Taking as the boundary condition that the parallel component of $\vec{B} = \mu_o \vec{H}$ be continuous at either surface, show that within the superconductor, $\vec{B} = \mu_o H_o \dfrac{\cosh(y/\lambda)}{\cosh(d/\lambda)}\hat{z}$, and the corresponding current density is

$$\vec{J} = \frac{H_o}{\lambda}\frac{\sinh(y/\lambda)}{\cosh(d/\lambda)}\hat{x}.$$

(b) The magnetization at a point within the slab is $\vec{M} = \vec{B}/\mu_o - \vec{H}_o$. Show that the net magnetization (averaged over the thickness of the slab) is

$$\langle \vec{M} \rangle = -\vec{H}_o \left\{ 1 - \frac{\lambda}{d}\tanh\left(\frac{d}{\lambda}\right) \right\}.$$

(c) Show that when the slab is much thicker than the penetration length ($d \gg \lambda$), the susceptibility approaches that of a perfect diamagnet,

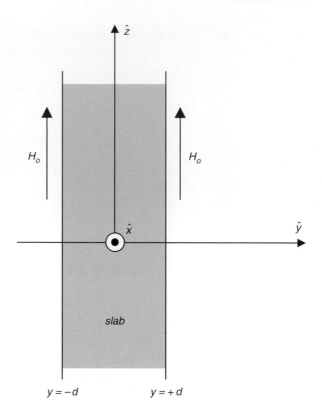

Figure 18.19

while when the slab is thinner than the penetration length ($d \ll \lambda$), the susceptibility vanishes like $\chi = \partial\langle\vec{M}\rangle/\partial H_o \approx -\frac{d^2}{3\lambda^2}$.

18.6. The thermodynamic state of a superconductor is determined by both the temperature and magnitude of an applied magnetic field, H. Assuming the volume remains constant, changes in the Helmholtz free energy density are given as $d\tilde{F} = -\tilde{S}dT - \mu_o M dH$, where $\tilde{S}$ is the entropy density. The coexistence curve, which separates the superconducting (S) and normal (N) phase, is given by the curve in Fig. 8.4.

(a) Show that for any point on this coexistence curve, the slope of the curve is given by

$$\frac{dH_c}{dT} = \frac{1}{\mu_o}\frac{\tilde{S}_N - \tilde{S}_S}{M_S - M_N}.$$

(b) Given that the normal phase has negligible ($\chi_m \approx 10^{-6}$) diamagnetism, while the superconducting state behaves as a perfect diamagnet, show from the result of part (a) above that the entropy is discontinuous by an amount

$$\tilde{S}_N - \tilde{S}_S = -\mu_o H_c \frac{dH_c}{dT},$$

and that the latent heat per volume, $\tilde{L} = T\Delta\tilde{S}$, when the transition occurs in a finite field, is

$$\tilde{L} = -\mu_o T H_c \frac{dH_c}{dT}.$$

(c) Show that when the transition occurs at zero field (at T_c) the specific heat exhibits a discontinuity given by

$$\frac{C_N - C_S}{V} = -\mu_o T \left(\frac{dH_c}{dT}\right)^2.$$

Suggested reading

Almost any standard Solid State textbook will include a chapter devoted to superconductivity. The first three listed do a good job of discussing the topic. Some of the seminal papers on the subject are listed for those interested.

N. A. Ashcroft and N. D. Mermin, *Solid State Physics* (Holt, Rinehart and Winston, New York, 1976).

C. Kittel, *Introduction to Solid State Physics*, 8th Ed. (John Wiley and Sons, 2005).

W. Buckel, *Superconductivity: Fundamentals and Applications* (VCH, New York, 1991).

P. Hofmann, *Solid State Physics* (Wiley-VCH, Weinheim, 2008).

H. K. Onnes, *Comm. Leiden* **120b** (1911).

W. Meissner and R. Ochsenfeld, *Naturwissenschaften* **21**, 787 (1933).

F. London and W. London, *Z. Phys.* **96**, 359 (1935).

J. Bardeen, L. N. Cooper and J. R. Schrieffer, *Phys. Rev.* **108**, 1175 (1957).

V. L. Ginzburg and L. D. Landau, *Zh. Eksp. Teor. Fiz.* **20**, 1064 (1950).

Appendix: Toolbox

Most of the material covered in this textbook can be understood without any additional prerequisite coursework other than that usually afforded by an introductory physics course that includes modern physics topics (harmonic oscillators and particle-in-a-box). In some places, concepts are drawn liberally from thermodynamics and statistical mechanics and considerable use is made of complex notation for describing waves, as well as considerable emphasis on Fourier transforms (albeit mostly at a conceptual level). While it is hoped that the reader is already well acquainted with these concepts, this is likely not to be the case for every reader and as an aid, this "toolbox" is included as a handy appendix to some of the relevant mathematical and theoretical background that appears within the textbook.

A.1 Complex notation

Complex notation is a very useful form of shorthand for dealing with the manipulation of trigonometric functions and is often used to describe propagating waves. A complex number is very much like a two-dimensional vector; it has two components, one that is "real" and the other that is "imaginary". One way in which we convey the "complex" aspect of the complex number z is to write it out as the vector sum

$$z = x + iy, \tag{A1.1}$$

where x represents the magnitude of the real component, y the magnitude of the imaginary component and $i \equiv \sqrt{-1}$. These associations are often expressed using "Re[]" and "Im[]" as

$$\begin{aligned} x &= \text{Re}[z] \\ y &= \text{Im}[z] \end{aligned} \tag{A1.2}$$

Since the complex number z is identical to a two-dimensional vector, an alternative means for describing z is to use a form of polar coordinates, as illustrated in Fig. A1.1. In polar coordinates, the components of z can be expressed as

$$\begin{aligned} x &= \text{Re}[z] = A \cos \phi \\ y &= \text{Im}[z] = A \sin \phi \end{aligned} \tag{A1.3}$$

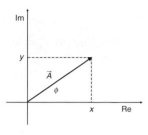

Representation of a complex number as a vector in either Cartesian or polar coordinates. The horizontal axis represents the real part of the complex number and the vertical axis represents the imaginary part.

Now here is where the shorthand feature begins. There is an important relationship between the mathematical properties of the exponential function and those of the two trigonometric functions, known as *Euler's equation*:

$$e^{i\phi} = \cos\phi + i\sin\phi \qquad (A1.4)$$

Returning to our graphical representation of the complex number z in Fig. A1.1, we see that we can then express any complex quantity as,

$$z = x + iy = A\cos\phi + iA\sin\phi = Ae^{i\phi}. \qquad (A1.5)$$

A.1.1 Trigonometric identities

To demonstrate how this shorthand notation can simplify trigonometric manipulations, consider the trigonometric identity

$$\cos(a + b) = \cos a \cos b - \sin a \sin b.$$

To prove this identity using geometry is painful, but it is remarkably easy to prove using our shorthand notation. We see that the left-hand side is just the real part of the corresponding exponential,

$$\cos(a + b) = \text{Re}\left[e^{i(a+b)}\right],$$

which can be expanded using the Euler formula as,

$$e^{i(a+b)} = e^{ia}e^{ib} = (\cos a + i\sin a)(\cos b + i\sin b)$$
$$= \cos a \cos b + i\cos a \sin b + i\sin a \cos b - \sin a \sin b.$$

Taking only the real part of this last expression, we obtain the desired identity.

A.1.2 Other items

A.1.2.1 Complex conjugate

The complex conjugate of z is denoted by an asterisk, z^*, and is obtained merely by replacing every occurrence of i in z by $-i$,

$$z = Ae^{i\phi} \Rightarrow z^* = Ae^{-i\phi} \qquad (A1.6)$$

A.1.2.2 Modulus

The modulus of a complex number is a representation of its vector magnitude and is obtained by square rooting the result of the product of z and its complex conjugate. The result is always a purely real number,

$$|z| = \sqrt{zz^*} = A \qquad (A1.7)$$

A.1.2.3 Angular equivalents

A little reflection on the polar diagram in Fig. A1.1 reveals that, for certain angles ϕ the exponential returns either a purely real or purely imaginary value,

$$\mathrm{e}^{\pm in\pi} = \begin{cases} +1 & \text{if n is even} \\ -1 & \text{if n is odd} \end{cases} \quad \text{and} \quad \mathrm{e}^{\pm i\pi/2} = \pm i. \qquad (A1.8)$$

An important item to remember with complex notion is that in many publications the shorthand is used without the "Re[]" and "Im[]" indicated. In these instances, the convention is that the final result of the manipulation is obtained by taking the real part of the corresponding complex quantity.

A.2 Wave notation

A large portion of this textbook is devoted to the scattering of waves by matter, so let us now clarify how a propagating wave is described using our complex notation shorthand. For a simple one-dimensional wave of frequency f and wavelength λ traveling in the positive x direction, the wave function is generally given as,

$$\psi(x,t) = A\cos(kx - \omega t + \delta), \qquad (A2.1)$$

where $k = 2\pi / \lambda$, $\omega = 2\pi f$ and δ is an arbitrary phase angle. This wave then travels with a speed $v = f\lambda = \omega / k$. In the complex notation (with the Re[] being suppressed or omitted) this wave is written as,

$$\psi(x,t) = A\mathrm{e}^{i(kx - \omega t + \delta)} = A\mathrm{e}^{i\phi}, \qquad (A2.2)$$

where ϕ is the total phase angle of the wave at position x and time t.

Next, we extend this notation to three-dimensional waves that are known as *plane waves*. These are waves that travel along a fixed direction of propagation parallel to the three-dimensional wave vector $\vec{k}$ and display a common value at all points in any plane perpendicular to $\vec{k}$. For this to happen the total phase angle, ϕ, must be common for every point in the plane at any given instant in time. This then means that the spatial portion of the total phase angle at any point in the plane is a constant, or that

$$\phi_r = \vec{k} \cdot \vec{r} = \text{constant},$$

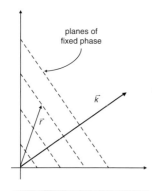

Figure A2.1

A plane wave is defined by planes of common phase. The location of these planes is determined by the projection $(\vec{k} \cdot \vec{r})$ of the position vector onto the wave vector.

where $\vec{r}$ is the position vector that describes any arbitrary point in space, as shown in Fig. A2.1. Hence the overall wave function describing a plane wave traveling along the direction $\vec{k}$ at a speed given by $v = f\lambda = \omega / k$, would be,

$$\psi(\vec{r},t) = A\mathrm{e}^{i(\vec{k} \cdot \vec{r} - \omega t + \delta)}. \qquad (A2.3)$$

A.3 Fourier analysis

A repeated theme found throughout the textbook is the notion that information concerning the structure or dynamics of a material is related by Fourier transformations to corresponding information in the scattering patterns. To develop some understanding behind the transformation, we begin with a simple example of Fourier analysis.

Suppose a musician plays a single sustained note on a clarinet, which is then recorded. If we play back the recorded signal into a strip chart recorder or other device that could produce a plot of the sound pressure versus time, we would see something like that shown in Fig. A3.1a. The $P(t)$ graph consists of a repeating pattern with an internal structure. The internal structure arises from the resonant nature of the clarinet which, in addition to the fundamental tone, ω, contains additional overtones at higher frequencies such as 2ω, 3ω, 4ω, etc., in varying amplitudes, as illustrated in Fig. A3.1b. The same tone played on, say, an oboe, would appear with a different internal structure because the amplitudes of the various overtones are different.

Since the $P(t)$ in the first figure is just the result of mixing together of the fundamental and overtones for a given frequency ω, each with different amplitudes, we could synthesize the same $P(t)$ using several electronic frequency generators (at ω, 2ω, 3ω, etc.) each adjusted to amplitudes, $P(\omega)$, that correspond to those in the second figure. This is the fundamental idea behind a Fourier transformation: both $P(t)$ and $P(\omega)$ contain the *same amount of information* and both provide equivalent descriptions of a clarinet playing a sustained note.

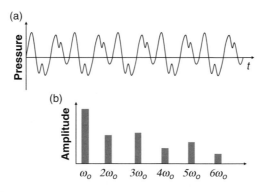

Figure A3.1 Example of sound produced by a sustained note played by a musical instrument. (a) The time-dependent pressure signal as recorded by a detector exhibiting a repeated pattern. (b) Illustration of how the corresponding Fourier components might appear as harmonics of differing magnitude.

A.3.1 Fourier series

Our example above illustrates that any periodic waveform, $f(t)$, can be reconstructed by an appropriate combination of single frequency waves of prescribed amplitudes,

$$f(t) = \frac{A_o}{2} + \sum_{m=1}^{\infty} A_m \cos(m\omega t) + \sum_{m=1}^{\infty} B_m \sin(m\omega t), \qquad (A3.1)$$

known as a *Fourier series*. All we require then are the set of amplitudes, A_o, A_m and B_m. The recipe for obtaining these unknown amplitudes relies on the particular *orthogonality* properties of the two trigonometric functions that emerge when integrated over a complete cycle, $T = 2\pi / \omega$:

$$\int_0^T \sin(m\omega t)\, \mathrm{d}t = \int_0^T \cos(m\omega t)\, \mathrm{d}t = \int_0^T \sin(m\omega t)\cos(n\omega t)\, \mathrm{d}t = 0, \qquad (A3.2)$$

and,

$$\int_0^T \sin(m\omega t)\sin(n\omega t)\, \mathrm{d}t = \int_0^T \cos(m\omega t)\cos(n\omega t)\, \mathrm{d}t = \frac{T}{2}\delta_{mn}, \qquad (A3.3)$$

where the Kronecker delta function is defined by,

$$\delta_{mn} \equiv \begin{cases} 0 & \text{if } m \neq n \\ 1 & \text{if } m = n. \end{cases} \qquad (A3.4)$$

Applying these to Eq. (A3.1), the amplitudes are then given by,

$$A_o = \frac{2}{T}\int_0^T f(t)\mathrm{d}t,$$

$$A_m = \frac{2}{T}\int_0^T f(t)\cos(m\omega t)\mathrm{d}t, \qquad (A3.5)$$

$$B_m = \frac{2}{T}\int_0^T f(t)\sin(m\omega t)\mathrm{d}t$$

A3.2 Fourier transforms

As an important example of using Fourier series, consider the square wave shown in Fig. A3.2a that is periodic but non-harmonic. Following the above recipe, one obtains,

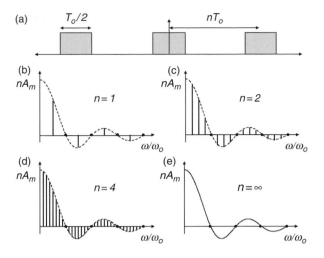

Top figure (a) shows the repeated pulse pattern occurring as a function of time. Lower figures (b through e) show the corresponding Fourier components. As the interval between pulses increases with increasing n, a greater density of Fourier components is required to describe the waveform approaching a continuum in the limit that the waveform assumes the properties of a single pulse.

$$f(t) = \frac{1}{2n} + \frac{1}{n}\sum_{m=1}^{\infty} \operatorname{sinc}\left(m\frac{\pi}{2}\frac{\omega}{\omega_o}\right)\cos(m\omega t), \qquad (A3.6)$$

where sinc $x \equiv \sin x/x$. Now consider what happens as the "pulse" width, $T_o/2$, remains fixed, while the period between pulses increases as n is systematically increased from 1 to 2 to 4. The progression is shown in Figs. A3.2b through A3.2d for the corresponding coefficients, A_m. What do we see? As $n \to \infty$, we (a) lose the sense of periodicity in $f(t)$ and develop only a single, very narrow, pulse occurring at $t = 0$; (b) increase the number of terms needed in the summation and develop an increasingly more "dense" set of frequencies, and (c) approach a continuum of A_ms, illustrated in Fig. A3.2e, that are better expressed as a *function* of ω, $A(\omega)$. In this continuum limit, the Fourier series is replaced by the *Fourier transform*:

$$f(t) = \int_0^{\infty} A(\omega)\cos(\omega t)d\omega + \int_0^{\infty} B(\omega)\sin(\omega t)d\omega, \qquad (A3.7)$$

whose amplitude functions, $A(\omega)$ and $B(\omega)$ are given by,

$$A(\omega) = \frac{1}{\pi}\int_{-\infty}^{\infty} f(t)\cos(\omega t)dt$$

$$\qquad\qquad\qquad\qquad (A3.8)$$

$$B(\omega) = \frac{1}{\pi}\int_{-\infty}^{\infty} f(t)\sin(\omega t)dt.$$

As an example of the Fourier transform, let's look at a simple square wave pulse,

$$f(t) = \begin{cases} 1 & \text{for} \quad -\frac{T_o}{4} < t < \frac{T_o}{4} \\ 0 & \text{otherwise} \end{cases}.$$

Firstly, as we have defined the time origin, this $f(t)$ is an even function. Therefore, there are no sine contributions and so $B(\omega) = 0$. This leaves us with only the Fourier cosine transform,

$$A(\omega) = \frac{1}{\pi} \int_{-\infty}^{\infty} f(t) \cos(\omega t) dt = \frac{1}{\pi} \int_{-T_o/4}^{+T_o/4} \cos(\omega t) dt.$$

This can be shown to reduce to,

$$A(\omega) = \frac{T_o}{2\pi} \text{sinc} \left(\frac{\omega T_o}{4} \right),$$

which is just seen to be the continuum result corresponding with the discrete situation in Eq. (A3.6) and illustrated in Fig. A3.2.

Although the Fourier transforms above are developed in reference to a time-dependent function $f(t)$ synthesized by the time-dependent oscillating functions $\sin(\omega t)$ and $\cos(\omega t)$, we could have equally well developed the Fourier transform to describe space-dependent functions, $f(x)$, described by the space-dependent oscillating functions $\sin(kx)$ and $\cos(kx)$, for which,

$$f(x) = \int_0^{\infty} A(k) \cos(kx) dk + \int_0^{\infty} B(k) \sin(kx) dk, \qquad (A3.9)$$

where,

$$A(k) = \frac{1}{\pi} \int_{-\infty}^{\infty} f(x) \cos(kx) dx$$

$$B(k) = \frac{1}{\pi} \int_{-\infty}^{\infty} f(x) \sin(kx) dx. \qquad (A3.10)$$

In either case, information contained in space (x) or time (t) is transformed into an equivalent measure of information contained in wave vector (k) or angular frequency (ω), respectively.

A.3.3 Fourier transforms expressed in complex notation

To show how these Fourier transforms will appear when complex notation is used, we begin by substituting our expressions for $A(k)$ and $B(k)$ (Eq. (A3.10)) back into Eq. (A3.9) to obtain,

$$f(x) = \frac{1}{\pi} \left\{ \int_{k=0}^{\infty} \cos(kx) \left[\int_{x'=-\infty}^{\infty} f(x') \cos(kx') dx' \right] dk \right.$$

$$\left. + \int_{0}^{\infty} \sin(kx) \left[\int_{x'=-\infty}^{\infty} f(x') \sin(kx') dx' \right] dk \right\},$$

which, after some rearranging, becomes,

$$f(x) = \frac{1}{\pi} \left\{ \int_{k=0}^{\infty} \int_{x'=-\infty}^{\infty} f(x')[\cos(kx')\cos(kx) + \sin(kx')\sin(kx)] dx' dk \right\}.$$

(A3.11)

Now we recall the following trigonometric identity:

$$\cos[k(x' - x)] = \cos kx \cos kx' + \sin kx \sin kx',$$

which allows Eq. (A3.11) to be rewritten as,

$$f(x) = \frac{1}{\pi} \left\{ \int_{k=0}^{\infty} \left[\int_{x'=-\infty}^{\infty} f(x') \cos(k(x' - x)) dx' \right] dk \right\}.$$

(A3.12)

Since the quantity inside the square brackets is an even function of k, we can replace the limits of the integration over dk to run from $k = -\infty$ to ∞ instead of 0 to ∞ giving,

$$f(x) = \frac{1}{2\pi} \left\{ \int_{k=-\infty}^{\infty} \left[\int_{x'=-\infty}^{\infty} f(x') \cos(k(x' - x)) dx' \right] dk \right\}.$$

(A3.13)

Next we add zero to this result, but in a creative way! Consider the quantity,

$$\frac{i}{2\pi} \left\{ \int_{k=-\infty}^{\infty} \left[\int_{x'=-\infty}^{\infty} f(x') \sin(k(x' - x)) dx' \right] dk \right\}.$$

(A3.14)

Since the quantity inside the square brackets is an odd function of k, the first half of the integral running from $k = -\infty$ to 0 must exactly cancel with the last half from $k = 0$ to $+\infty$. Thus the entire quantity in Eq. (A3.14) is identically zero. If we add this 'zero' onto our last result (Eq. (A3.13)),

$$f(x) = \frac{1}{2\pi} \left\{ \int_{k=-\infty}^{\infty} \left[\int_{x'=-\infty}^{\infty} f(x') \cos(k(x' - x)) dx' \right] dk \right\}$$

$$+ \frac{i}{2\pi} \left\{ \int_{k=-\infty}^{\infty} \left[\int_{x'=-\infty}^{\infty} f(x') \sin(k(x' - x)) dx' \right] dk \right\},$$

we can rearrange to obtain,

$$f(x) = \frac{1}{2\pi} \left\{ \int\limits_{k=-\infty}^{\infty} \left[\int\limits_{x'=-\infty}^{\infty} f(x')[\cos(k(x'-x)) + i\sin(k(x'-x))] dx' \right] dk \right\},$$

which, with Euler's relationship becomes,

$$f(x) = \frac{1}{2\pi} \left\{ \int\limits_{k=-\infty}^{\infty} \left[\int\limits_{x'=-\infty}^{\infty} f(x')e^{ikx'} dx' \right] e^{-ikx} dk \right\} = \frac{1}{2\pi} \left\{ \int\limits_{k=-\infty}^{\infty} F(k)\, e^{-ikx} dk \right\}.$$

Thus the Fourier transform can be expressed in complex notation as,

$$f(x) = \frac{1}{2\pi} \int\limits_{k=-\infty}^{\infty} F(k)\, e^{-ikx} dk, \tag{A3.15}$$

and the inverse Fourier transform that provides the amplitudes as,

$$F(k) = \int\limits_{x=-\infty}^{\infty} f(x)\, e^{+ikx} dx. \tag{A3.16}$$

The functions $f(x)$ and $F(k)$ are referred to as Fourier transform *pairs* because, as we have emphasized earlier, each contains identical amounts of information. A similar set of Fourier transform pairs can be developed for time-dependent functions as,

$$f(t) = \frac{1}{2\pi} \int\limits_{\omega=-\infty}^{\infty} F(\omega)\, e^{-i\omega t} d\omega$$

$$F(\omega) = \int\limits_{t=-\infty}^{\infty} f(t)\, e^{+i\omega t} dt. \tag{A3.17}$$

As an example of the Fourier transform, let us evaluate the Fourier transform of the Gaussian function, $f(x) = e^{-x^2/a^2}$:

$$F(k) = \int\limits_{x=-\infty}^{\infty} f(x)\, e^{+ikx} dx = \int\limits_{x=-\infty}^{\infty} e^{-x^2/a^2}\, e^{+ikx} dx.$$

We use the Euler formula to separate this into real and imaginary parts as

$$F(k) = \int\limits_{x=-\infty}^{\infty} e^{-x^2/a^2} (\cos kx + i\sin kx)\, dx$$

$$= \int\limits_{x=-\infty}^{\infty} e^{-x^2/a^2} \cos kx\, dx + i \int\limits_{x=-\infty}^{\infty} e^{-x^2/a^2} \sin kx\, dx.$$

In the second integral, the integrand is an odd function of x that must vanish when integrated over all x. In the first integral, the integrand is an even function of x and for this integral we can replace the limits as

$$F(k) = \int_{x=-\infty}^{\infty} e^{-x^2/a^2} \cos kx\, dx = 2 \int_{x=0}^{\infty} e^{-x^2/a^2} \cos kx\, dx = \sqrt{\pi} a e^{-k^2 a^2/4}.$$

Note that the Gaussian function is somewhat unique in that its Fourier transform is itself another Gaussian function.

A.3.4 Extension of Fourier transforms to higher dimensions

In three dimensions, the above spatial Fourier transforms become:

$$f(x,y,z) = \frac{1}{(2\pi)^3} \int_{k_x=-\infty}^{\infty} \int_{k_y=-\infty}^{\infty} \int_{k_z=-\infty}^{\infty} F(k_x,k_y,k_z)\, e^{-i(k_x x + k_y y + k_z z)} dk_x dk_y dk_z,$$

$$F(k_x,k_y,k_z) = \int_{x=-\infty}^{\infty} \int_{y=-\infty}^{\infty} \int_{z=-\infty}^{\infty} f(x,y,z)\, e^{+i(k_x x + k_y y + k_z z)} dx\, dy\, dz.$$

If we think of k_x, k_y and k_z as the coordinates of some "k-space" similar to x, y and z in real space, then we could tidy up these last two expressions to read,

$$f(\vec{r}) = \frac{1}{(2\pi)^3} \int_{-\infty}^{\infty} F(\vec{k})\, e^{-i(\vec{k}\cdot\vec{r})} d^3\vec{k}$$

$$F(\vec{k}) = \int_{-\infty}^{\infty} f(\vec{r})\, e^{+i(\vec{k}\cdot\vec{r})} d^3\vec{r},$$

(A3.18)

where $d^3\vec{r}$ and $d^3\vec{k}$ represent volume elements in real space and k-space, respectively.

A.4 The Dirac delta function

There is an important but improper function known as the *Dirac delta function* that appears frequently in the textbook. The Dirac delta function is defined by two properties. Firstly, it is a function of some variable, x, such that

$$\delta(x - x_o) \equiv \begin{cases} 0 & \text{if } x \neq x_o \\ \infty & \text{if } x = x_o. \end{cases}$$

(A4.1)

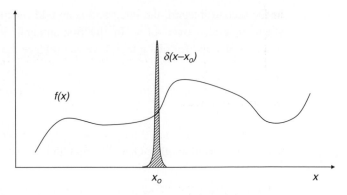

Pictorial representation of a Dirac delta function (shaded peak) together with an arbitrary function $f(x)$.

Secondly, the area under the curve of $\delta(x-x_o)$ is normalized to unity,

$$\int_{-\infty}^{\infty} \delta(x - x_o)\mathrm{d}x = 1. \tag{A4.2}$$

Although the function is ill-defined, we can visualize it as looking something like a very sharp spike that occurs at $x = x_o$ and falls rapidly away to zero as we move away from x_o, as illustrated in Fig. A4.1.

Since the Dirac delta function is zero at points other than x_o, it also follows that the integral (Eq. (A4.2)) need not extend all the way to infinity, but merely needs to encompass the point x_o,

$$\int_{x_o-\varepsilon}^{x_o+\varepsilon} \delta(x - x_o)\mathrm{d}x = 1.$$

Similarly, when multiplied with any arbitrary function $f(x)$, the result is non-zero only in the vicinity of x_o, where the Dirac delta function is non-zero. Hence, integrating the product,

$$\int_{-\infty}^{\infty} f(x)\delta(x - x_o)\mathrm{d}x \cong f(x_o) \int_{-\infty}^{\infty} \delta(x - x_o)\mathrm{d}x = f(x_o),$$

just returns the arbitrary function evaluated at x_o.

The Dirac delta function is readily extended to three dimensions as,

$$\delta(x - x_o)\delta(y - y_o)\delta(z - z_o) \equiv \begin{cases} \infty & \text{if } x = x_o \text{ and } y = y_o \text{ and } z = z_o \\ 0 & \text{otherwise} \end{cases},$$

with,

$$\int_{x=-\infty}^{\infty} \int_{y=-\infty}^{\infty} \int_{z=-\infty}^{\infty} \delta(x - x_o)\delta(y - y_o)\delta(z - z_o)\mathrm{d}x\,\mathrm{d}y\,\mathrm{d}z = 1,$$

and,

$$\int\limits_{x=-\infty}^{\infty} \int\limits_{y=-\infty}^{\infty} \int\limits_{z=-\infty}^{\infty} f(x,y,z)\delta(x-x_o)\delta(y-y_o)\delta(z-z_o)\mathrm{d}x\,\mathrm{d}y\,\mathrm{d}z = f(x_o,y_o,z_o),$$

or, more compactly,

$$\int\limits_{-\infty}^{\infty} f(\vec{r})\delta(\vec{r}-\vec{r}_o)\mathrm{d}^3\vec{r} = f(\vec{r}_o). \tag{A4.3}$$

One common application of the Dirac delta function, used extensively in the textbook, is the definition of a discrete particle number density, $n(\vec{r})$:

$$n(\vec{r}) = \sum_{i=1}^{N} \delta(\vec{r}-\vec{r}_i), \tag{A4.4}$$

whose validity is seen by integrating over all the N particles as,

$$\int\limits_{\text{particles}} n(\vec{r})\mathrm{d}^3\vec{r} = \sum_{i=1}^{N} \int\limits_{\text{particles}} \delta(\vec{r}-\vec{r}_i)\mathrm{d}^3\vec{r} = \sum_{i=1}^{N} 1 = N.$$

A.4.1 Dirac delta functions and Fourier transforms

Recall the one-dimensional Fourier transform pairs (Eq. (A3.15) and Eq. (A3.16)) we introduced earlier. When we substitute the second equation into the first we obtain,

$$f(x) = \frac{1}{2\pi} \int\limits_{k=-\infty}^{\infty} \left[\int\limits_{x'=-\infty}^{\infty} f(x')\mathrm{e}^{+ikx'}\mathrm{d}x' \right] \mathrm{e}^{-ikx}\mathrm{d}k$$

$$= \int\limits_{x'=-\infty}^{\infty} \left[\frac{1}{2\pi} \int\limits_{k=-\infty}^{\infty} \mathrm{e}^{-ik(x-x')}\mathrm{d}k \right] f(x')\mathrm{d}x'.$$

Apparently, for this to be meaningful in the light of Eq. (A4.3), the quantity in square brackets must be equivalent to the one-dimensional Dirac delta function! Thus we find that we can express the Dirac delta function in our complex notation as,

$$\delta(x-x') \equiv \frac{1}{2\pi} \int\limits_{k=-\infty}^{\infty} \mathrm{e}^{-ik(x-x')}\mathrm{d}k. \tag{A4.5}$$

Furthermore, since x and x' merely represent dummy variables, we could readily exchange them such that $\delta(x-x') = \delta(x'-x)$. As a consequence, the Dirac delta function is

$$\delta(x) = \frac{1}{2\pi} \int_{k=-\infty}^{\infty} e^{-ikx}dk = \frac{1}{2\pi} \int_{k=-\infty}^{\infty} e^{+ikx}dk. \tag{A4.6}$$

That is, the Fourier transform of unity is just the Dirac delta function! Conversely, the inverse Fourier transform of a Dirac delta function,

$$F(k) = \int_{x=-\infty}^{\infty} \delta(x)e^{+ikx}dx = e^0 = 1, \tag{A4.7}$$

is unity. These two Fourier transform pairs (Eq. (A4.6) and Eq. (A4.7)) appear frequently in the textbook.

A.5 Elements of thermodynamics

A.5.1 First and second laws

Thermodynamics was developed to explain the gross behavior of systems containing large numbers of particles. At equilibrium, such a system is described by an equation of state, which is a function of all the relevant state variables (e.g. pressure, temperature, number density, etc.) of the form,

$$f(P, T, n, \cdots) = 0. \tag{A5.1}$$

A familiar example is the ideal gas law for a system of non-interacting point particles,

$$n - P/k_B T = 0, \quad \text{or} \quad n = P/k_B T.$$

Beyond this, the bulk of thermodynamics boils down to two fundamental laws that govern how a system behaves during processes in which one or more state variables change. The first law of thermodynamics basically states that energy is neither created nor destroyed during any process but only changes form. For a system of particles at equilibrium, there is some quantity of internal energy, U. This energy is stored in the system of particles in the form of energy that each particle may have individually (such as kinetic energy) and potential energy stored in the form of interactions between the particles or with an external field. In order for this internal energy to change, energy must be introduced or withdrawn from the system and there are two main mechanisms for this transfer. Either some form of mechanical work dW can be performed on the system or some quantity of heat dQ can be transferred to the system. Thus the *first law* is formally stated as

$$dU = dQ + dW. \tag{A5.2}$$

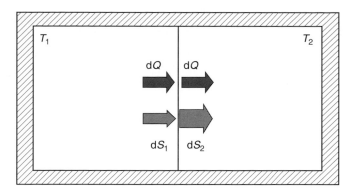

An isolated container with two systems in thermal contact. For every small exchange of heat moving from the warmer system (1) to the cooler system (2), a smaller amount of entropy exits system 1 than enters system 2. Although heat is conserved in the process, new entropy is created.

Now there are actually a number of ways in which work could be performed on the system. Most common is the work done by pressure when a system increases or decreases in volume,

$$\mathrm{d}W = -P\mathrm{d}V. \tag{A5.3}$$

But a similar type of work is performed in, say, a magnetic system whenever the magnetization, M, increases in the presence of a fixed external field, H,

$$\mathrm{d}W = H\mathrm{d}M. \tag{A5.4}$$

Unlike work, heat is a bit of an oddity in that whenever it flows into or out of a system it carries along with it an amount of stuff called "entropy", S. For example, when two systems at different temperatures are placed into thermal contact, as illustrated in Fig. A5.1, heat flows from the warm system (at T_1) into the cooler system (at T_2). However, for each amount of $\mathrm{d}Q$ that leaves the warm system, its entropy decreases by an amount $\mathrm{d}S_1 = -\mathrm{d}Q/T_1$, while the same amount of heat entering the cooler system causes its entropy to increase by $\mathrm{d}S_2 = \mathrm{d}Q/T_2$. For the two systems as a whole, the total entropy has increased because of the heat flow. Although heat is neither created nor destroyed, this entropy stuff appears to develop from nowhere. Indeed, it is this tendency for entropy to be magically produced during processes that forms the *second law* of thermodynamics: "Entropy tends to increase".

The above discussion allows us to express the first law somewhat differently, in the form of what is often called the *thermodynamic identity*

$$\begin{aligned}\mathrm{d}U &= T\mathrm{d}S - P\mathrm{d}V \quad \text{(fluids, gases)} \\ \mathrm{d}U &= T\mathrm{d}S + H\mathrm{d}M \quad \text{(magnetic).}\end{aligned} \tag{A5.5}$$

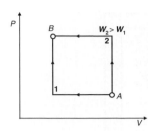

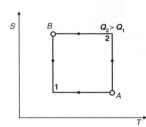

Figure A5.2

The amount of work performed or heat exchanged during a thermodynamic process depends on the path taken. In the upper figure, we see that the work performed on the system is larger in going from point A to point B along path 2 than along path 1. Similarly, in the lower figure, we see that the heat entering the system is larger in going from point A to point B along path 2 than along path 1.

A.5.2 The free energies

During its development, thermodynamics has spawned a number of special functions of the state variables known as thermodynamic potentials or *free energies*. The most common examples include the enthalpy,

$$H = U + PV,\qquad\qquad(A5.6)$$

the Helmholtz free energy,

$$F = U - TS,\qquad\qquad(A5.7)$$

and the Gibbs free energy,

$$G = U - TS + PV.\qquad\qquad(A5.8)$$

Like potentials, the utility of these functions lies not so much in the actual values they have, but in the amount by which they change when a system undergoes a process from one set of state variables to another. For example, when a system moves from state 1 to state 2, the enthalpy change is

$$\Delta H = H_2 - H_1 = (U_2 - U_1) + (P_2 V_2 - P_1 V_1).$$

It is important to notice that the change in the quantity PV in the second term is NOT the work performed during the process. As is illustrated in Fig. A5.2, both the work and the heat are path dependent quantities.

Another endearing feature of these thermodynamic potentials emerges when they are combined with the thermodynamic identity in Eq. (A5.5). As an example, consider the enthalpy whose incremental change is given by

$$dH = dU + PdV + VdP.$$

Upon introducing the thermodynamic identity, this reduces to

$$dH = TdS + VdP \Rightarrow H(S,P).\qquad\qquad(A5.9)$$

This result tells us two things. Firstly that H is really only a function of two state variables, $H(S,P)$, and secondly, that for a process occurring at a fixed pressure, the change in H equals the heat transfer. By a similar means, one can show that for the other thermodynamic potentials,

$$dF = -SdT - PdV \Rightarrow F(T,V),\qquad\qquad(A5.10)$$

$$dG = -SdT + VdP \Rightarrow G(T,P).\qquad\qquad(A5.11)$$

A.5.3 Free energy and the second law

In our earlier example, we observed that when two systems at different temperature are allowed to exchange energy, one loses entropy while the

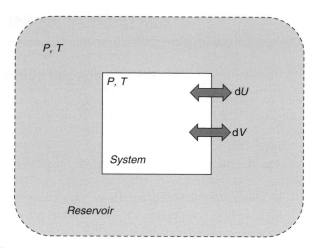

A single system is in thermal and mechanical contact with a reservoir (representing the rest of the universe). During exchanges of energy or volume, the second law requires that the total entropy of the two systems combined either remains unchanged or increases.

other gains entropy. Furthermore, the gain by the one exceeds the loss from the other and their approach to equilibrium is accompanied by a net production of entropy. This is the second law that applies to the two systems as a whole when they are isolated from the rest of the universe. But suppose we are interested in just a single system whose only contact is that with the rest of the universe, or, more reasonably, in contact with a sizeable subsection of the universe, which we might describe as a "reservoir". What does the second law imply about how this system approaches equilibrium?

Imagine our system is at the same temperature and pressure as its surroundings and is able to exchange both energy and volume, as illustrated in Fig. A5.3. The second law implies that,

$$dS_{TOT} = dS + dS_R = dS + \left(\frac{dU_R}{T} + \frac{P}{T} dV_R \right) \geq 0,$$

where the thermodynamic identity of Eq. (A5.5), has been applied for the reservoir. Now because the source of any dU_R or dV_R comes from the system, it reasons that $dU_R = -dU$ and $dV_R = -dV$, so that,

$$dS_{TOT} = dS + \left(\frac{-dU}{T} - \frac{P}{T} dV \right) = -\frac{1}{T} \left(dU - TdS + PdV \right) = -\frac{dG}{T}.$$

$$(A5.12)$$

Here we see another useful quality of the free energy. In any process, the demand by the second law for net production of entropy translates into a corresponding demand for minimization of the free energy of the system alone.

A.6 Statistical mechanics

A.6.1 Microstates and macrostates

Classical thermodynamics offers little in the way of interpreting what this entropy stuff really is. To obtain a better sense of it, one needs to explore the foundations of *statistical mechanics*, wherein the thermodynamics is built from the ground up, by examining the multitude of possible configurations that a system of particles might assume. Each of these configurations is known as a *microstate* and any subset of these that shares a common collective property (e.g. total energy) is known as a *macrostate*. As a more concrete example, consider a simple two-state model of $N = 3$ magnetic moments placed in a magnetic field, as illustrated in Fig. A6.1. By two-state, we mean that the moments are restricted to be either up or down with respect to the field direction and the orientation of each moment contributes to the total energy an amount $E = 0$, if the moment is opposed to the field and $E = \varepsilon$ when it is

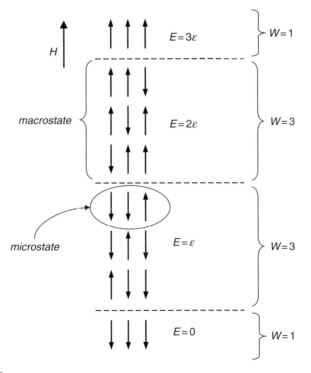

Figure A6.1 A system consisting of three magnetic moments interacting with an external magnetic field, H. There are a total of eight distinct microstates derived from the possible ways in which the three moments can be oriented. Of these, there are four possible macrostates associated with the net magnetization or total energy.

aligned with the field. From Fig. A6.1 we see that for this system there are a total of eight microstates and four allowed macrostates, defined by their common total energy. For each of the macrostates characterized by N_u upward moments, there are

$$W(U = N_u \varepsilon) = \frac{N!}{N_u!(N - N_u)!},$$ (A6.1)

possible microstates, and in statistical mechanics the entropy associated with any given macrostate is

$$S = k_B \ln W,$$ (A6.2)

where $k_B = 1.38 \times 10^{-23}$ J/K is the Boltzmann constant.

A.6.2 The Boltzmann factor

The fundamental assumption of statistical mechanics is that the system is equally likely to be in any one of the microstates corresponding to a given macrostate, but that the probability that the system is in a given macrostate is proportional to a so-called Boltzmann factor,

$$P(E) \propto e^{-E/k_B T}.$$ (A6.3)

This expresses the notion that the probability of a given macrostate of energy E depends upon the available thermal energy, $k_B T$, present in the system and increases with increasing temperature. The Boltzmann factor is particularly useful in conjunction with the *partition function*,

$$Z = \sum_i e^{-E_i/k_B T},$$ (A6.4)

which is a sum of Boltzmann factors, one term for each and every microstate. The partition function allows for proper normalization of the probability in Eq. (A6.3), and the average value of any quantity of interest, X, can be expressed as

$$\langle X \rangle = \sum_i X_i P(E_i) = \frac{1}{Z} \sum_i X_i e^{-E_i/k_B T}.$$ (A6.5)

By way of an example, we can determine the average energy of our two-state model, which would be given by,

$$U = \langle E \rangle = \frac{1}{Z} \sum_i E_i e^{-E_i/k_B T}$$

$$= \frac{3\varepsilon e^{-3\varepsilon/k_B T} + 3 \cdot 2\varepsilon e^{-2\varepsilon/k_B T} + 3 \cdot 1\varepsilon e^{-1\varepsilon/k_B T}}{e^{-3\varepsilon/k_B T} + 3e^{-2\varepsilon/k_B T} + 3e^{-1\varepsilon/k_B T} + 1} = k_B T^2 \frac{\partial \ln Z}{\partial T}.$$ (A6.6)

A.7 Common integrals

$$\int_{-\infty}^{+\infty} e^{-ax^2} dx = \sqrt{\frac{\pi}{a}} \tag{A7.1}$$

$$\int_{-\infty}^{+\infty} x^{2n} e^{-ax^2} dx = \frac{1 \cdot 3 \cdot 5 \cdots (2n-1)}{2^n a^n} \sqrt{\frac{\pi}{a}} \tag{A7.2}$$

$$\int_{-\infty}^{+\infty} x^{2n+1} e^{-ax^2} dx = 0 \tag{A7.3}$$

$$\int_{0}^{+\infty} x^n e^{-ax} dx = \frac{n!}{a^{n+1}} \tag{A7.4}$$

$$\int_{0}^{+\infty} \frac{x^{p-1}}{e^x - 1} dx = \Gamma(p)\zeta(p), \quad p > 0 \tag{A7.5}$$

where $\Gamma(p) \equiv \int_{0}^{+\infty} x^{p-1} e^{-x} dx$, and $\zeta(p) \equiv \sum_{n=1}^{\infty} n^{-p}$.

For n a positive integer: $\Gamma(n+1) = n\Gamma(n) = n!$

Selected values of interest include:

$$\Gamma(1/2) = \sqrt{\pi} \qquad\qquad \Gamma(3/2) = \sqrt{\pi}/2$$

$$\zeta(1) = \infty \qquad\qquad \zeta(3/2) = 2.612$$

$$\zeta(2) = \pi^2/6 = 1.645 \qquad\qquad \zeta(5/2) = 1.341$$

$$\zeta(3) = 1.202 \qquad\qquad \zeta(7/2) = 1.127$$

$$\zeta(4) = \pi^4/90 = 1.082 \qquad\qquad \zeta(9/2) = 1.055$$

$$\int_{0}^{+\infty} \frac{x^3}{e^x - 1} dx = \frac{\pi^4}{15} \tag{A7.6}$$

$$\int_{0}^{+\infty} \frac{x^4 e^x}{(e^x - 1)^2} dx = \frac{4\pi^4}{15} \tag{A7.7}$$

$$\int_{0}^{+\infty} \frac{x^2 e^x}{(e^x + 1)^2} dx = \frac{\pi^2}{6} \tag{A7.8}$$

$$\int_{-\infty}^{+\infty} \frac{x^2 e^x}{(e^x + 1)^2} dx = \frac{\pi^2}{3} \tag{A7.9}$$

$$\int_{-\infty}^{+\infty} \frac{e^x}{(e^x + 1)^2} dx = 1 \tag{A7.10}$$

Glossary

A	lattice spacing
$\vec{a}_1$	unit lattice vector
$\vec{a}_2$	unit lattice vector
$\vec{a}_3$	unit lattice vector
A	area
A, B	elastic wave amplitudes of diatomic modes
$\vec{A}$	vector potential
b	renormalization scale factor
$\vec{b}_1$	unit reciprocal lattice vector
$\vec{b}_2$	unit reciprocal lattice vector
$\vec{b}_3$	unit reciprocal lattice vector
B_{12}, B_{21}	stimulated absorption and emission coefficients
$\vec{B}$	magnetic induction (field)
$C(\vec{r}_1, \vec{r}_2)$	correlation function
$C(\vec{r}_1, \vec{r}_2)$	density–density correlation function
C, C_1	elastic spring constant
C_p	specific heat at constant pressure
C_V	specific heat at constant volume
d	spatial dimension
d_{hkl}	plane separation distance
D	diffusivity or diffusion coefficient
D_f	fractal dimension
D_V	kinematic viscosity
D_T	thermal diffusivity
e	fundamental charge
e^{-2W}	Debye–Waller factor
E	energy
E_g	energy gap
E_F	Fermi energy
ΔE	crystal energy bandwidth
$\vec{E}_o$	amplitude of incident electric field
$\vec{E}_S$	amplitude of scattered electric field
f	force

$f(\vec{q})$	form factor
F	Helmholtz free energy
F_o	generalized force or field
$\tilde{F}$	free energy density
g	splitting factor
$g(r)$	pair distribution function (isotropic)
$g(\vec{r}_1, \vec{r}_2)$	pair distribution function (non-isotropic)
$g(E)$	density of states
$g_{TLS}(E)$	density of two-level systems of energy E
$g(K)$	density of states
G	shear modulus
G	Gibbs free energy
$G(r)$	van Hove correlation function
$G(\vec{r}, t)$	van Hove space-time correlation function
$\vec{G}_{hkl}$	reciprocal lattice vector
h	Miller index
h	magnetic field scaled to critical point
$\hbar$	1.0545×10^{-34} Js
H_c	coercivity
H_C	critical field
H_{c1}	lower critical field
H_{c2}	upper critical field
$\vec{H}$	magnetic field
$\vec{H}_{int}$	molecular or mean field
I	current
I_S	scattered intensity
I'_S	normalized scattering intensity
I_{SC}	short circuit current
$\vec{j}_Q$	heat flow per cross-scctional area
J	shear compliance
J	total atomic angular momentum quantum number
J	current density
$J_{ex}(r_{ij})$	exchange integral
$\vec{J}$	total atomic angular momentum vector
k	Miller index
k	8.99×109 Nm^2/C^2
k_B	Boltzmann constant, 1.38×10^{-23} J/K
K	phonon or electron wave number
K_{max}	Debye cutoff wave number
K_F	Fermi surface wave number

$\vec{k}_i$	incident wave vector
$\vec{k}_s$	scattered wave vector
l	Miller index
l_{mfp}	mean free path
L	atomic orbital angular momentum quantum number
L	length
L_f	latent heat of fusion
$\vec{L}$	atomic orbital angular momentum vector
m	mass
m	magnetization scaled to critical point
m^*	effective mass of electron
m_J	magnetic quantum number
m_o	electron mass
m_p	proton mass
M	number of Rouse beads in a chain
M_R	remanence
M_S	saturation magnetization
$\vec{M}$	magnetization (per volume)
n	electron density in conduction band
$n(\vec{r})$	local number density
$n_e, n_e(\vec{r})$	electron number density
n_C	critical density
n_F	Fermi energy level
$n_{\mathrm{ph}}(E)$	number density of phonons of energy E
n_s	cluster mass distribution
n_S	density of Cooper pairs
$\langle n \rangle$	global average number density
$\langle n \rangle$	Planck distribution or occupancy
$\langle n(r) \rangle_{\mathrm{excl}}$	local number density excluding central
$\Delta n(\vec{r}, t)$	local density fluctuation
$\langle \Delta n^2 \rangle$	mean square density fluctuation
$\hat{n}$	liquid crystal director
N	number of monomers in a polymer chain
N	number of particles
N_e	number of electrons
N_{ex}	number of excitable electrons
N_{TLS}	constant density of two-level systems
p	hole density in valence band
p	fraction of occupied sites
p	pressure scaled to critical point

p^*	fixed point
p_1, p_2	probability of TLS level occupancy
p_c	percolation threshold
p_{ij}	lattice sum terms
p_o	dipole moment amplitude
Δp	relative deviation from percolation threshold
$\vec{p}$	momentum
P	pressure
P	probability a site belongs to spanning cluster
P	Kronig–Penney barrier strength
$P(\vec{r})$	probability for separation $\vec{r}$
$P_{FD}(E)$	Fermi–Dirac distribution function
$\vec{q}$	scattering wave vector
Q	heat
r_{ij}	particle separation
r_o	nearest neighbor separation
Δr	particle separation
$\langle r^2 \rangle$	mean squared displacement
$\vec{r}_1$	position vector
$\vec{r}_2$	position vector
R	gas constant, 8.314 J/K
$R(t)$	time-dependent response function
$R^*(\omega)$	frequency-dependent response function
R_G	radius of gyration
$\vec{R}_i$	basis vector
R_{LP}	Landau–Placek ratio
R_o	amplitude of response function
R_s	cluster radius of gyration
$\langle R \rangle$	root-mean-squared displacement
s	cluster mass
s_{max}	largest cluster mass
$\vec{s}$	electron spin
S	entropy
S	average cluster mass
S	atomic spin quantum number
$S(\vec{q})$	static structure factor
$S(\vec{q}, t)$	dynamic structure factor
$S(\vec{q}, \omega)$	dynamic structure factor
$S_X(\omega)$	power spectrum
$\vec{S}$	atomic spin vector

t	temperature scaled to critical point
Δt	relative deviation from critical temperature
T	temperature
T_c	critical temperature
T_c	Curie temperature
T_f	melting/freezing temperature
T_F	Fermi temperature
T_g	glass transition temperature
T_K	Kauzmann temperature
$\vec{T}$	translation vector
$u(r_{ij})$	pair potential
$u_K(x)$	modulation of Bloch wave function
$\vec{u}(\vec{r}, t)$	fluid velocity
U	internal energy
v	volume scaled to critical point
Δv	relative deviation from critical volume
V	volume
$V(r)$	potential energy
V_o	steady-state junction potential
V_{OC}	open circuit voltage
W	multiplicity
W	work
W	amorphous barrier distribution width
X	number ratio of liquid to gas molecules
ΔX	generalized deformation
Z	valence
α	lattice angle
α	Madelung constant
α	specific heat critical exponent
α_P	isobaric expansivity
β	lattice angle
β	stretching or Kohlrausch exponent
β	order parameter exponent
χ_m	magnetic susceptibility
χ_S	bulk modulus
χ_T	isothermal compressibility
δ	critical isotherm exponent
δ_{ij}	Kroenecker delta function
ε	van der Waals energy scale
ε_o	permittivity of free space

ε_r	dielectric constant
ϕ	work function
ϕ	order parameter
ϕ_i	phase angle
ϕ_K	relative phase angle of Kth Rouse mode
$\phi_{q,K}$	polarization angle
$\phi_R(t)$	pulse response function
Φ	magnetic flux
Φ_o	quantum of magnetic flux (fluxon) 2.07×10^{-15} Tm^2
γ	lattice angle
γ	shear strain
γ	surface tension
γ	susceptibility/compressibility exponent
γ	average cluster mass exponent
γ	ratio of C_P/C_V
Γ	Brillouin linewidth
$\Gamma(\vec{r})$	moment–moment (spin–spin) correlation function
Γ_{12}, Γ_{21}	stimulated absorption and emission rates
η	viscosity
η_S	shear viscosity
η_V	bulk viscosity
κ_{th}	thermal conductivity
λ	molecular field strength
λ	wavelength
λ	penetration length
μ	diffusivity exponent
μ_B	Bohr magneton, 9.27×10^{-24} Am^2
μ_o	permeability of free space
$\vec{\mu}$	magnetic moment
$\langle\vec{\mu}\rangle$	average magnetic moment
ν	correlation length exponent
θ	scattering angle (half angle)
θ	phase angle
θ_D	Debye temperature
ρ	orbital radius
ρ	mass density
ρ	resistivity
ρ_K	density of allowed modes in K-space
σ	van der Waals lengthscale
σ	conductivity

σ	shear stress
σ	largest cluster mass exponent
τ	relaxation time
τ	cluster mass distribution exponent
τ_{coll}	lifetime between collisions
υ	velocity
υ	speed of sound
υ	specific volume
υ_F	Fermi velocity
υ_g	group velocity
υ_o	speed of sound in Debye limit
υ_{drift}	drift speed
υ_{therm}	thermal speed
ω	angular frequency
ω	spin renormalization exponent
ω_B	Brillouin frequency shift
ω_K	angular frequency of Kth phonon mode
ω_o	resonant frequency
$\Delta\omega$	Larmor frequency
ξ	correlation length
ξ	Ginzburg–Landau coherence length
ξ_o	BCS intrinsic coherence length
ψ	wave function
Ψ	wave function
ζ	drag coefficient

References

Aharony, A., Alexander, S., Entin-Wohlman, O. and Orbach, R. (1987) "Scattering of fractons, the Ioffe-Regel criterion, and the 4/3 conjecture," *Phys. Rev. Lett.* **58**(2), 132–135.

Alexander, S., Laermans, C., Orbach, R. and Rosenberg, H. M. (1983) "Fracton interpretation of vibrational properties of cross-linked polymers, glasses, and irradiated quartz," *Phys. Rev. B* **28**(8), 4615–4619.

Anderson, P. W. (1958) "Absence of diffusion in certain random lattices," *Phys. Rev.* **109**(5), 1492–1505.

Ashcroft, N. W. and Mermin, N. D. (1976) *Solid State Physics*, New York, Holt, Rinehart and Winston.

Bardeen, J., Cooper, L. N. and Schrieffer, J. R. (1957) "Theory of superconductivity," *Phys. Rev.* **108**, 1175–1204.

Berne, B. J. and Pecora, R. (1976) *Dynamic Light Scattering*, New York, John Wiley and Sons.

Blakemore, J. S. (1974) *Solid State Physics*, 2nd Ed., Philadelphia, W. B. Saunders Co.

Boese, D. and Kremer, F. (1990) "Molecular dynamics in bulk cis-polyisoprene as studied by dielectric spectroscopy," *Macromolecules* **23**, 829–835.

Bondi, M. A., Forrester, A. T., Garfunkel, M. P. and Satterthwaite, C. B. (1958) "Experimental evidence for and energy gap in superconductors," *Rev. Mod. Phys.* **30**(4), 1109–1136.

Born, M. and Wolf, E. (1997) *Principles of Optics*, 6th Ed., Cambridge, Cambridge University Press.

Buckel, W. (1991) *Superconductivity: Fundamentals and Applications*, New York, VCH.

Callister, W. D., Jr. (2000) *Science and Engineering: An Introduction*, 5th Ed., New York, John Wiley & Sons.

Chaikin, P. M. and Lubensky, T. C. (2003) *Principles of Condensed Matter Physics*, New York, Cambridge University Press.

Chandler, D. (1987) *Introduction to Modern Statistical Mechanics*, New York, Oxford University Press.

Cullity, B. D. (1978) *Elements of X-ray Diffraction*, 2nd Ed., Reading, MA, Addison-Wesley.

Decker, D. L., Mapother, D. E. and Shaw, R. W. (1958) "Critical field measurements on superconducting lead isotopes," *Phys. Rev.* **112**(6), 1888–1898.

Dimon, P., Sinha, S. K., Weitz, D. A., *et al.* (1986) "Structure of aggregated gold colloids," *Phys. Rev. Lett.* **57**(5), 595–598.

Dugdale, J. S. and Gugan, D. (1962) "The effect of pressure on the electrical resistance of lithium, sodium and potassium at low temperatures," *Proc. Royal Soc. (London)* **A270**, 186–211.

Efros, A. L. and Shklovskii, B. I. (1976) "Critical behavior of conductivity and dielectric constant near the metal-non-metal transition threshold," *Phys. Status Solidi* **B76**(2), 475–485.

Elliott, S. R. (1990) *Physics of Amorphous Materials*, 2nd Ed., New York, John Wiley and Sons.

Farnoux, B., Boue, F., Cotton, J. P., *et al.* (1978) "Cross-over in polymer solutions," *J. de Physique* **39**, 77–86.

Finney, J. L. (1970) "Random packings and the structure of simple liquids. I. The geometry of random close packing," *Proc. Roy. Soc. (London)* **A319**, 479–493.

Gangopadhyay, S., Elminyawi, I. and Sorensen, C. M. (1991) "Optical structure factor measurements of soot particles in a premixed flame," *Appl. Optics* **30**(33), 4859–4864.

Gefen, Y., Aharony, A. and Alexander, S. (1983) "Anomalous diffusion on percolating clusters," *Phys. Rev. Lett.* **50**, 77–80.

Graebner, J. E. and Golding, B. (1986) "Phonon localization in aggregates," *Phys. Rev. B* **34**, 5788–5790.

Graebner, J. E., Golding, B. and Allen, L. C. (1986) "Phonon localization in glasses," *Phys. Rev. B* **34**, 5696–5701.

Greenler, R. (1980) *Rainbows, Halos, and Glories*, New York, Cambridge University Press.

Griffiths, D. J. (1999) *Introduction to Electrodynamics*, 3rd Ed., New Jersey, Prentice Hall.

Guggenheim, E. A. (1945) "The principle of corresponding states," *J. Chem. Phys.* **13**, 253–261.

Hansen, J. P. and McDonald, I. R. (1986) *Theory of Simple Liquids*, 2nd Ed., New York, Academic Press.

Havlin, S. and Ben-Avraham, D. (1987) "Diffusion in disordered media," *Adv. Phys.* **36**(6), 695–798.

Hecht, E. and Zajac, A. (1974) *Optics*, Reading, MA, Addison-Wesley.

Henniger, E. H., Buschert, R. C. and Heaton, L. (1967) "Atomic structure and correlation in vitreous silica by X-ray and neutron diffraction," *J. Phys. Chem. Solids* **28**(3), 423–432.

Henry, W. E. (1952) "Spin paramagnetism of Cr^{3+}, Fe^{3+} and Gd^{3+} at liquid helium temperatures and in strong magnetic fields," *Phys. Rev.* **88**(3), 559–562.

Hofmann, P. (2008) *Solid State Physics*, Weinheim, Wiley-VCH.

Ioffe, A. F. and Regel, A. R. (1960) "Non-crystalline, amorphous, and liquid electronic semiconductors," *Prog. Semicond.* **4**, 237–291.

Jackson, L. C. (1936) "The paramagnetism of the rare earth sulphates at low temperatures," *Proc. Royal Soc. (London)* **48**, 741–746.

Jan, N., Hong, D. C. and Stanley, H. E. (1985) "The fractal dimension and other percolation exponents in four and five dimensions," *J. Phys. A: Math. Gen.* **18**, L935–L939.

Jan, N., Hong, D. C. and Stanley, H. E. (1983) *Handbook of Chemistry and Physics*, 64th Ed., Boca Raton, CRC Press.

Kauzmann, W. (1948) "The nature of the glassy state and the behavior of liquids at low temperatures," *Chem. Rev.* **43**(2), 219–256.

Kittel, C. (2005) *Introduction to Solid State Physics*, 8th Ed., Hoboken, NJ, John Wiley and Sons.

Kittel, C. and Kroemer, H. (1980) *Thermal Physics*, 2nd Ed., San Francisco, W. H. Freeman and Co.

Kronig, R. de L. and Penney, W. G. (1931) "Quantum mechanics of electrons in crystal lattices," *Proc. Royal Soc. (London)* **A130**(814), 499–513.

Kubo, R. (1966) "The fluctuation-dissipation theorem" *Rep. Prog. Phys.* **29**, 255–284.

Landau, L. D. (1965) *Collected papers of L. D. Landau*, New York, Gordon and Breach.

Lien, W. H. and Phillips, N. E. (1964) "Low-temperature heat capacities of potassium, rubidium, and cesium," *Phys. Rev.* **133**, A1370–A1377.

London, F. and London, W. (1935) "The electromagnetic equations of the supraconductor," *Proc. Royal Soc. (London)* **A149**(866), 71–88.

MacDonald, D. K. C. and Mendelssohn, K. (1950) "Resistivity of pure metals at low temperatures I. The alkali metals," *Proc. Royal Soc. (London)* **A202**, 103–126.

Maxwell, E. and Lutes, O. S. (1954) "Threshold field properties of some superconductors," *Phys. Rev.* **95**(2), 333–338.

Maxwell, E. (1952) "Superconductivity of the isotopes of tin," *Phys. Rev.* **86**(2), 235–242.

McCrum, N. G., Read, B. E. and Williams, G. (1991) *Anelastic and Dielectric Effects in Polymeric Solids*, New York, Dover Publications.

Meissner, W. and Ochsenfeld, R. (1933) "Ein neuer effect bei eintritt der supraleitfahigkeit," *Naturwissenschaften* **21**(44), 787–788.

Moran, F. J. and Maita, J. P. (1954) "Electrical properties of silicon containing arsenic and boron," *Phys. Rev.* **96**(1), 28–35.

Mott, N. F. (1990) *Metal-Insulator Transitions*, 2nd Ed., London, Taylor and Francis.

Nigh, H. E., Legvold, S. and Spedding, F. H. (1963) "Magnetization and electrical resistivity of gadolinium single crystals," *Phys. Rev.* **132**(3), 1092–1097.

Omar, M. A. (1975) *Elementary Solid State Physics*, Reading, MA, Addison Wesley.

Onnes, H. K. (1911) "On the sudden change in the rate at which the resistance of mercury disappears," *Comm. Leiden* **124c** (1911).

Ornstein, L. S. and Zernike, F. (1914) "Accidental deviations of density and opalescence at the critical point of a single substance" *KNAW Proc. Akad. Sci. (Amsterdam)* **17**, 793.

Phillips, W. A. (1987) "Two-level states in glasses," *Rep. Prog. Phys.* **50**, 1657–1708.

Schroeder, D. V. (2000) *An Introduction to Thermal Physics*, New York, Addison, Wesley and Longman.

Seddon, J. M. (1990) "Structure of the inverted hexagonal phase and non-lamellar phase transitions of lipids," *Biochemica et Biophysica Acta* **1031**(1), 1–69.

Shaw, R. W., Mapother, D. E. and Hopkins, D. C. (1960) "Critical fields of superconducting tin, indium, and tantalum," *Phys. Rev.* **120**(1), 88–91.

Silfvast, W. T. (2004) *Laser Fundamentals*, 2nd Ed., New York, Cambridge University Press.

Sorensen, C. M., Oh, C., Schmidt, P. W. and Rieker, T. P. (1998) "Scaling description of the structure factor of fractal soot composites," *Phys. Rev. E* **58**(4), 4666–4672.

Stanley, H. E. (1971) *Introduction to Phase Transitions and Critical Phenomena*, New York, Oxford University Press.

Stauffer, D. (1985) *Introduction to Percolation Theory*, Philadelphia, Taylor and Francis.

Strobl, G. (2004) *Condensed Matter Physics*, Berlin, Springer-Verlag.

Susman, S., Volin, K. J., Price, D. L., *et al.* (1991) "Intermediate-range order in permanently densified vitreous SiO_2: A neutron-diffraction and molecular-dynamics study," *Phys. Rev. B* **43**, 1194–1197.

Temkin, R. J., Paul, W. and Connell, G. A. N. (1973) "Amorphous germanium II. Structural properties," *Adv. Phys.* **22**(5), 581–641.

Thermophysical Properties of Matter, edited by Y. S. Touloukian (1970), New York, Plenum Press.

Thomas, J. E. and Schmidt, P. W. (1963) "X-ray study of critical opalescence in argon," *J. Chem. Phys.* **39**, 2506–2516.

Townsend, P. and Sutton, J. (1962) "Investigation by electron tunneling of the superconducting energy gaps in Nb, Ta, Sn, and Pb," *Phys. Rev.* **128**(2), 591–595.

van de Hulst, H. C. (1957) *Light Scattering by Small Particles*, New York, John Wiley and Sons.

Weitz, D. A. (2004) "Packing in the spheres," *Science* **303**, 968–969.

Weitz, D. A. and Oliveria, M. (1984) "Fractal structures formed by kinetic aggregation of aqueous gold colloids," *Phys. Rev. Lett.* **52**(16), 1433–1436.

Weitz, D. A., Huang, J. S., Lin, M. Y. and Sung, J. (1985) "Limits of the fractal dimension for irreversible kinetic aggregation of gold colloids," *Phys. Rev. Lett.* **54**(13), 1416–1419.

Whitesides, G. M. and Grzybowski, B. "Self-assembly at all scales," (2002) *Science* **295**, 2418–2421.

Zallen, R. (1983) *The Physics of Amorphous Materials*, New York, John Wiley and Sons.

Zarzycki, J. (1991) *Glasses and the Vitreous State*, New York, Cambridge University Press.

Index

Anderson transition, 241–242
anomalous diffusion, 307–310

basis set, 5
BCC (body centered cubic), 12–13
Bethe lattice, 303
Bloch waves, 218–220
Bohr magneton, 54
Boltzmann factor, 381
Boltzmann superposition, 253
bonds
    covalent, 38–41
    hydrogen, 41–42
    ionic, 38
    metallic, 42
    van der Waals, 36–38
Bragg scattering, 88–89
    phonons, 168, 191
Bragg scattering
    electrons, 220
Bravais lattices, 10
    2D nets, 9
Brillouin function, 58
Brillouin scattering, 148, 180
Brillouin zone, 163
    and Bragg scattering, 168
    and electrons, 227, 230
    and phonons, 188, 191
    definition, 167
    higher zones, 168
Brownian motion
    diffusion coefficient, 143
    photon correlation spectroscopy, 144
    random walk, 142
    systemic drag, 257–258

chalcogenide, 30
chemical potential, 207, 275
close packed
    ellipsoids, 27–28
    hexagonal, 16
    random, 26–27
coexistence, 268, 269, 286
coherence length, 355–356, 357
cohesive energy, 42
    ionic crystals, 45–46

van der Waals crystal, 44
compressibility
    isothermal, 103, 150, 277, 281, 321
conventional cell, 10
coordination
    and radial distribution function, 25
    number, 14
correlation function, 52
    density–density, 79
    spin–spin, 62
    van Hove space, 104
    van Hove space-time, 142
correlation length, 53, 62, 279, 282, 297, 299,
        327, 329
critical
    density, 273
    exponents, 286, 326
        mean field, 303, 321–324
        percolation, 297
    opalescence, 279
    Ornstein–Zernike theory, 279–282
    phenomena, 276–278
    point, 270
CRN (continuous random network),
        28–29
cross section, scattering, 71–72
cubic lattice, 12–15
    BCC, 12–13
    FCC, 12–13
    simple cubic, 12
Curie law, 60
Curie temperature, 60, 61
Curie–Weiss law, 60
cyclic boundary conditions,
        169–170

Debye
    approximation, 187
    model, 184, 186–187
    temperature, 189
Debye–Waller factor, 178
density fluctuations
    in liquids, 141
density of states
    electron, 211, 228, 242
    fractal, 198

phonon, 187, 195, 200
  two-level systems, 194
diamagnetism, 55–57
diffusion
  anomalous, 307–310
  coefficient, 143, 144
  electron, Drude model, 203
  equation, 143
  mean free path, 190, 192
  thermal, 189
  thermal coefficient, 148
dipole scattering, 70
Dirac delta function, 373
director, 126
dispersion relation
  free electron, 210
  Kronig–Penney model, 223
  magnon, 172
  phonon, diatomic, 174
  phonon, monatomic, 165
dissipation, 256–257
DLA (diffusion limited aggregation), 117
DLCA (diffusion limited cluster aggregation), 117
  soot, 118–119
Dulong–Petit law, 183
dynamic structure factor, 139–142

Einstein model of specific heat, 184–186
Einstein relation, 145
electrical conductivity
  and band occupation, 230–231
  and electron scattering, 213–214
  Drude model of, 203–204
  in the free electron model, 212–213
ellipsoids
  packing efficiency, 27–28
energy bands, 223–230
enthalpy, 378
equation of state, 376
  Ising model, 319
  scaled forms, 319–321
  van der Waals, 316
equipartition theorem, 183, 202
Ewald–Oseen extinction theorem, 105, 214
exchange interaction, 61, 171, 283–284, 314
exclusion principle, 37, 39, 205

FCC (face centered cubic), 12–13
Fermi energy, 207, 212
Fermi–Dirac energy distribution, 207
ferromagnetism, 60
  hysteresis curve, 63–64
  order parameter, 283
  phase diagrams, 284–285
  spin waves, 171–173
field emission, 210

finite-sized scaling, 298–300
first law of thermodynamics, 267, 376
five-fold symmetry, 9
fixed points, 302
fluctuation–dissipation theorem, 257–261
fluctuations
  in amorphous matter, 102
  in liquids, 145
  in order parameter, 313
  near critical point, 278–279
form factor
  atomic, 75
  cell, 83, 89
Fourier
  analysis, 367
  series, 368
  transforms, 368–373
fractals, 117–118
  Sierpinski gasket, 117, 300
free energy, 378–379

gelation, 31, 305–307
Gibbs free energy, 267, 378
Ginzburg–Landau
  coherence length, 355–356
  equations, 354
  mean field theory, 349–351
glass transition, 150–151
  Kauzmann paradox, 151–153
  structural relaxation, 153–154
Guinier scattering regime, 119–121
gyration
  radius of, 120

Hall effect, 214–216
  and holes, 233
  Hall coefficient, 215
hard sphere fluid, 26
harmonic oscillator, 185
HCP (hexagonal close packed), 16, 27
heat capacity, 182–183
  amorphous matter, 193–195
  Debye model, 186–187
  Dulong–Petit, 183
  Einstein model, 184–186
  electronic contribution, 207–210
  two-level systems, 193–195
Helmholtz free energy, 324, 378
hexagonal lattice, 16
homogeneous functions, 294, 330
Hookian solid, 247
hydrodynamics
  hydrodynamic modes, 145
  Rayleigh–Brillouin spectrum, 148–150
hydrophilic and hydrophobic, 129
hysteresis, 63–64

inelastic scattering
  by phonons, 179–180
intermediate-range order, 105–106
Ioffe–Regel criterion, 197
Ising model, 314, 318
  critical exponents, 324
  equation of state, 319
  Kadanoff renormalization, 327–330

Kadanoff renormalization, 327–330
Kauzmann paradox, 151–153
Kronig–Penney model, 221–225

Landau theory
  mean field, 324–326
  superconductors, 349–351
Landau–Placek ratio, 150
Larmor frequency, 56
lattice, 4
  Bravais lattices, 9, 10
  conventional cell, 10
  primitive cells, 6
  symmetry, 8–9
lattice vibrations
  Debye approximation, 187
  Debye model, 186–187
  dispersion relation, 165, 174
  phonon density of states, 187
  waves, 164–165
Laue conditions, 85
Lennard-Jones potential, 37, 314
lever rule, 271–272, 277
linear response, 247
lipids, 129
liquid crystals, 124–129
  LCD display, 126
  lyotropic, 129–131
  microemulsions, 130
  nematic phase, 126
  smectic phase, 128
  thermotropic, 124–129
liquids
  and the glass transition, 150–151
  density fluctuations, 141
  hydrodynamic modes, 145–150
  Newtonian, 247, 248
  polymer, 154
  Rayleigh–Brillouin spectrum, 148–150
  structure factor, 97
long-range order, 20, 30
lower critical field, 351, 357–359

Madelung constant, 46
magnetic moment, 53–55
magnons, 171–173
  dispersion relation, 172
  specific heat contribution, 200

mean field, 304, 313
  critical exponents, 321–324
  interaction, 313–315
  Ising model, 318–319
  Landau theory, 324–326
  theory, 313–315
  van der Waals, 130, 316–318
micelles, 130
microemulsions, 130, 131
Miller indices, 86–87
molecular field, 60, 61, 315, 318
Mott transition, 226

Navier–Stokes equations, 146
nematic liquid crystal, 126
Newtonian fluid, 247
nuclear magneton, 55
nucleation
  barrier, 287
  critical radius of, 275
  heterogeneous, 287
  homogeneous, 274–276

opalescence. *See* critical opalescence
order parameter, 272–274
  exponent, 322
  fluids, 273
  in percolation, 296
  magnetic, 283
orientational order, 62
orientational ordering, 51
  correlation function, 52
  correlation length, 53
Ornstein–Zernike theory, 279–282

pair distribution function
  defined, 22–25
  radial distribution function, 25
  related to structure factor, 97
  self-similar matter, 112–114
paramagnetism, 57
partition function, 381
penetration length, 338–339, 357
percolation
  bond percolation, 289
  correlation length, 297
  exponents, 297
  site percolation, 289
  spanning cluster, 292
  threshold, 292
phase diagrams
  coexistence curve, 269
  critical point, 270
  fluids, 269–272
  lever rule, 271–272
  magnetic, 284–285
  triple point, 270

phase transition
first order, 274
second order, 274, 276, 279, 293, 313
phonons, 170–171
inelastic scattering by, 177–180
localization, 193, 197–199
Umklapp collisions, 191
photoelectric effect, 210
photovoltaic cell, 240–241
pn-junction, 237–240
polymer
coil structure, 110–115
dynamics
reptation model, 157–159
Rouse modes, 154–157
entropic spring, 154–155
pair distribution function, 113
Porod scattering law, 124
power spectrum, 258
primitive cells, 6
pulse response function, 253

Raman scattering, 180
random walk, 110–112, 281
diffusion, 142 144
self-avoiding, 114–115
Rayleigh scattering, 100–101
blue sky, 100
fiber attenuation, 104
Rayleigh–Benard convection, 131
Rayleigh–Brillouin spectrum, 148–150
RCP (random close packed), 26–27
reciprocal space
Fourier transform, 86
lattice, 85–86
lattice vectors, 86
related to Bragg scattering, 88–89
relaxation
polymer, 159
polymers, 157
structural, 153
viscoelastic, 248–249
renormalization
Kadanoff (Ising model), 327–330
triangular lattice, 300–303
response function, 251, 253–256
in fluctuation–dissipation, 258
rotational invariance, 22

scaling
finite-sized, 298–300
law of corresponding states, 285–286, 319–321
relations, 297–298, 309, 312, 330
scattering
Bragg, 88–89
by fractals, 118
by liquid crystals, 126

by phonons, 177–180, 191
cross section, 71–72
in crystals, 83–84
in glass, 97–100
in liquids, 97, 141
lengthscale, 77–79
multiple, 193, 197, 220
of electrons, 213–214
Rayleigh, 100–101
wave vector, 74
second law of thermodynamics, 267, 377, 379
self-assembly, 131–132
self-avoiding walk, 114–115
self-similarity, 111–112, 116, 279, 289, 296, 299, 300, 308, 327
fractals, 117–118
semiconductors
acceptor levels, 236–237
donor levels, 236–237
doped, 235
holes, 232–233
intrinsic, 233–235
conductivity, 235
n-type, 236
photovoltaic cell, 240–241
pn-junction, 237–240
p-type, 236
short-range order, 26, 30, 97–100
Sierpinski gasket, 117, 300, 327
small angle scattering
SAXS, SANS, 105–106
smectic liquid crystal, 128
soot, 118–119
spin waves. See magnons
spin–spin correlation function, 62, 328
structure factor
amorphous matter, 97
and density–density correlation function, 79–80
and G(r), 103
dynamic, 141
fractals, 118
in crystals, 83–84
missing reflections, 89–91
self-similar matter, 113
static, 77
superconductors
coherence length, 355–356
Cooper pairs, 346–348
critical field, 340
discovery, 334
energy gap, 343
flux quantization, 348
high Tc, 360
isotope effect, 344–345
London equation, 338
Meissner effect, 335–337

superconductors (cont.)
    penetration length, 338–339
    specific heat, 340–343
    Type II, 351–352
        critical fields, 356–360
        Ginzburg–Landau theory, 352–356
    vortex state, 352
supercooling/heating, 274–276
surface tension, 275
susceptibility
    Curie–Weiss law, 60
    diamagnetic, 56
    divergence, 285, 323, 325, 331
    paramagnetic, 58
symmetry
    amorphous, 21–22
    breaking, 50, 51
    crystal, 8–9
    five-fold, 9
    operations, 9

thermal conductivity
    crystals, 192
    defined, 189–191
    electronic contribution, 217
    glasses, 193
    two-level systems, 195–197
thermal expansion, 181, 278
thermal fluctuations, 102, 145, 258, 265, 279
thermionic emission, 210
thermotropic liquid crystals, 125–129
Thomson scattering, 71
translation vector, 4, 10, 83, 177

translational invariance, 22
triple point, 270
two-level systems, 194–197, 342

Umklapp process, 191
unit cell, 6, 10
universality, 285–286, 303, 324
upper critical field, 351, 359–360

van der Waals
    bonds, 36–38
    equation of state, 316–318
van Hove
    correlation function, 104
    discontinuities, 230, 242
    space-time correlation function, 142
viscoelastic relaxation, 154, 248–249,
        254
    dissipation, 256–257
    Rouse modes, 154–157
viscosity
    bulk, 146
    near gel point, 306
    shear, 145, 248
Voronoi polyhedra, 27

Wigner–Seitz cell, 7, 27, 168
work function, 210

X-ray
    cross section, 71
    diffraction, 89
    scattering from glasses, 97–100